OWNERSHIP OF KNOWLEDGE

INSIDE TECHNOLOGY

Edited by Wiebe E. Bijker and Rebecca Slayton

A list of books in the series appears at the back of the book.

OWNERSHIP OF KNOWLEDGE
BEYOND INTELLECTUAL PROPERTY

edited by Dagmar Schäfer, Annapurna Mamidipudi,
and Marius Buning

THE MIT PRESS
CAMBRIDGE, MASSACHUSETTS
LONDON, ENGLAND

© 2023 Massachusetts Institute of Technology

This work is subject to a Creative Commons CC-BY-ND-NC license.

Subject to such license, all rights are reserved.

Subject to such license, all rights are reserved.

Co-funded by the ERC project "Before Copyright: Printing privileges and the politics of knowledge in early modern Europe," funded by the European Union. Views and opinions expressed are however those of the author(s) only and do not necessarily reflect those of the European Union or the European Research Council. Neither the European Union nor the granting authority can be held responsible for them.

This book was set in Stone Serif and Futura by Westchester Publishing Services. Printed and bound in the United States of America.

Library of Congress Cataloging-in-Publication Data

Names: Schäfer, Dagmar, editor. | Mamidipudi, Annapurna, editor. | Buning, Marius, 1979– editor.
Title: Ownership of knowledge : beyond intellectual property / edited by Dagmar Schäfer, Annapurna Mamidipudi, and Marius Buning.
Description: Cambridge, Massachusetts : The MIT Press, [2023] | Series: Inside technology | Includes bibliographical references and index.
Identifiers: LCCN 2022038290 (print) | LCCN 2022038291 (ebook) | ISBN 9780262545594 (paperback) | ISBN 9780262374637 (epub) | ISBN 9780262374644 (pdf)
Subjects: LCSH: Knowledge management. | Intellectual property.
Classification: LCC HD30.2 .O926 2023 (print) | LCC HD30.2 (ebook) | DDC 658.4/038—dc23/20220811
LC record available at https://lccn.loc.gov/2022038290
LC ebook record available at https://lccn.loc.gov/2022038291

10 9 8 7 6 5 4 3 2 1

CONTENTS

List of Figures   vii
Acknowledgments   ix

    Ownership of Knowledge: Introduction   1
    Dagmar Schäfer and Annapurna Mamidipudi

1   **Excavations of Knowledge Ownership: Theoretical Chapter**   15
    Annapurna Mamidipudi and Dagmar Schäfer

I   MUTUAL CONDITIONING

2   **Intellectual Property with Chinese Characteristics**   47
    Cynthia Brokaw

3   **Teaching Intellectual Property: Constructing the Historical Narrative of Intellectual Property in University Textbooks**   91
    Marius Buning

II   THE THREE PRACTICES: PERFORMANCE, USE, NAMING

4   *Raga* **and the Problem of Ownership: Knowledge and Culture in Carnatic Music**   121
    Annapurna Mamidipudi and Viren Murthy

5   **Imitating Crackles: Material Mimesis in Stones and Textiles**   153
    Marjolijn Bol

6   **Educational Inequities and the Distribution of Technical Knowledge: Three Instruments**   181
    Amy E. Slaton

III   THE THREE DOMAINS: SOCIETY, ECONOMY, EPISTEMOLOGY

7   **An Aesthetic of Knowledge: Relations and the Documentation of Traditional Knowledge in Papua New Guinea**   219
    James Leach

8   Names for Work: Crafts, Bureaucracy, and Law in Yuan and Ming China (Thirteenth–Seventeenth Century)   251
Dagmar Schäfer

9   Ownability, Ownership, Knowledge, and Genetic Information in the United States   293
Myles W. Jackson

IV   THE ROLE OF SCHOLARSHIP

10   Objects, Knowledge, and Museums: Reflections on the Endangered Material Knowledge Programme   319
Lissant Bolton

11   A Reader's Guide to *Ownership of Knowledge*: Diagrammatic Chapter   343
Vivek S. Oak, Jörn Oeder, and Annapurna Mamidipudi

Contributors   363
Index   367

LIST OF FIGURES

3.1 Semiotic square outlining the various relationships between IP and the PD   105

5.1 Flask (*bianhu*), Eastern Zhou dynasty, Warring States period   154

5.2 The Warwick Castle table, attr. to Baldassare Artima and Diacinto Cawcy, England   155

5.3 Quench-cracked quartz crystals   159

5.4 Imitation gems colored with copper green and alkanet (red dye) dissolved in hot resin   160

5.5 Quilt, Coromandel Coast (made for the European market)   164

5.6 Silk fragment, lampas weave, Italy   165

5.7 Block-printed linen, woven (originally black), Germany   166

5.8 Selendang (shawl) or belt, Java   169

11.1 The spatial and temporal splits and how they are fixed   345

11.2 Defining first- and second-order relations through practice-material instantiation relations   347

11.3 The grid for splitting   349

11.4 Fixing the split   351

11.5 The combinations of material instantiations (MI) for each case   352

11.6 Axes of tacit/explicit knowledge, and alienable/inalienable ownership imposed by the science-law relation on the grid   353

11.7 Proliferating the quadrants with the WBO notations signifying fixing of ownables and knowables   354

11.8 Overlaying the quadrants and notation with cases to analyze hierarchy of knowledge ownership   355

11.9 Locating Myles Jackson's case in the grid   357

11.10 Locating James Leach's case in the grid 359
11.11 Locating Lissant Bolton's case in the grid 360
11.12 Locating Amy Slaton's case in the grid 361

ACKNOWLEDGMENTS

This book has its origin in a conversational bike ride along Lake Michigan in May 2016, when two of the editors were spending time at the Neubauer Collegium for Culture and Society. The bikes were rented and our thoughts easily shared as we stopped frequently to accommodate the pace of one who was not as fit as the other. Our generous host Jacob Eyferth allowed us time to enjoy open-ended conversations with Don Harper, BuYun Chen, Kaijun Chen, Yuhang Li, and Soumhya Venkatesan, among others, and the very enthusiastic group of graduate students we met at the history seminar. Their questions are in this book.

"Strong reasons make strong actions" (Shakespeare, *King John*). Galvanized by the issues raised, we pursued them further in two conferences at the Max Planck Institute for the History of Science, Berlin (MPIWG). The first, in November 2016, probed "Ownership of Knowledge." The follow-up conference in August 2018 forged a path "Beyond Intellectual Property." In addition to the authors of this book, conference participants included Martha Buskirk, Hyo Yoon Kang, Alain Pottage, Jérôme Baudry, Graeme Gooday, Stathis Arapostathis, Luca Molà, Catherine Fisk, Berris Charnley, Mario Biagioli, Sean Bottomley, Giorgio Riello, Sarah van Beurden, Jung Lee, and Peng-Sheng Chiu. We thank all of these colleagues for sharing their work and thoughts with us. In 2018, Marius Buning joined the editorial team.

"The deep waters will have their ferry boats and tall mountains have passes that can be crossed" (Sun Wukong, *Journey to the West*)—with knowledgeable guides to show the way. Our guides included Dorothy Ko, Jose Bellido, Véronique Pouillard, Jaya Remond, Hansun Hsiung, Pamela O. Long, Shiv Viswanathan, Sumithra Vasudev, Uzramma, Satish Poludas, Gauri Nori, Ellen Harlizius-Klück, Klaus Staubermann, Christian Götter, Wiebe Bijker, Marianna Szczygielska, Noa Hegesh, Lisa Onaga, Tamar Novick, Wilko Graf von Hardenberg, and Chun Xu. Thanks also to the participants of the colloquia of Department III (Artifacts, Action, Knowledge) at the MPIWG between 2018 and 2019, who read and commented on earlier versions of parts of this book, as did independent translator and researcher Xiujie Wu. The inevitable formalities and practicalities were overcome with the help of Helen Rana and Michael Thomas Taylor, who provided aid that ran the gamut of editing from content to mechanical, together with the Department III editorial staff, Gina Grzimek and Melanie Luise Glienke, supported by librarian Cathleen Paethe, and the student assistants Spencer Forbes, Jing-Shin (Anita) Lin, Paul Kaemmerer, Wiebke Weitzmann, and Yi Zheng. The anonymous reviewers and the editorial board of the Inside Technology series spurred us on to cross the finish line of this challenging project.

Saint Kabir, the fifteenth-century weaver poet of Varanasi, wrote, "I set out in search of red, I became red myself," telling us that seeking knowledge is a journey inside ourselves. We do not claim ownership of our publication, because insight should not be owned. We do, however, take responsibility for the word we have coined to give this insight space to grow and have an impact in the real world:

kn/own/able.

OWNERSHIP OF KNOWLEDGE: INTRODUCTION

Dagmar Schäfer and Annapurna Mamidipudi

For many centuries, and still today, humans across the world have transformed the skins of dead animals into leather. Individuals develop and learn or teach tanning. Societies provide tools and production sites for making leather, or they might promote its trade. Now read this passage again, and see how in each of these activities, identifying how certain aspects of tanning can be known presupposes something being owned. Developing and learning or teaching tanning, for instance, implies notions of how knowledge is gained—and thus owned, and given away—through sharing. The reverse is also evident. Identifying which parts or aspects of tanning can be owned—flaying, preparing the hide, managing the production site or the end product rather than the working body or the tanning mixture—presets how tanning can be known. This mutual conditioning of knowing and owning, and how knowledge ownership is constructed around it, is the topic of this book.

Knowledge and property are important themes in historical, sociological, and anthropological research, and in recent decades a growing body of literature has come to investigate the long, and ongoing, history of their checkered relationship. Scholars of science, technology, medicine, and law have all come to emphasize knowledge as the *sum* of human understanding and the many *forms* it can take, and they have shed light on its ownership as possession by law, as well as in acts of social sharing. This book consolidates these two strands of research by approaching knowledge ownership as a complex process in which the first moment of having intrinsically anticipates owning, and in which ownership comprises enactments of possession that include, from the first grasp of understanding, all kinds of dispossession. Such dispossession may consist of ignorance by law, social silencing, outright denunciation, or nonactions such as the acceptance of material decay.

If we look at knowledge ownership in this comprehensive way, we can see how the actions that our modern world insists on distinguishing as knowing and owning actually collapse into each other when society, for instance, spatially distances itself from the bloody flaying of carcasses by moving the sites of slaughter away from human settlements, not *wanting* to know the unpleasant and polluting effects of the trade. Regulating the sites of production *conditions* the possibilities of owning such knowledge. Knowing and owning also collapse into each other and are mutually conditioned when formulas for tanning hides are devised and either shared or guarded, when the end product alone is preserved, or when recipes are memorized through chants, and then later protected by law. Practitioners who know with their bodies may achieve a high status in some societies, whereas other communities cast out such experts because of the smell that sticks to their skins, even though this stench is essential to both knowing and owning the task. Each of these acts in the process of tanning facilitates possibilities of knowing and owning, or makes it impossible for knowledge to be known or owned.

Working from this viewpoint on the mutual conditioning of knowing and owning, this book builds on scholarship that has foregrounded the roles of practices and materiality.[1] Historical, sociological, and anthropological studies of technology have convincingly put materials and practices on par with words as statuses of knowledge.[2] In line with such scholarship, we suggest that all statuses of knowledge are equally concerned with owning and thus with acts of *kn/own/ing*. We critically engage with those strands that have long attempted to address injustices and imbalances of knowing and owning in our past and modern worlds, which might include key moments such as the work of Karl Marx, for whom the sciences and technology and alienated labor facilitated capitalist disappropriation;[3] Michel Foucault's notion of "naming and power";[4] and more recent attempts that have looked at "the ways in which we know and represent the world (both nature and society)."[5] Inasmuch as knowledge is not a thing but a process realized in action, we suggest that its ownership is equally at stake at every moment; what can be known, the *knowable*, always defines what can be owned, the *ownable* (and vice versa). They are not separate, and there is no sequence of cause and effect. Only *kn/own/ables* exist, and hence when actors attempt to distinguish what can be known, they always also manipulate what can be owned (and vice versa). We refer to this as a mutual conditioning that produces knowables and ownables.

The analytical focus of this book is to explore how kn/own/ables are intrinsic to all negotiations of knowledge ownership: when scientists examine genes and make them intellectual property; when a tenth-century Chinese carver produces or destroys woodblocks; or when a banana becomes the tool to explain the laws of physics to underprivileged students in US schools. In this context we propose technology as a useful heuristic to shed light on how an emphasis on law and science has given preference to words as a way to own knowledge, thus centering debates on knowledge as (legal) property and on possessing knowledge as a form of intellectual achievement. Since we flag our approach to knowledge ownership in this book as one informed by technology studies, authors of the case studies reflect on the past and present of the development, circulation, and exchange of knowledge, and on its appropriation or decay, down to the role of current scholarship on science, technology, society, and law in such debates. We have arranged case studies to highlight four crucial tactics actors use to enable or disable knowledge ownership in attempting to split the kn/own/able—and thereby also create imbalances and inequalities, in the past and in our contemporary world. Focusing on these tactics, four corresponding sections investigate: (1) how different cultures and societies approach kn/own/ables in various ways and thus manipulate their mutual conditioning to set the structural premises for knowledge ownership; (2) how actors in these societies rely on or are constrained by these premises when employing the three practices of knowledge ownership—namely, performance, use, and naming; (3) how this consequently allows actors to apply these practices differently in different domains of society, economy, and epistemology to own knowledge; and (4) how scholarship interferes in a contemporary world where science and law exist.

Each individual case unfolds the rich ways in which actors, past and present, manifest knowledge ownership. In sum, these arguments spurred us to develop a theoretical framework for analyzing knowledge ownership, which we offer to readers in chapter 1. Since that framework developed directly out of the discussions that produced this scholarship, we have opted to locate the more in-depth summaries of the following case studies in that theoretical chapter rather than in this introduction, as would be customary for such a volume. In this introduction, we instead invite the reader to explore how kn/own/ables govern all knowledge ownership, at all times, and how this knowledge ownership is continuously manipulated—including how such manipulations set the premise for science and law today. This approach makes visible the possibilities—and benefits—that come from the work of scholars engaging in and with knowledge ownership. The book thus ultimately also aims to shed light on scholarship and its use as a powerful tool in the politics of knowledge, property rights, and knowledge ownership.

THREE PRACTICES AND THE PREMISES OF KNOWING AND OWNING

For historians of science and technology, anthropologists, and sociologists, what we identify as the three major practices of knowledge ownership—*performance*, *use*, and *naming*—resonates with research on different ways of knowing or owning in practices, in bodies, in materials, or through codification. Let us return for a moment to our introductory example of tanning to illustrate how we approach these three practices of knowledge ownership as kn/own/ables and how the mutual conditioning of knowing and owning plays out in different scenarios. In *performing* tanning, the tanner at once knows and *has* (owns) that part of tanning knowledge. In pointing to bodily performance as practice, sociological, anthropological, and historical scholarship into the twenty-first century has addressed both the act of knowing and owning. Yet, it has examined these acts only as separate instances,[6] whereas in fact, knowing and owning happen simultaneously. Scholars have recognized that an agency of knowledge can be inscribed into an object—for instance, Bruno Latour acknowleged this in the case of using scientific instruments.[7] *Using* makes an inscription that entails both knowing and owning. When the tanner uses the knife, they know and own the use of the knife; when they wield the knife, they know and own through performance. When the tanner knows a recipe, they have (own) the knowledge that is *named*.[8] We can see how each practice grasps an instance of a specific part of knowledge that can be known and owned: the kn/own/able. We can also see how our description has highlighted certain moments with regard to specific material instantiations. In performance as practice, knowing and owning are collapsed in the tanner's *body*. Use as practice collapses knowing and owning into the knife *object*. Naming as practice collapses knowledge and *word*.

We introduce a new terminology here—of kn/own/ables, mutual conditioning, and the three practices—to highlight the process of dis-/enabling that governs all ownership of knowledge. All three of these practices are practices of kn/own/ing, which past and present actors mobilize to produce knowables and ownables and distinguish them from each other. We describe such dis-/enabling as processes of *decision-making* to pinpoint one consequence of this mutuality: that actors condition possibilities of knowing and owning irrespective of how and when they make knowledge or ownership explicit, from the first performance, use, or naming, as well as in the refusal to perform, use, or name, or in the act of forgetting. Because of the mutual conditioning that governs all these practices, actors can also own knowledge in other ways, such as by ignoring or acknowledging only the knowing or owning part in each of these practices, and/or by silencing or emphasizing one or the other practice; or by reconsidering what instantiation, material or immaterial, matters most.

Listing these three practices in reverse order to how most scholars are accustomed serves as a reminder that regional and epistemic histories have prioritized codification—defined occasionally quite narrowly as the systematic, rule-based organization of knowledge in writing—over practices and materials when it comes to knowledge ownership, whereas in fact, every knowledge ownership is always a result of all three practices working *in combination*. We can see the mutual conditioning and combined usages of all three practices governing all negotiations of knowledge ownership, past and present; and we find a rich variety of blends of how actors silence or emphasize one or the other of these practices for knowledge ownership. An early third-century Chinese philosophical text, for instance, showcases the cook Ding performing the knowledge he owns while his lord, Wenhui, learns the Way (*dao* 道) only when Ding describes his performance verbally. Cook Ding impresses his lord by cutting an ox up in perfect rhythm and time, "as if performing the [sacred] dance of the mulberry grove (*sanglin* 桑林) in a proper blend with the dancing rhythms of [the music of] the capital cantata (*jing shou* 經首)."[9] Ding knows through his body's performance and through his use of the knife. Grasping the entire animal's body, he sees the spaces between the flesh and bones. He does not operate with his eyes alone. Instead, understanding the Way lies in his whole being. His "faculties knowing where to stop (*guan zhi zhi* 官知止)" is a form of owning that he further enforces by rejecting Lord Wenhui's attempt to *name* his proficiency as *skills* (*ji* 技)—which would denote lesser abilities—and by *naming* it instead as "behavior guided by divine forces (*shen yu xing* 神欲行)."[10] In contrast to Ding's complex claiming of knowledge and its ownership, Lord Wenhui admits that his understanding and approach to the highest form of *zhi* 知 (i.e., *Erkenntnis* as knowledge and cognition), as quoted above, relies on Cook Ding's explanations: "By listening to Cook Ding's *words*, now finally I can learn the Way to lead a life in sound health."[11]

In this example, Cook Ding and Lord Wenhui negotiate knowledge ownership by claiming different combinations of knowing and owning the Way—through words, bodily performance, and the use of the knife as a tool. But whereas Ding accesses the Way through all three practices, the king achieves it through naming alone. How then do we understand the historical and social differences that exist in practices of knowledge ownership and the manner by which they come to be authorized?

THE KING'S WAY AND THE ROLE OF DOMAINS

This story of cook and lord brings up a third point in the mutual conditioning that this book explores, which we address as the historically changing role of three *domains*—namely, society, economy, and epistemology—for the authorization of knowledge ownership practices. Like the three practices, this trio of domains has been central in scholarship on knowing, especially as the history of technology and science and technology studies (STS) about the validation of knowledge ownership look out from our world (of science and law) to others.

When viewed from such "other" worlds, the roles of these three domains for knowing and owning are highly variable. In third-century China, cook and lord both negotiate the kn/own/able, without breaking apart knowing and owning. Both rely on the fact that the practices of knowledge ownership they invoke can be authorized in any one of these domains. Cook Ding, for instance, owns his knowledge in the sense that he can decide whether to *share* such enactments with others, or not. He controls his relationships with others, so he knows and owns in the domain of society. He also owns by understanding—or to put it more literally, by grasping—the *use* of the knife. Thus, he makes a living and owns his knowledge in the domain of economy. And Cook Ding, like Lord Wenhui, also knows and owns by explaining and *naming* his knowledge, thus operating in the domain of epistemology. In the third century, for both lord and cook, society, economy, and epistemology were all *equally* authoritative for claiming knowledge ownership.

Certainly, in the reality of third-century China, power relations mattered in this complex negotiation of knowledge ownership, too. Lord Wenhui ultimately wielded the political power and social status to "own" Cook Ding and thus control his body and work. However, Cook Ding and Lord Wenhui's story is informative in unexpected ways when considering that Wenhui is a king seeking advice about good government. It is thus the cook who, in aligning all three practices, *has* the Way and educates Wenhui. The lord makes no effort to mirror and *have* Cook Ding's comprehensive—embodied, physical, and mental—knowledge of the Way. He is content with owning the Way through description alone—and because his practice is governing, not cooking.

Many early Chinese texts of this period depict lords and kings being challenged by practitioners' complex ways of knowing and owning, and learning from them: "Yi Yin was a bartender, Tai Gong a butcher, Guanzi produced leather, Boli Xi was a slave clerk. But when within the four seas there was disorder, they stood up to be the teachers of the age."[12] Inasmuch as power hierarchies—politics—came to play an increasingly dominant role in knowledge ownership in China during the subsequent centuries, however, elites past and present have emphasized such early examples of complex negotiations around knowledge ownership in favor of the "king's way." A pattern emerged here as successive generations of scholars accordingly came to interpret the practitioners' status not only as a social failure but also as politically powerless in their own time, thereby downplaying that such early sources indeed depict practitioners as wise men, while also suggesting instead that practitioners were wise men humbly remaining in the low status of practitioners until needed.[13] Such interpretations of an increasingly scholarly literati-elite throughout the centuries *also* came to highlight one singular practice of knowing and owning—to wit, naming—and constituted the *explicit* product of the mind, words, as being the highest epistemological standard.

While this Chinese approach to a king's way is a historically distinct case (and should not be essentialized as cultural), it showcases that in many cases, past and present, power is given to one way of owning and knowing (i.e., a king's way), and that elite actors also predominantly lessen the role of performance and use as knowledge and ownership practice, assigning naming a key role to control knowledge socially and politically and reap its economic benefits.

MANIPULATING OWNERSHIP OF KNOWLEDGE: LAW AND SCIENCE AS REFERENCE POINTS

One important difference becomes evident when comparing the abilities and possibilities of owning knowledge in third-century China and in our contemporary world: our world is bound to two reference points in a way that the cook and lord were not. Since (at least) the mid-nineteenth century, law has come into the picture and science has dominated debates on knowledge and epistemology all around the globe. Law matters because, even as modern states rely on it to equalize the power relation between the lord and the cook's ability to *know*, its introduction has meant that the king's way of *owning* knowledge, naming, has become the major reference point for owning *all* knowledge. This means that with the increased efforts to protect all kinds of knowledge *by* law, actors have made naming mandatory for owning *all* parts of knowledge as property, because law requires knowledge to be named—for instance, as performance or use—in order to qualify as knowledge that can be legally owned.[14] Science matters because modern law (i.e., as performed in courts and referenced in public debates) adopts scientific discourse as the highest standard for the ownership of knowledge—both when it constructs modern copyright from a historical concern about scientific publishing and even more so when it asserts nature or natural as different from man-made and distinguishes conditions, materials, and artifacts as criteria for defining what can be patented or branded, or called an invention, innovation, or discovery.[15]

In our theoretical chapter, we trace in greater detail how the fixing of owning by naming to the discourse of law, as we find in knowledge politics and scholarly debates, has favored science even over technology as the major reference point or standard for ways of knowing. In this introduction, however, we need only note that we consider it crucial to understand that these developments have led actors in the contemporary world to lose sight of the kn/own/able and the mutual conditioning of knowing and owning, and to begin to own knowledge by highlighting or ignoring the three practices' roles as *either* knowing *or* owning. For this reason, too, actors in our modern world can *manipulate* knowledge ownership in a new fashion. We deliberately use the word *manipulate* here to point to political action when actors attribute certain practices of owning and knowing only to certain domains—and then draw new relationships between knowing and owning *as if* they were not mutually conditioned.

This can be seen, for instance, in a recent case related to Chinese leather tanning practices. In 2007, the international *China Daily News* English website celebrated the excavation of a tannery site in Rome, "the largest ever found."[16] At that time, China, which historically defines itself as a culture rooted in silk, had not yet taken full ownership of its fur and leather traditions as aspects of its cultural heritage or intellectual property. When the article was published, a negative view of leather as a technology "made in China" with a high degree of pollution dominated the daily newspaper landscape in the People's Republic of China. Wenzhou, an important center for industrial leather production since at least 1898, subsequently resituated its factories, forced major stakeholders to invest in improving their cleaning facilities, and suspended more than one hundred companies in 2007 alone.[17] At the same time, though, the leading citizens of Wenzhou attempted to improve the reputation of leather technology by reframing it as knowledge and enthusiastically engaging in China's launch of a national scheme to promote, and an international scheme to protect, intangible cultural heritage (ICH) (*feiwuzhi wenhua yichan* 非物質文化遺產).[18] After the *China Daily News* article was published, three specific tanning techniques were added to the list of Chinese ICH.[19]

In this example, actors manipulate the ownership of knowledge by shifting how tanning is known or owned between different domains, while keeping in view that science and law have now become the gold standard for making ownership claims. The tanner as craftsman practiced knowledge socially as part of a particular community. Politicians claimed this as a knowledge practice to be owned as culture—indeed, as a form of culture that could no longer be defined as the proprietary knowledge of an individual. In order to enable legal ownership of such knowledge, as something not separable from the community, it was regulated as intangible cultural heritage. Through a series of naming actions, tanning was thus transformed from being knowledge that was performed and owned by singular bodies into a good owned by a collective as national asset. Yet, the part of tanning that was undesirable remained unnamed, and society thus also exercised ownership by sidelining the tanner's body along with his knowledge. This knowledge gained economic credence as a technology that a master could perform and own legally as an industry, while the act of tanning itself, again inseparable from the body of the tanner, had to be performed as labor, in the economy. This example further highlights how consumers matter, in that they take ownership of tanned leather products as traditional craft, as local or global, and/or as the result of highly polluting technology or sustainable new sciences.

THE SCHOLARLY CONTRIBUTION

Our footnoting in the above passage flags the ways in which historical, anthropological, and sociological scholarship (and sciences examining the past) were actively involved in these debates in attempting to promote or protect craft knowledge. Our intention is to highlight that it was not only politicians who occasionally mobilized academic studies to verify "Chinese" crafts or ancient sciences. Contemporary physical material scientists, for instance, sequenced the chemical composition of a tanning solution used by one of Wenzhou's companies against that used by another and unlocked recipes for tanning obtained from archaeological excavations on ancient sites—which were then given credence as historical material or even a trace of ancient scientific knowledge that was worth owning culturally and could be owned legally, too.

This interference of scholarly academic work—its active *and* implicit repercussions—is thus where our interest in mutual conditioning coalesces with our practical concerns about the current scholarly debates on knowledge equality or epistemic injustices and the role of scholarly work in the politics of knowledge ownership. In analyzing knowledge, scientists as well as historians, anthropologists, and sociologists become part and parcel of manipulations of knowledge ownership, too—sometimes deliberately, but often implicitly through the nature of their research methods or interests. This can happen, for instance, as simply as in a desire to elucidate the "science" behind the material manipulations of what others name craft, or by analyzing texts as epistemological objects and denoting performance as something identified as *tacit*—that is, as not normally explained in words or formula. In the Chinese politics of treating tanning as cultural heritage and industry, science and technology matter as the highest standard of knowledge for making it possible to transfer the legal ownership of a chemical composition's sequencing, even as society as a whole ignores the actors who performed this practice or distances itself from its other aspects, such as flaying the animal and tanning the hide.

Scholars moreover affect the possibilities of owning knowledge when they object to notions of indigeneity and of traditional or local knowledge, as well as when they subtly shift attention to the importance of such practices as scientific or assign agency to materials and animals by calling them ontological. What is even more important to highlight here—because it remains much more implicit—is that (our) scholarship *always* and *inevitably* interferes with knowledge ownership, even if (we as) scholars do not wish to do so, and that such interferences arise often long *before* politics come into play. This happens both when a scientist takes interest in one or the other material of an archaeological excavation and when a scholar in historical, anthropological, or sociological research *equates* historical or cultural practices of *owning* knowledge to contemporary notions of science and law, thereby relegating the preparation of the animal skin—and all the skills and knowledge required for it—entirely to the realms of economy and society, outside law and science. And it also happens when they *refuse to equate* historical or cultural practices of *owning* knowledge to contemporary notions of science and law and thereby also refuse to illustrate alternatives.

This is not least, then, a book with a reflexive scholarly aim: to shift the perspective within the academy regarding its own role in manipulating the relation between knowing and owning. If law is about justice, then the responsibility of a historian, sociologist of knowledge, or anthropologist is to reveal tensions in our normative frameworks of producing ownership of knowledge that come from privileging one practice of knowledge ownership over another. This is the core politics of the book.

Notes

1. Wiebe E. Bijker, Thomas P. Hughes, and Trevor Pinch, eds., *The Social Construction of Technological Systems: New Directions in the Sociology and History of Technology* (1987; repr., Cambridge, MA: MIT Press, 2012). For approaches to sciences, see Bruno Latour, *Science in Action: How to Follow Scientists and Engineers through Society* (Milton Keynes, UK: Open University Press, 1987); Donald MacKenzie and Judy Wajcman, eds., *The Social Shaping of Technology* (Buckingham, UK: Open University Press, 1985).

2. For representative approaches to practices from the viewpoint of technology and anthropology, see, e.g., Nelly Oudshoorn and Trevor Pinch, *How Users Matter: The Co-construction of Users and Technology* (Cambridge, MA: MIT Press, 2003). Stephen Shapin and Simon Schaffer have foregrounded practices in the sciences. Shapin and Schaffer, *Leviathan and the Air-Pump: Hobbes, Boyle, and the Experimental Life* (Princeton, NJ: Princeton University Press, 1985). For an approach concentrating on words and semantics instead, see, e.g., Karine Chemla and Jascques Virbel, eds., *Texts, Textual Acts and the History of Science* (Cham, Switzerland: Springer, 2015), which, following speech act theory, identifies the structuring of words and information in texts from mathematics to zoology.

3. Karl Marx tackled alienated labor in *Ökonomisch-philosophische Manuskripte aus dem Jahre 1844* (Berlin: Contumax-Hofenberg, 2017). In his later work he described such alienation in various instances, such as the manner in which society becomes increasingly mediated by technology with large-scale industry. Karl Marx, *Grundrisse der Kritik der politischen Ökonomie* (*Rohentwurf*, "raw draft," 1857–1858; first published 1939–1941, Marx-Engels-Lenin Institute, Moscow; repr. 1953, Dietz Verlag, Berlin, GDR; new ed. 1983, MEW 42, 47–768, Dietz Verlag; and MEGA II.1–2, 49–747). For an example of later debates, see Sean Sayers, *Marx and Alienation: Essays on Hegelian Themes* (London: Palgrave, 2011).

4. For Foucault's naming function as a means of giving forms to modes of knowledge, see Michel Foucault, "The Subject and Power," *Critical Inquiry* 8, no. 4 (1982): 777–795, esp. 781.

5. Sheila Jasanoff, *States of Knowledge: The Co-production of Science and the Social Order* (London: Routledge, 2004), 2.

6. Andrew Pickering, "From Science as Knowledge to Science as Practice," in *Science as Practice and Culture*, ed. Andrew Pickering (Chicago: University of Chicago Press, 1992), 1–26. For a recent review of this scholarship and its relation to STS studies, see Simon A. Cole and Alyse Bertenthal, "Science, Technology, Society, and Law," *Annual Review of Law and Social Science* 13 (2017): 351–371, https://doi.org/10.1146/annurev-lawsocsci-110316-113550.

7. Latour, *Science in Action*.

8. Propositional, codified, explicit knowledge that can be named in words. Harry Collins, *Tacit and Explicit Knowledge* (Chicago: University of Chicago Press, 2010).

9. *Zhuangzi* 莊子, chap. 3, our translation. Zhuangzi lived 369–286 BCE. Both dance forms/music pieces appear in the classics, but it is mostly later thinkers who attributed them to eras of high antiquity when dance and music were used as a cosmological tool to enhance their ritual authority. For a standard interpretation of Zhuangzi, see Chen Guyin 陳鼓應, *Zhuangzi jinzhu jinyi* 莊子今註今譯 (Beijing: Zhonghua Shuju, 1983).

10. *Zhuangzi*, chap. 3.

11. *Zhuangzi*, chap. 3, our translation. Multiple different translations exist, many highlighting different implications—a point that we also address in the theoretical chapter of this book. For the current standard translations, see Wm. Theodore de Bary and Irene Bloom, eds., *Sources of Chinese Tradition*, 2nd ed. (New York: Columbia University Press, 1999), 1:103–104.

12. He Guanzi 12:83/1–4, quoted with minor adjustments from Carine Defoort and Ho-kuan tzu, *The Pheasant Cap Master (He guan zi): A Rhetorical Reading* (Albany: State University of New York Press 1997), 129. We have used the literal translation "teachers of age/teachers of the generations" here to indicate to the importance of these men as transmitters of knowledge rather than their roles as political advisers, as most later Chinese comments emphasize.

13. For the role of craftsmanship in pre- and early imperial state literature and philosophy, see Michael Puett, *The Ambivalence of Creation: Debates concerning Innovation and Artifice in Early China* (Stanford, CA: Stanford University Press, 2001).

14. It is important to note here that the issue is not that performance and use are not protected or governed by law; they are. What is missing is that they do not act as practices of knowledge ownership; instead, law protects them as ways of knowing.

15. Mario Biagioli and Marius Buning, "Technologies of the Law/Law as a Technology," *History of Science* 57, no. 1 (2019): 3–17. See also Mario Biagioli, "Nature and the Commons: The Vegetable Roots of Intellectual Property," in *Living Properties: Making Knowledge and Controlling Ownership in the History of Biology*, ed. Jean-Paul Gaudillière, Daniel J. Kevles, and Hans-Jörg Rheinberger (Berlin: Max Planck Institute for the History of Science, 2009), 241–250.

16. See "Archaeologists Excavate Ancient Tannery," *China Daily*, last modified August 1, 2007, http://www.chinadaily.com.cn/world/2007-08/01/content_5447206.htm. The same story also made

it on *CTV* News; see "Archaeologists Excavate Ancient Tannery," *CTV News*, last modified August 1, 2007, https://www.ctvnews.ca/archeologists-excavate-ancient-tannery-1.251027. Some websites from the United States such as *Washington Post*, *NBC News*, and Sott.net featured the news in sections such as "Tourism" or "News of the World," but not in the general news sections.

17. Most articles mention 126 companies. Not all pursued tanning, though. In Shuitou 水头镇, Wenzhou 温州, 162 leather companies were suspended. For a summary, see "Liangge 'lianxü sannian' zheshe Shuitou zhibian 两个'连续三年'折射水头之变" [Two "consecutive three years" reflect the changes in Shuitou], Pingyang xinwen wang 平阳新闻网, last modified December 31, 2009, http://py.66wz.com/system/2009/12/31/010162648.shtml. While several articles are no longer available, websites feature initiatives of local officials. See also "Zhong wuran qiye weihe guan er bu ting?" 重污染企业为何关而不停？ [Why were the enterprises with heavy pollution not completely shut down?], Sina, last modified August 14, 2007, http://news.sina.com.cn/c/2007-08-14/112912382093s.shtml. For an overview of Wenzhou's campaign since this period, see Wang Jinyan 王金燕, "Wenzhou feiwuzhi wenhua yichan de zhishi chanquan baohu yanjiu" 温州非物质文化遗产的知识产权保护研究 [Research on intellectual property protection of Wenzhou intangible cultural heritage], *Zhejiang gongmao zhiye jishu xueyuan xuebao* 4 (2011): 47–51. The notion of Wenzhou's "long tradition" in leather production was invoked repeatedly in newspaper and scientific articles since at least 2001. See Li Minxiao 李民校, "Wenzhou bei guanming 'Zhongguo xiedu'" 温州被冠名"中国鞋都" [Wenzhou was called "the capital of shoes in China"], *Xibu pige* 8 (2001): 9.

18. Originally, ICH was set up to help protect diverse local customs. By 2007, Chinese attempts increasingly targeted arts and crafts. Shadow puppet theater was added to the Chinese National List of ICH in 2006; see Chinese National List of ICH—First Batch 国家级非物质文化遗产名录——第一批, Sequence no. 235, item no. IV-91, Chinese Cultural Studies Center, accessed April 23, 2020, https://www.culturalheritagechina.org/national-list-first-batch. In 2011 it was also inscribed on the UNESCO Representative List of the Intangible Cultural Heritage of Humanity; see "Chinese Shadow Puppetry," UNESCO ICH, accessed April 23, 2020, https://ich.unesco.org/en/RL/chinese-shadow-puppetry-00421. An extraordinary session was held in 2007 in Beijing. The documents were not made public.

19. No. 894–897, accessed April 23, 2020, http://www.ihchina.cn. In this first inclusion, China's view of leather favored "minority" practices. In the following decade, it shifted to also include "Chinese" knowledge.

Bibliography

Bary, Wm. Theodore de, and Irene Bloom, eds. *Sources of Chinese Tradition*. 2nd ed. Vol. 1. New York: Columbia University Press, 1999.

Biagioli, Mario. "Nature and the Commons: The Vegetable Roots of Intellectual Property." In *Living Properties: Making Knowledge and Controlling Ownership in the History of Biology*, edited by Jean-Paul Gaudillière, Daniel J. Kevles, and Hans-Jörg Rheinberger, 241–250. Berlin: Max Planck Institute for the History of Science, 2009.

Biagioli, Mario, and Marius Buning, "Technologies of the Law/Law as a Technology." *History of Science* 57, no. 1 (2019): 3–17.

Bijker, Wiebe E., Thomas P. Hughes, and Trevor Pinch, eds. *The Social Construction of Technological Systems: New Directions in the Sociology and History of Technology*. 1987. Reprint, Cambridge, MA: MIT Press, 2012.

Chemla, Karine, and Jacques Virbel, eds. *Texts, Textual Acts and the History of Science*. Cham, Switzerland: Springer, 2015.

Chen Guyin 陳鼓應. *Zhuangzi jinzhu jinyi* 莊子今註今譯. Beijing : Zhonghua Shuju, 1983.

Cole, Simon A., and Alyse Bertenthal. "Science, Technology, Society, and Law." *Annual Review of Law and Social Science* 13 (2017): 351–371

Collins, Harry. *Tacit and Explicit Knowledge*. Chicago: University of Chicago Press, 2010.

Defoort, Carine, and Ho-kuan tzu. *The Pheasant Cap Master (He guan zi): A Rhetorical Reading*. Albany: State University of New York Press 1997.

Foucault, Michel. "The Subject and Power." *Critical Inquiry* 8, no. 4 (1982): 777–795.

Jasanoff, Sheila. *States of Knowledge: The Co-Production of Science and the Social Order*. London: Routledge, 2004.

Latour, Bruno. *Science in Action: How to Follow Scientists and Engineers through Society*. Milton Keynes, UK: Open University Press, 1987.

Li Minxiao 李民校. "Wenzhou bei guanming 'Zhongguo xiedu'" 温州被冠名"中国鞋都" [Wenzhou was called "the capital of shoes in China"]. *Xibu pige* 8 (2001): 9.

Marx, Karl. *Ökonomisch-philosophische Manuskripte aus dem Jahre 1844*. Berlin: Contumax-Hofenberg, 2017.

Oudshoorn, Nelly, and Trevor Pinch. *How Users Matter: The Co-construction of Users and Technology*. Cambridge, MA: MIT Press, 2003.

Pickering, Andrew. "From Science as Knowledge to Science as Practice." In *Science as Practice and Culture*, edited by Andrew Pickering, 1–26. Chicago: University of Chicago Press, 1992.

Puett, Michael. *The Ambivalence of Creation: Debates concerning Innovation and Artifice in Early China*. Stanford, CA: Stanford University Press, 2001.

Sayers, Sean. *Marx and Alienation: Essays on Hegelian Themes*. London: Palgrave, 2011.

Shapin, Stephen, and Simon Schaffer. *Leviathan and the Air-Pump: Hobbes, Boyle, and the Experimental Life*. Princeton, NJ: Princeton University Press, 1985.

1

EXCAVATIONS OF KNOWLEDGE OWNERSHIP: THEORETICAL CHAPTER

Annapurna Mamidipudi and Dagmar Schäfer

Let us begin by restating the basics of our proposition: knowledge is an activity that always involves the body and the mind and, hence, its ownership can be authorized by manipulating the different practices of *use*, *performance*, and *naming*. This chapter builds on the preliminary thoughts outlined in the introduction and offers a theoretical framework for historians, sociologists, and anthropologists who wish to excavate the explicit effects and the subtle implications of the *mutual conditioning* of knowing and owning and to trace the many processes through which knowledge is made property.

We call our approach to the analysis of knowledge ownership an excavation, picking up from Michel Foucault's archeology of knowledge. Like Foucault, we approach knowledge as an understanding that is justified and ascertained by experience. But because our interest differs from that of Foucault, so too does our starting point. For Foucault, knowledge was inseparable from discourse. He thus started with "words" to trace the set of "things said" about knowledge in all its interrelations and transformation, defining *knowledge* as "the space in which the subject may take up a position and speak of the objects with which he deals in his discourse . . . knowledge is also the field of coordination and subordination of statements in which concepts appear, and are defined, applied and transformed . . . lastly, knowledge is defined by the possibilities of use and appropriation offered by discourse."[1] Unlike in the Foucauldian world, however, in the real world, knowledge and discourse do not entirely coincide. In the formation of knowledge ownership in the real world, power lies exactly in the processes that make knowledge a discursive matter or not: using or not using, performing or not performing, saying or not saying can all be acts of owning knowledge. Precisely because knowledge ownership is established through processes of distinction, we use the heuristics of technology studies. Or, seeing how knowledge equally concerns matter, body, and mind,[2] we could say that we suggest applying the methods of archeology throughout our investigations in order to see that *objects*, *bodies*, and *words* are all equally relevant *material instantiations* of knowledge ownership.

In our introduction we coined the term *kn/own/able* as the yet-to-be-distinguished knowable and ownable that marks our entry point. In this theoretical chapter we first show how to set up the excavation site, how to use tools, and the ways to investigate and identify the stratigraphy, the find and the "fill." As we are digging into the same soil where others before us have worked, the second part of this chapter traces how we have come to understand these issues through the case studies of the book's individual chapters. In a third section we offer a selective reading for those who are interested in seeing how technology studies and scholarship on knowledge, economy, and law have critically engaged with the quasi-ontological status that has been given to some surrogates or proxies for knowledge—as things that can be owned as "product" (in economics) or "property" (in law) in our modern world. In scholarship or scientific research, as in public speech, this quasi-ontological status pertains to the ownership of responsibility for words that is attributed to an author or subject.

THE MAKING OF THE SITE: FLAGS AND MAPS

Archaeologists follow leads. While in the early days, digs mainly targeted solid artifacts as "evidence," researchers nowadays start by flagging pertinent points of intervention and then mapping out the field. When one is interested in the processes that make knowledge ownership possible or not, it is important to begin with a yet-to-be-opened ground that holds bodies, objects, and words, as the material instantiations of the practices for owning knowledge that guide our investigations: *performance*, *use*, and *naming*. Similar to how archeologists carefully identify the find, stratigraphy, and soil that has to be moved away as "fill," the roles or meanings of objects, bodies, and words must be understood by a combined analysis of the properties of each trace and their relative spatial positioning. The researcher of knowledge ownership, like the archeologist-to-be, may want to be cautious about "the use of the word 'natural' as a synonym for the undisturbed subsoil."[3] What has perished and withered away or left an empty spot has effects; absences matter as much as presences, or different evidences have different shelf lives; some leads such as words or objects are persistent, while bodies fade, and part of knowledge can be lost; even if words and objects survive, knowledge must be regained in each body, over and over.[4] Knowledge ownership is therefore bound to the nature of the body—and the body learns, shares, forgets, dies. This is to say, in other words, that kn/own/ables are socially generated and have material effects. They can be turned into other kn/own/ables or even not-kn/own/ables too, with time.

Flags are our tool of choice to mark how, in any scholarly analysis of knowledge ownership, space and time matter. Anthropology, the various historical disciplines, and science and technology studies (STS) have all highlighted ethical concerns about the analyst's role and have studied the material manifestation of decision-making, as actants or as ontological effects with regards to the politics of knowledge.[5] Flagging can indeed be understood as a form of "judgment" (*Urteil*) in a Kantian sense that emphasizes the ethics and effects of such an intervention by the archeologist, and hence also displays the values the archeologist holds.[6]

When taking not only knowledge but also its ownership into account, we suggest seeing flags as a decision-making tool that works in two directions: they pierce down to specific traces of moments generated by past actors and thereby elevate a specific moment of the analysts' knowledge-making. Flags set a target or define where evidence could be, even if this knowledge is intuitive or the evidence is not yet known. In our excavations, they mark how one *word* is differentiated and named as concept from the multiple others; how one *object* turns into a model when being inscribed with particular functions that determine its *use*; how one *body* becomes expert over others who mainly practice or work, in *performance*. That each flag can be moved reflects that the expectations of the analyst are not always met and that part of the analysis is to align analyst expectations with the yet-to-be-discovered fact. Despite such flexibility, however, the placement has material effects. When it pierces through layers of time, flagging moves other materials and creates new "false" layers that were not previously there. With each flag, not one but many claims are made. With many flags, a field site emerges. And at this point, we suggest that the excavator pause for a first survey of how the analyst demarcates the terrain through their intervention, thereby also determining what *can* become kn/own/able and what remains not-kn/own/able.

At the end of this book, Vivek S. Oak, Jörn Oeder, and Annapurna Mamidipudi offer a diagrammatic, hands-on guide to analyzing the different ways by which terrains have been demarcated in a global world of knowledge ownership. Here we elaborate how individuals and cultures have always designated spaces/areas and/or moments that should or could not (yet) be known and owned—not in the sense of a flat, two-dimensional dichotomizing principle; rather, there is a gradual progression in which actors (1) peel or core, (2) partition, or (3) slice through the three-dimensional globe that is knowledge ownership. Tracing these techniques further allows one to question the relative size of kn/own/able and not-kno/wn/able within the globe.

The peeling and coring is about the dividing lines between the kn/own/able and what is not (yet) known. It is always possible to push the dividing lines to the extreme inner or outer rim, which reflects a world in which everything can be kn/own/able or, inversely, in which no knowledge is actually ownable at all. When that happens, we can see how, in producing the kn/own/able, actors also produce the not-kn/own/able—there is no moment when a kn/own/able exists by itself. Lines of demarcation between such spherical layers can be blurry, leaving unclear the boundaries between that which is known as un-kn/own/able or entirely un-kn/own/ed. In cases where knowledge was not yet known but its ownability was considered relevant, it has always been expressed by individuals as uncertainty or claimed as magic. This made knowledge something owned or not ownable to all by naming it, for example, as sacred or profane. Others have asserted that knowing is mainly a human capacity (different from, e.g., nonhuman agency) and hence ownable only by individuals or social organizations. The excavator, too, causes such a distinction in demarcating their interests and what—for now—cannot or will not be touched or might be left to be found in the future (e.g., an HIV vaccine).

The partitioning addresses the fact that kn/own/ing becomes expressive through three different practices (use, performance, naming) and is instantiated through the body, object, and word as kn/own/able—and conversely, in the ways that the not-performing body, a knife not used, or a thought not named cannot be kn/own/ed. Unlike in the Foucauldian world, in the real world these three material instantiations of kn/own/ables are predominantly associated with three areas of human activity: the performing body with *society*, the naming of words with *epistemology*, and objects and their use with *economy*. Or we could say that this association of practices and instantiations of not-kn/own/ables or kn/own/ables to such areas creates what we call *domains* of knowledge ownership. Actors have always prioritized some associations between practices, instantiations, and domains over others to define what could be legitimately known and owned where and when—deciding that bodies signify social sites; or that objects portend property in markets; or that an accumulation of texts indicates archives or libraries of knowing. This fixation of a tripartite structure can also be seen as an artifact of the present intellectual formation of the social sciences and the humanities and their categories of science, law, and property. For the analyst, it is important to see that any legitimizing association between the three instantiations and three domains is variable and also not necessarily mutually exclusive: bodies can identify burial sites (social), but also markets (economic); objects can signify markets (economic) and be seen to embody knowledge (epistemological) or culture, or constitute a factor in our environment.

Other domains are therefore also possible. But inasmuch as society, economy, and epistemology are the triptych (i.e., the core and two ends) of all knowledge ownership, we suggest using them as a heuristic scaffolding to understand how the third practice of slicing is regularly employed to create flattened worlds where the kn/own/able can be split into knowable and ownable that live in different domains. To understand the power of a flattened view—of slicing through complex global worlds of knowledge ownership—it helps to keep in mind that any denial of one practice, instantiation, or domain as kn/own/able or not-kn/own/able, or any ignorance toward the *inseparability* of knowing and owning, is an exertion of power that constitutes a new regime of knowledge ownership.

The cases in our book illustrate plural regimes of knowledge ownership, past and present. In some, actors like Cook Ding, decide to operate within a regime in which all kn/own/ables have to be synchronized; in others, only one domain can have kn/own/ables, such as the social sphere (Leach). Or we see that kn/own/ables are defined cyclically (Mamidipudi and Viren), or discontinuously (Jackson, Slaton), or in multiples (Bol). For all cases in our book such issues apply to varying degrees; yet for all cases, at one point, actors give power to one particular historical combination of peeling/coring, partitioning, and slicing. In this moment, this combination becomes central for knowledge ownership, as one moment of knowledge ownership is given power over another.

This is also where the fourth dimension—that is, the role of time—for staking knowledge ownership claims comes into play. In fixing their view on a certain moment, actors are able to highlight one or the other practice, instantiation, or domain as a case of either "knowing" or "owning," or both. Actors can then apply a linear model of time to establish a temporal or hierarchical causality between practices, instantiations, and domains of either knowing or owning—also inasmuch as some instantiations of kn/own/ing last longer in time than others; words and objects persist, even when the body and the practice perish.

There is a wide (though not universal) acceptance across time and cultures that, because of such perishing and persistence, some relations between practice, instantiations, and domains are somewhat "natural"—more "permanent" and "stable"—and therefore "more" legitimate than others. For instance, many cultures have come to prioritize the association between words and naming, assuming that knowing persists through texts and can be more easily retrieved, whereas the knowledge of use or performance perishes with the body. Emphasis is then placed on the idea that bodily knowledge or knowing of performance and use always *needs to be* repeated/retrieved in order to be owned. In fact, though, all knowing relies on constant training—so that naming, for instance, means speaking things out, or performing reading and using writing. Others have come to identify knowing and owning as separate acts by saying that owning an object is not concerned with knowing and thus is solely about property, whereas in fact, all ownership affects knowing, as the actor can hinder other people from using and knowing this object.[7]

Such temporal distinctions hinge on the legitimizing function that domains have for certain practices of knowing and owning or their instantiations. For instance, epistemology is the domain where words that name things or actions can be legitimately owned as knowledge; in this domain, in order to own actions performed by one's body as knowledge, the action still needs to be named. Here a distinction emerges between first- and second-order relationships that creates a space allowing/enabling knowledge ownership to be manipulated and performance to be defined as not-knowing and ownable only in society; or for objects to be owned and traded in the marketplace; or for the assumption that, in order to own the knowledge of using such objects, epistemology counts, and hence this use still has to be named. In regimes of ownership where one order of relationship is placed higher in the hierarchy than others, knowledge ownership claims are inherently less easily manipulated, even in the other domain.

Because equality in knowledge ownership matters to us as scholars, we propose approaching every step of this excavation as a process of *manipulation* that is likely to or can facilitate or impede knowledge ownership. The flags with which we began this chapter pierce through various layers of the stratigraphy by fixing a singular moment. So we, too, fix moments as scholars. Giving valence to one moment of knowing, for instance, already interferes with the example we gave in our introduction, namely Cook Ding's way of owning knowledge—that is, "the Way." This "Way" owns knowledge indivisibly through naming a word for its practice, performing it through Ding's body, and using an object—a knife. This is because "the Way" is an ongoing process of acquiring knowledge that also includes the yet-to-be-learned, the yet-to-be-known; for Ding, such a singular moment cannot exist in the first place. Rather, his knowing is always in the liminal space between the production of the kn/own/able and the not-kn/own/able.

In affixing the moment to an actor (which can be a single body or a group), the flag can also be seen as an act of enclosure or individuation. As the cases in our book exemplify, actors in the past and present have come to manipulate knowledge ownership by emphasizing the material instantiations of the three practices—word, body, object—in their first-order relation to domains, and then by acting as if all other knowing and owning comes second—or is even ill-placed or illegitimate. For example, evidence of writing and the presence of books can identify an area as a library or center of learning. As the practice is then defined depending on the domain of knowing, "writing" a text (as practice of naming)—on child-rearing, for example—is legitimized as knowable and ownable. It then becomes easier to make all other evidence for knowing by other practices look out of place—for example, in the claim that a child's crib as the *object used* for child-rearing no longer furnishes legitimate evidence for owning knowledge in the domain of epistemology, because in this domain the use of the crib has to be described in order to count as knowing. Libraries may collect formulas that have been written down, or they may only sample extractions of different chemical components.

Cultures may decide it is not the word but performance that constitutes the primary evidence creating the domain of epistemology—and thus the only legitimate order. As the cases examined in this book illustrate, actors used different temporalities, created and adhered to different orders as "our laws." But then too, all these cases show that at some point, capitalism and Eurocentric legal frameworks have been/are interfering in how we have come to think about such knowledge ownership claims.

We can see that what historians, sociologists, and anthropologists have variously attempted to address as "ensembles" or "assemblages" or "complexities" of knowledge *do not stand outside*, but instead constitute the very core of knowledge ownership regimes. In the same way that the mutuality between knowing and owning is addressed and operationalized in the course of a dig, distinctions are not only made with regard to find, stratigraphy, and soil or fill; rather, analysts detect and interpret patterns of relationships between subjects of research interests based on spatial and temporal relations. Historians and sociologists studying knowledge, technology, and sciences do not need reminding of this; approaches such as actor-network theory (ANT) and the social construction of technology (SCOT) have emphasized the social character of all knowing—and drawn attention to the need to study the sociotechnical, sociomaterial complexes of knowledge, and so on at work.[8]

ANT, SCOT, and assemblage studies have been useful for explaining networks and how the interaction between the whole and its parts "either constrains or enables the parts that compose it to act differently."[9] And yet, from the viewpoint of knowledge ownership, all such research suffers from a major flaw, because its analytical starting point lies after the moment when the kn/own/able is divisible across the social, technical, and material. Such research attempts to "reassemble" what seems to have been broken apart, but in effect could not, cannot, and has not been broken apart to begin with: the kn/own/able. This creates an imbalance.

For us, it is crucial to point out that the most forceful act of manipulation is making invisible how the kn/own/able is always at work, as it allows actors to *condition* knowing by defining owning and to condition owning by defining knowing. Other concepts that attempt to address this relation in terms of simultaneity have chosen terms such as *coproduction* and *coevolution*. However, knowledge ownership regimes are constituted not because knowledge and ownership are produced simultaneously, but rather because once knowledge is defined, the conditions for ownership are determined; the process then ceases to be either open-ended or productive. In a broken world, such conditioning has substantial effects when knowing and owning are split: experimenting in a laboratory produces science, whereas tanning skins in ancient Chinese workshops primarily constitutes culture. In order to then see tanning as knowledge worth knowing and owning individually or as a patentable process, it has to be *re*-defined, *re*-conditioned, or *re*-assembled. In this, the social constitutes the basic prerogative not only for constructing knowledge—as many studies have emphasized—but *always* for knowledge ownership as well; individuals and collectives own by deciding to "share" or not to "share" with their bodies, by way of objects, or through words. All cases in our book illustrate these concerns, and we have arranged them to highlight particular points of these dynamics. Each of the four sections contains up to three cases illustrating variations in the execution of specific techniques.

THE SITE PLAN: READING OUR CASES

In studying worlds of knowledge ownership, we suggest placing emphasis first on the *process* used—by actors and analysts—to manipulate different variables and relations. This focus is reflected in the organization of our book into four sections that highlight four such crucial strategies employed by past and present actors: (1) how actors have historically attempted to operationalize *mutual conditioning* to disable or enable knowing and owning; (2) how actors utilize different combinations of word, body, and object, thereby employing (or neglecting) these three things variously as material instantiations of owning knowledge through the *practices* of performance, use, or naming; (3) the ways in which actors prioritize one practice of knowledge ownership over another by emphasizing a relation to the *domains* of society, economy, or epistemology, in order to authorize owning; and (4) how *scholarship* inevitably manipulates knowledge ownership, inasmuch as "knowledge" is its very business—especially as we enter a world of accreted practices and domains in which science represents the highest standard of knowing and "certain" (i.e., modern, liberal, etc.) laws of property dominate ownership.

In reading through all cases, it also becomes apparent that the four strategies are mostly employed in conjunction, while no strategy is inevitable or its effects completely irreversible. As we briefly touched on in our introduction, though, past and present scholarly influences have meant that some practices—such as naming in words—have historically and regionally been given more room—or significance—than others. In unveiling actors' views, the analytical view also often detects the emergence of knowables that *seem* to be distinct from ownables, whereas in fact, at each level and in every moment, what actually produces regimes of ownership are always kn/own/ables.

Part I: Mutual Conditioning
One way that actors, past and present, regulate the ownership of knowledge is by addressing the structural premises for knowledge ownership and "determining" the

mutual conditioning of knowing and owning. As stated earlier, we introduce mutual conditioning to show how identifying which aspect is known always presets the conditions for how it can be owned, and vice versa.

The two cases in part I of this book, "Mutual Conditioning," tackle the history of book publishing, a topic that is at the center of past and present debates around regimes of knowledge ownership, their practices, and their technologies. These cases pinpoint two extreme ends of book publishing as such a tactic. We can imagine them as an effort to divide up the grounds and award to this partitioning the authoritative role in deciding what is only knowable or only ownable. In chapter 2, Cynthia Brokaw shows how societies decide to favor a logic that does not allow individual knowledge to be owned as economic property; she thus connects to a growing body of literature on other temporalities illustrating similar concerns.[10] Chinese authors promoted an idea or piece of knowledge around what today might be called "moral rights" of ownership of the knowledge, but they were not able to claim "material" or "legal" rights as their European colleagues could in the sense of the ownership of copyright or the right to file a patent, or "profitable ownership" when printers claimed possession of the woodblocks or, in the case of a movable press, the setting up of production facilities. In elucidating how the history of teaching intellectual property in the twenty-first century is based on texts delivered by legal professionals rather than historians, Marius Buning reveals in chapter 3 how time-bound intellectual property actually is. And yet even today, historians and politicians rely on this specific moment of fixation when they chart the field of knowledge ownership. Both cases make clear the role that scholars play in manipulating these relations by fixing knowing and owning through mobilizing concepts.

Part II: The Three Practices: Performance, Use, Naming

A second important technique that is observable in different times and contexts is actors' focusing, in their manipulations, on the authority of the three practices for knowing or owning. In our introduction to this book, we briefly outlined how those belonging to the "elite" of any given society have often prioritized naming and research from among the many possible ways of manipulating knowledge ownership throughout history, even as they engaged with varied combinations of the three techniques—naming, performance, and use.

The three cases in this second section foreground these three different practices of authorizing knowing and owning. The study by Annapurna Mamidipudi and Viren Murthy lays out the ways in which ownership is embedded in practice. Discussing four very different performers, in chapter 4 the authors exemplify how the knowledge of Carnatic music is owned in the performance practices of contemporary classical musicians in South India who have historically used, rejected, or relied on "Western" or "colonial" standards of naming to own their performance as knowledge. Mamidipudi and Murthy emphasize that such knowledge remains valid only when performed by a particular community.

In chapter 5, Marjolijn Bol stresses the agency given to materials in their "use" as the major subject and validation of knowledge ownership. In a consumer market, discourses about what is authentic or fake condition notions of ownership and knowledge theft. Bol points to histories of "material mimesis," such as crystals made to look like more precious gems, and Dutch batiks that were not recognized by Javanese customers as equivalent to the original resin-printed fabrics from their own island, because they ironically judged these products based on a material characteristic that local dyers had been trying to avoid. Later, these Dutch copies found new acceptance in African markets precisely because of this particular process of fabricating the cloth and using the raw materials.

A dominance of naming governs the modern educational system, which functions as a sorting mechanism, as Amy Slaton shows in chapter 6. In a classroom exercise designed to teach a scientific method, students are asked to draw on a graph paper a set of points that they can use to name the unknown object in a box as a banana. This paper bears the name of the student who produced it, who can now own that knowledge, and those who don't have their name on such a paper cannot own that knowledge. This naming functions as a sorting mechanism for knowers and not-knowers. Once thus labeled, the two classes of students are set on markedly divergent trajectories of knowledge ownership.

In each of these three cases, actors rely on the mutual conditioning of knowing and owning through all three practices, but they regulate knowledge ownership by flagging one dominant practice.

Part III: The Three Domains: Society, Economy, Epistemology

In the third section of the book, we tackle the domains of society, economy, and epistemology, which have consequences for how knowledge ownership is enabled or disabled. Some actors, for instance, regulate knowledge ownership by using the authority that one practice has over different domains to enable ownership.

In chapter 7, James Leach describes how Reite people, in Papua New Guinea, assert kinship relations as a purpose in itself. For them, actors cannot own knowledge, but only the social relationship that allows rights of performance. Reite people use knowledge ownership not to produce property, but to create community. The dominant domain in each regime depends on which element is valued as the authoritative form, such as society valuing the body, or epistemology valuing the word. Thus, in performing their knowledge, *knowing* Reite bodies own their knowledge authoritatively in the domain of *society*. Or we could say that in this community, kn/own/ables exist only through bodies that know and own relationships as knowledge.

While the history that Dagmar Schäfer examines in chapter 8 seems to mainly concern naming, this practice was used specifically in legal practice as a means to access and control the labor of craftspeople and their institutions—in this case, their households. It was obvious to imperial authorities in China that craftsmen's bodies owned knowing. Hence, in order to put this labor to use, imperial clerks manipulated its ownership by *naming* or *refusing to name* the craftsmen's expertise as subject to state appropriation via tax. This strategy attempted to claim dominion over *performance* as the authoritative mode of owning knowledge. In this way, Chinese literati effectively denied such expertise any visibility in their political economy *and* epistemology—and thus the scholar whose profession it is to name turns into the major gatekeeper for knowledge ownership.

Science and modern law come into the picture in chapter 9 when Myles Jackson tackles how scientists and politicians have changed possibilities of knowing and owning genes and the products that are derived from such knowledge. The case follows a shift in scientific practice—from a context in which the knowability of genes depends on laboratory practices and the "wet" science of molecular biology, to one in which describing this gene mainly happens in a computer sequence. It shows how a translation of material practices in the sciences is made into a *naming* practice in the domain of epistemology, the products of which then become property by law as "knowledge," in the form of patents. Then law is used to operationalize owning over knowing, thereby foregrounding *economy*.

In all three cases, the kn/own/able is present as the body. Actors authorize knowledge ownership by foregrounding practices within specific domains or shifting them through domains (which we can also imagine as a slicing through layers that creates different cross sections).

Part IV: The Role of Scholarship

Our fourth and last section of the book highlights the role of our own scholarship. Given that knowledge ownership is a process and dynamic, the integrity of the kn/own/able is at stake once practices and domains have been separated and have solidified as either knowing or owning. This has consequences especially for science and law, "the two institutions that, perhaps more than any other, are responsible for making order, and guarding against disorder, in contemporary societies."[11] As in archeology, we are thus left with standing ruins and fields featuring both complete vessels and shards affected by weather, water, and the surrounding soil, which constitute the hard facts of our time.

In our final case study, in chapter 10, Lissant Bolton highlights the material side of such ossifying acts and the challenges museums face in making objects of the Vanuatu "knowable." When something becomes a public object for display, it is no longer "ownable" by the Vanuatu. She explores museums as places in which the multiple instances of regulating the relationship between knowing and owning become politically fraught. She suggests it is important to "disentangle" the political strategies inherent in reparations for colonial wrongs and identity formation through ownership from the work of acknowledging and recording the richness of human knowledge and practice.

In fact, though, all our case studies illustrate how scholarship affects and is part of ownership debates, sometimes making it, as Myles Jackson's study illustrates, even a matter of court debates. Other cases are more subtle, such as when Amy Slaton follows how learners are sorted in the US educational system. In being named as knowers and not-knowers, these learners emerge as "underprivileged," and thus unfit for further scholarship, or "meritorious," defined as achieving validation of their knowledge only through a staging of "apparent" mastery. This case also elucidates how analytics matter with regard to language as the major tool that scholars use to express their knowledge. Additionally, Slaton's case, like Bolton's, is a sharp nod to the politics of exclusion, of constructing the not-knowable—without exposing it as problematic—in the body of the non-knower. Epistemic hierarchies reinforce social sorting, turning teaching in educational institutions from the making of knowers into its opposite: the making of people deemed to be incapable of knowing, and a class of non-knowers.

The sum of our cases shows the implication of scholarly decision-making that produces knowers and intervenes to turn people into not-knowers, in the past and present. Near to the present we can see how the topic of knowledge ownership is limited by our fixation on a particular moment dominated by the language of science and Western legal frameworks. Such cases show how deeply fragmented the shards are, as the vase can neither be recovered nor easily pieced back together again. Hence, either the ownership of knowledge remains incompletely addressed or one is left with a remainder that does not fit.

Taken together, the cases in this book show how *mutual conditioning* holds power at all times, even when actors are not aware of it. The task of the researcher/analyst is to draw attention to the application of any such strategies as acts of *manipulation*. We suggest calling these acts manipulation because of their multiple political implications, which affect possibilities of owning past, present, or future knowledge, even *when they are not meant to*. Some examples of this are museums being unable to prevent the decay of materials due to a lack of funding decades earlier; knowledge being lost because of wars, disasters, changing values, or human ignorance; and musicians in India wishing to "protect" music as the heritage of their caste rather than as practice. Other examples are contemporary ideals reflecting on historical standards or those of one geographical region reflecting its particular experience onto the other. In the same way that our book's chapters on underprivileged students in the United States and on Vanuatu objects lead us back into our modern world, they also especially drive home that in this reading we—the scholars and analysts—are in a position of great responsibility when dealing with questions of knowledge ownership in our analysis across time and place. Scholars act in the domain of epistemology, and thus when we identify practices and situate them in domains, as social or scientific, economic or intellectual, we draw boundaries of knowability and ownability with lasting consequences.

With this notion in mind, we now wish to shift gears and, in the final section of this chapter, offer our readers a reflection on how scholarship has both engaged with and contributed to past and current understandings of knowledge and its ownership as product or property. The following historiography is an invitation to a critical reading of this scholarship. First, we explain how debates in scholarship and knowledge politics have offered a basis to separate concerns about knowledge from its effects on owning. As both such scholarship and knowledge politics grew increasingly uncomfortable with this separation, they then attempted to fix it. Attempts were made, on the one hand, to elucidate variations of knowledge and ownership regimes across regions and times, and on the other, to address multiple disciplines such as history, sociology and anthropology, philosophy, politics, sciences, technology, and law in cross-, inter- or transdisciplinary forms.

Into the twenty-first century, these approaches have created a curious conundrum with effects in real-life politics: we have both come to *fragment* knowledge into a plurality of "kinds" of knowledge, and then also attempted to legally own these fragments, each on its own or all together. Socially and politically, we have then come to address the injustices and imbalances created by enabling the ownership of such fragments by attempting to once again validate all parts of knowing and authorize all different forms of owning—always by making them all equally legible to law and science.[12] This ultimately means acting as if throwing together all these shards into one container would restore the broken vase. A substantial number of scholars have attempted to *fix* the world by reconnecting the world epistemologically in varied pairings of coproduction, coevolution, and so on across the three sides of culture/science, individual/collective, and body/brain divides—as if the patched vase would equal the original. And as scholars ourselves, we can only pick and choose parts of the fragmented knowledge . What is called for, however, is acknowledging that what cannot be broken apart to begin with does not need to be fixed. Rather, it needs to be acknowledged—as the kn/own/able.

THE SEPARATION OF KNOWING AND OWNING: ALIENABLE AND TACIT

It is curious—even ironic—that the standpoints that have allowed us to ignore the inextricable linkage between forms of owning and knowing in legal and scientific practice, or in economic and social life, are the very ideas that have attempted to emphasize this relationship: the ideas of alienation and of tacitness.[13] Many other terms could be noted, such as *estrangement*; or *distancing*; or *implicit*, *local*, and *indigenous*. But it is these two ideas that have become subject to a range of histories of knowing and owning and have been mobilized in new ways, and if not together, then at least with mutually constitutive effects, for a new global hegemony of the ownership of knowledge. Over the twentieth century, in an astonishingly uninterrupted line of development, the endeavor of pluralizing knowledges became a fix for what had been broken apart by debates concerned with the difference between socioeconomic orders of capitalism and communism, and thus with the general tension between private or collective owning and using.

While it would indeed be hard to pin down a starting point for the separation of knowing and owning as such, important anchoring points have been the works of Karl Marx as well as those of Michael and Karl Polanyi and their debates about capital and society, and knowledge and ownership.[14] An introduction to the theoretical framework of this book can hardly to do justice to the many steps of this development, or to the diverse viewpoints held in the many scholarly disciplines that have been involved in these debates, which range so widely as to encompass philosophy and engineering science. But it is possible to pinpoint some crucial junctures in these scholarly perspectives and how they have contributed to a solidification of knowledge ownership regimes.

Karl Marx is important because he connected alienation to ownership in his arguments about work, capital, and society, thus bringing economy, social order, and laws to the forefront of debates concerning knowledge. This connection paved the way for future approaches that acted *as if* the self and the fruits of one's labor—that is, the means of production and their product—were or *can* ever be fully separated, whereas in fact they *cannot*. Ownership became property. In due course, the areas to which alienation was applied expanded. Once connected mainly to "lowly" labor, its segmentation, and the loss of human autonomy to market forces, the concept became an issue of importance to various realms of knowing, including the relation of science to different political, social, and value systems. The most telling assessment here might be that of the sociologist Robert K. Merton, who was heavily engaged in debates about science and its history. In 1957, he critically noted that the "scientific worker" was subject to market forces in the same way as the blue-collar worker, "as the scientist is *separated* [our italics] from his technical equipment—after all, the physicist does not ordinarily own his cyclotron. To work at his research, he must be employed by a bureaucracy with the laboratory resources."[15] Merton suggested that both craft labor and science were ruled by the same mechanisms of alienation and owned in economy. But this ignored that a laborer would have been unable to own their knowledge (and possess it) in the same way that a scientist of Merton's day could have (by means of their scholarship or grants, for instance).

An important avenue leading up to such junctures over the twentieth century is found in the attempts of different groups of scholars, politicians, and entrepreneurs to reassess and upgrade work within a globalizing world and economy. Many of them did so by reconnecting work to knowledge. This group includes political figures such as the Indian lawyer and leader M. K. Gandhi (1869–1948) and his championing of hand spinning; the chairman of a committee that drafted the Japanese constitution in ordinary language, Yuzo Yamamoto, and his efforts to protect Japanese cultural heritage in 1949; and the British chemist and diplomat Joseph Needham and his interest in the Chinese roots of modern sciences.[16] For scholars of the mid-twentieth century, such efforts were a motivation to refine their own analyses of knowledge, more carefully differentiating between its forms by distinguishing, for instance, between what can or cannot be abstracted. It is at this juncture that actors started to distinguish between knowledge that is inalienable and labor that is valued in the domain of economy, or between knowledge that is inalienable and science that is valued in the domain of epistemology. In the latter case, actors favored the category of "tacit." Tacit became an epistemological value, and inalienability became inextricably linked to the economic realm, to livelihoods, and to exploitation.

The first proponent of "tacit" as a concept for understanding scientific knowledge, Michael Polanyi, suggested in the 1970s that science "is always more than we can tell." His aim was to highlight the relation between the individual and the collective in a way that would allow the freedom and rights of the individual to trump over those of collective bodies in particular nation-states.[17] Polanyi promoted the notion of "tacit" as a concept in order to emphasize the relational character of all knowing, and in making a shift from the collective to the individual to protect individual scientific freedom, he also addressed ownership. Since this implied that tacit knowledge has a social character, he associated the scientific enterprise with the master/apprentice relationship through which the neophyte becomes initiated into membership in the scientific community.

As soon as these terms—and their pairings of tacit/explicit, alienable/inalienable—entered the discourse, they were contested. New scholarship was developed aiming to show that those parts of knowing seen as "tacit" in the sense of being inseparable and inextricable from the individual body,[18] or that those parts of work and labor that were seen as inalienable from bodies, had value. The conclusion was thus that they, too, should be subjected to ownership. Over the twentieth century, tacitness took on the role of identifying the inalienable part of knowing and owning, and as a marker of how scholars, politicians, entrepreneurs, scientists, and practitioners grew uncomfortable with notions of "work" as "lowly" or science as a merely abstract intervention. Scholars increasingly pushed for an awareness of how all work and human practice entailed knowledge and was worthy of recognition, if not ownership. Building on Polanyi's case, Harry Collins showed in 1974 how tacit knowledge can be acquired only by an individual through social membership in the scientific community.[19] In other cases, as scholars showed, communities were formed that allowed collective tacit knowledge to be produced and shared. Especially historians of technology—in contrast, for instance, to those studying economic change—have brought to light how craftsmen and practitioners have protected their skills and shared them to create community, or how collectives have relied on the "sharing" of knowledge, ideas of differentiated labor, or skills, and not on individual ownership. Scholarship from this late twentieth-century period drastically shifted away from a view on the unifying features of a modern world of science and toward a growing interest in understanding differences, diversity, and variations across time and space.[20]

There is a vast body of literature unfolding different pasts or presents from which we could quote here, but a case in point for carrying such concerns over into the analysis of knowledge and its ownership in early modern Europe is Pamela O. Long's study, which suggested that while European guilds mobilized ideas of "'intellectual property' (a kind of intangible property) . . . with regard to material inventions,"[21] the concern of many craftsmen was actually openness and the transmission of skills—not its secrecy.[22] Studies about the sciences called attention to practices, showing how even the most abstract sciences in the West relied on materials, use, and performance, and not only on words and texts.[23] Approaches naturally diverged. Whereas French sociology discusses such issues as a question of political power, the study of technology more broadly attempts to empower this tacitness within the study of sociotechnical context as the power of society and materials. In all cases, analysts addressed power as *exercised* rather than possessed.[24] That is to say, as Philip Mirowski suggests: "Since tacit knowledge was intrinsically dispersed throughout the community, and could only be passed along piecemeal through a socialization process inculcating a particular personal commitment, there could never be any effective rationalization or codification of the process of research."[25]

There are other political implications that postcolonial studies have emphasized and that, again, have historical roots. One example is Indian skill and design, particularly in the field of textile production—which are now seen as having played an important role in raising the quality of European textile industries.[26] Attributing ignorance as the reason for the inalienability of artisans' knowledge from their bodies, the colonial British engineer Alfred Chatterton assumed that "the ordinary artisan . . . unacquainted with principles is therefore quite unable to explain why one way of doing a thing is better than another."[27] Craft skills that could not be codified in text were seen as offering no actual contribution to global knowledge and as not making the grade of "genuine knowledge" on all levels.[28] Ownership shifted as colonial authors codified this knowledge, which had been personal property of craft groups,[29] into surveys, gazetteers, and monographs. This turned proprietorial knowledge into a public good or common cultural heritage. In making private knowledge public, the colonial state was seen to represent an exercise of power that intervened in not just knowledge, but its ownership.[30]

What we can take from this survey is that scholars grew increasingly uncomfortable with such rifts between the different ways of knowing and owning and so attempted to fix them. Such efforts went beyond the colonial period and were more complex than pointing out that a powerful party is oppressing a weaker one; yet, this led to a discourse that seemed to compare or oppose sociotechnical differences as fundamental to a culture. It is no coincidence that debates around the ownership of craft were politicized in India even before Gandhi, and that the precursors of intangible cultural heritage (ICH) originate in Japan. From the East came a view on the practice of embodied craft knowledge as spiritual revelation embedded in oral and religious texts.[31] This speculative view provided a basis for the unity of culture as national heritage, conceptually opposing such unity characteristic of the East or Orient to a more enlightened and analytic Occident—or, the rest to the West; the premodern to the modern. This again contributed to efforts that constructed (and romanticized) the scholarly and political discourse on the native craftsperson as being seen to bear traditional knowledge steeped not only in "tacit" and "intangible" culture, but also in ideas about how that culture had to be owned, collectively or individually, as a public and common or commodifiable good. Preserving this knowledge meant preserving the body and bodies that inalienably carried it as unchanging culture; it meant labeling this knowledge as "traditional" knowledge as opposed to "modern" ways of knowing.[32] As a result, owning knowledge tacitly meant framing the knower not only as intrinsically traditional, but also as incapable of dealing with the abstractions brought about by "knowing" as an abstraction indicative of science or of various sciences.

In this sense, any epistemic category is simultaneously always a political one. And it is because of these developments, in legal practice and in scholarship, that our contemporary world has come to differ from the world of guilds, Carnatic music, or Chinese naming-as-owning practices in its view of ownership and knowledge. In our world, the ownership facilitated by claims of tacitness—and of tacitness as a characteristic of knowledge that should not prevent the knowledge from being ownable—has become a one-way street. Work cannot be owned as science or propositional knowledge, while when it comes to science, calling parts of knowledge "tacit" often means identifying ways of knowing that are inalienable, and that thus should not or *cannot* be owned as property. What is most important to understand, then, is that the scientist can claim tacit knowledge that is available only in the "social" domain, whereas the craftsman cannot claim explicit knowledge in the "scientific" domain because "tacit" is no longer seen as an inexplicable remainder but as the opposite of "explicit." Once again, the notion of alienable is opposed to inalienable components of knowledge, or for that matter, propositional concepts are opposed to material conditions, commensurable to incommensurable, or textual to embodied. This leads to a separation between science and culture, with some groups, regions, or nations having sciences and others culture. Knowledge that is not explicit, and that is inalienable from communities (such as the knowledge of craftspeople), is stuck in that it cannot be framed as science and can be owned only as culture. Knowledge in the forms of arts and crafts continues to fall under the category of intangible cultural heritage, suggesting that there is a historical bias toward comprehending knowledge that is inalienable from bodies as somehow innate—inexpressible by means of words—and thus unscientific.

Scholars have directly contributed to this development. They, as analysts single out *tacitness, making it relevant* as a specific form of knowledge that cannot be extracted and abstracted from the body or materials and therefore needs no verbal expression. The divide is enforced paving the way for tacitness being applied as a legally acceptable category for claiming ownership (for example, as ICH), as if the nomenclature of tacitness did not manipulate ownership claims to begin with. Scholarly work contributes to such manipulations whenever it treats science and law as *the* primary categories or as preset rather than bounded and distinct technologies of social, material, or epistemic ordering. As modes of socialization or belief systems,[33] science and law cannot exist by themselves—even as they develop distinct characteristics and rules.

Historical and sociological scholarship has highlighted, for instance, how different groups own(ed) or have been disowned and how they have attempted to protect knowledge by legal methods. But there are very few studies that have successfully attempted to show *and* emphasize that actors have had different approaches to how to *keep* knowledge and *not give it away*. It is only a minority of scholars who have contributed to refining a particular nomenclature of alienable/inalienable and tacit/explicit into one that would value all knowledge—brain and body—equally while simultaneously acknowledging differences. And whenever such scholarship has emerged, it has operated with a nomenclature of distinction and in efforts to equalize the field while emphasizing variety, thus concluding that there are different ways of knowing, such as everyday knowledge, science as practice, and so on.[34] Many of these efforts have critically engaged with dichotomies—local/global, indigenous/universal, traditional/modern, epistemological/ontological—and attempted to tip the balance by giving, if not more, than at least "equal" power or agency to objects, bodies, or things.[35] Yet, one important factor in these developments is that even in this critical engagement, scholars operate within categories that have emerged from a flattened view, in which knowing and owning have been separated. This is most evident in attempts by scholars to overcome the separation and reunite what had been broken apart, such as by emphasizing the category of technoscientific.[36]

There is a second line of scholarship that, in the awareness of the political implications of knowledge debates, then chose to focus its analytic interest on the relation between knowledge and ownership, suggesting among other things that they are coproduced or have coevolved.[37] This approach, we suggest, holds particular dangers because describing knowledge in terms of "co-"s (i.e., coming together, mutually, in common) suggests an open process, whereas we wish to emphasize, for instance, that scholars *precondition* the scope of owning when they focus on knowing. Another equally crucial factor is that the efforts of analysts in fact replicate, or repeat and resemble, those of the historical actors they are examining, in identifying explication (the ability to verbalize what one knows) as the heart of this debate, and law as the dominant way to manipulate ownership of knowledge. Explication, then, has become the widely accepted silver bullet for manipulating the relation between knowing and owning knowledge. The vast body of literature dealing with or using terms such as *tacitness*, or equivalences such as *intangible*, *local*, *indigenous*, and so on, reflects this problem in the struggle of scholars and the actors that they analyze to communicate all knowing, and to explain the equal importance of bodily performance alongside cognitive, chemical, or physiological processes in the brain and the body. Accepting the deficiency or difficulty of communication, these scholars attempt to use such terms as a vehicle for negotiating ownership claims by bracketing this difficulty. But in the process, they further substantiate language as *the* determinant of knowing and owning. Conversely, we could say that the many material and embodied ways of manipulating the relation between owning and knowing, such as those described in this volume by James Leach or Cynthia Brokaw, have never entered the debate, while the field of law studies remains attached to a rather formalistic or mechanistic stance on its responsibilities for dealing with knowledge that has not yet been made explicit by way of naming.[38]

In recent decades, philosophers, historians, sociologists, and anthropologists have begun critically reflecting on how they themselves have come to adopt an understanding of knowing and owning as a two-step process in which two different qualities are *defined*, rather than analyze the mutual conditioning between knowing and owning. In this understanding, we can identify two major strands of arguments, with the dominant one pointing to power hierarchies between knowledge of the mind and knowledge of the body, as *knowing* in contrast to *doing*. The first is owned as intellectual property and the second as labor. It is important to add to this discussion that superficial similarities in such power hierarchies exist across space and time, and they have allowed researchers to make generalizations when they analyze the methods of owning knowledge that we addressed in our introduction. Whether they have been located in China, India, or Europe, or whether they have identified knowledge as residing in the brain, the heart-mind, or the soul, elites have favored *naming* as their vehicle of knowing and owning, rather than an individual or collective body, or a hand performing menial labor or bringing tools to use.[39]

In the archeology of knowledge, as in attempts to dig out the material traces of human activities, it is "only too easy to invent new surfaces (and even new buildings!) halfway down a thick layer."[40] In the case of knowledge ownership, such layers have obscured the mutual conditioning of kn/own/ables, giving primacy to modern law as a way of owning knowledge. This made the endeavor of pluralizing knowledge appear to be a fix for what debates about capitalism, science, and modernity seem to have broken apart: the relation between body and mind. Yet, even as our modern sciences tell us that all knowledge is bodily and knows no such divide,[41] modern scholarship has accepted naming as a tactic to validate and own knowledge. And when knowledge is named as culture, it is a commons that cannot and must not be owned individually. Some sociologists and historians of science and knowledge, by contrast, have accepted that the relation between knowing and owning knowledge, as determined by science and law, does not consider the inalienable social and material components of knowledge—its tacit dimensions. They have thus looked into alternative practices of knowing by attempting to elevate "tacit" knowledge as knowledge worth knowing. Suggesting that visual language or material compositions can compensate for this lack of explicitness in knowledge, that they can become "things that talk,"[42] these scholars have attempted to validate the "rest" of the relation, explicitly giving a name to what we might call inalienably tacit. And yet, even given our media-enhanced world of sonic and film records, something always remains "unsaid" and thus underrepresented, creating an imbalance in our attention between explicit and tacit. Naming the aspect of knowledge that is "tacit" thus becomes an evaluation of not only what is possible to know or own but also at the same time what is *worth* knowing and owning *at all*. What remains as fill is untranslated materials in their muteness and bodies in their labor that by definition fall outside the scope of knowledge ownership. Tacitness—and any term that might be chosen to replace what cannot be made kn/own/able—thus always is and has been political at its core. It enhances some ownership claims and inhibits others. Epistemologically, it holds the future of knowledge ownership hostage.

IN CONCLUSION

Where then does our own scholarly analysis stand vis-à-vis this critique of the politics of epistemologies? Since equality matters in the ownership of knowledge, we suggest that the first step is to acknowledge that we, too, practice *naming*. We, too, prioritize one aspect of the kn/own/able or *how* one can kn/own it. It is our further understanding that all approaches to knowing are implicit prerogatives that *condition* owning. The use of materials entails not just knowing but also owning, and bodies that perform knowing become the means of owning knowledge. In defining categories of use, performance, and naming as ways of relating knowing and owning, we thus remain in the realm of language. The key difference is that we make visible the practice through which the relation between the two is materialized and acknowledge our responsibility in this process.

In sum, the contributions in this book suggest that as long as actors or analysts only try to fix one or the other side (i.e., knowing or owning) and ignore the conditioning, inequality will always persist within approaches to the ownership of knowledge—and in the results of these approaches. Achieving equality requires that we do more than fix the epistemic and/or legal side. We need to champion the indivisible kn/own/able. As scholars, we can contribute by making the kn/own/able visible, as it is apparent in and made available through different combinations of naming, performing, and use; by words, objects, and bodies; and in different domains such as society, economy, and epistemology. When a gene sequence is named to describe the underlying structure, it is only partially knowable. But because our modern laws consider this an acceptable form of knowledge, the use of the gene becomes completely ownable even if it is not completely knowable. As a weaver performs weaving on the loom, they own a conceptual principle through use, even if the piece of fabric is not completed yet. In bureaucracies, protocols can routinely produce particular categories of expertise. Asking why China did not produce a paper practice of codification, or intellectual property, misses the point, whereas focusing on the conditioning unveils how differently Chinese authors throughout time wished to own their knowledge by writing or were disowned from their knowledge(s) because some considered naming the key for all kn/own/ing. These kinds of questions then allow us to examine, on equal grounds, how the regimes that value naming, such as science today, function as authoritative ownership of knowledge compared to those that value use, such as the making of printing blocks. We suggest that such comparisons require a shift in the values underlying our analytical understanding—and that pushing such ways of owning into the social or economic sphere is a manipulation of the relationship, too.

This book has brought together scholars who look beyond the ways in which boundaries have been drawn around what we have identified as kn/own/ables, in fields or concepts such as law, history, science, knowledge, and property. Taking interdisciplinary approaches, these scholars make visible the methodological values—and hence also the blind spots—that underlie individual disciplines. Such interdisciplinary encounters force methodological apparatuses that are generally stable to shift in order to accommodate what is newly learned. In a final note, we thus turn to our own trade to address the tool that historians, sociologists, and anthropologists best deploy: performing and using epistemologies. Identifying ourselves as analysts, we move our analysis along a timeline—always keeping sight of the moment in which ownership is inextricably attached to the knowable word/object/body; sometimes working forwards, other times backwards. Taking a *longue-durée* perspective on the making of epistemes and their normative effects allows us to capture moments of stabilization synchronously as moments of change, or as signified in rites of passage. Yet, transitions inevitably point to ruptures. At one moment in time, the body/object/word may constitute the relation between the knowable and the ownable. But in other moments, this relation is in a new state in which it vacillates between a new set of possible states in which the thing itself is ownable as knowledge, just as protocols and rules can both be codified in social arrangements and codified as knowledge.

Cognizant of these difficulties and their own responsibility in these power structures, some scholars have also become activists in attempting to empower those disowned by current law. Such research has aimed to widen our view of "soft" ownership, according to the standards of modern capitalism, toward further inclusion by law, representing the "hard" version of ownership. It has also paved the way for appropriating such knowledge while not always protecting those who know and own beyond the structures of *legal property rights*. In our world, *legal* knowledge ownership usually means control. And this control usually defines inclusion and exclusion of use or access, the right to exploit something for economic, professional, or any other legally recognized gain, and the right to buy and sell. Scholars inquiring about knowledge or ownership historically, sociologically, and anthropologically thus bear an enormous responsibility with their research. This is true across all these fields, whether scholars study the different ways in which genes have been known, and how this has mattered for further consideration by courts in the United States; whether they reveal that taxing crafts during Ming-era China differs substantially from contemporary notions of crafts within intangible heritage; or whether they emphasize that, to Indonesians, batik techniques for making cloth and the use of materials have been more important than any accurate replication of a given form or aesthetic. As each of these cases shows, a history of kn/own/ability enables us to live up to the responsibility that is incumbent on scholarship to contribute to, and promote, better practices of owning knowing and sharing it in today's world.

ACKNOWLEDGMENTS

The European Research Council project PENELOPE (ERC funding HORIZON 2020 number 682711) at the Deutsches Museum in Munich and the Deutsche Forschungsgemeinschaft (DFG) project 435681850 at the Technische Universität Berlin have supported Annapurna Mamidipudi's research into ownership of knowledge in craft and hand weaving. She is grateful to the PIs of both projects, Ellen Harlizius-Klück and Friedrich Steinle, for taking a leap of faith into the world of traditional Indian crafts and immersing themselves in understanding its naming, performance, and use in modern times.

Notes

1. Michel Foucault, *The Archaeology of Knowledge the Discourse on Language*, trans. A. M. Sheridan Smith (New York: Pantheon Books, 1972), 182–183.

2. For discussions on materials and cognition, see, e.g., Carl Knappett, *Thinking through Material Culture* (Philadelphia: University of Pennsylvania Press, 2005). For the role of things, see Timothy J. LeCain, *The Matter of History: How Things Create the Past* (Cambridge: Cambridge University Press, 2017). The role of embodied knowledge is highlighted in Joy Parr, *Sensing Changes: Technologies, Environments, and the Everyday, 1953–2000* (Seattle: University of Washington Press, 2010).

3. Philip Barker, *Techniques of Archeological Excavation*, 3rd ed. (London: Routledge, 2015), 111

4. Research in technology studies addresses the loss and the "regaining" of embodied knowledge in terms of knowledge-in-action and knowledge-in-use. This field focuses on the historical understanding of the nature of "knowledge" and how to research past and present knowledge generation. See, e.g., David Edgerton, "From Innovation to Use: Ten Eclectic Theses on the Historiography of Technology," *History and Technology* 16, no. 2 (1999): 1–26; Mikael Hård and Andrew Jamison, *Hubris and Hybrid* (London: Routledge, 2005), esp. part III. For related work in sociology, see John Law, ed., "Power, Action, and Belief: A New Sociology of Knowledge," special issue, *Sociological Review* 32, no. 1 (1984).

5. See, e.g., Annemarie Mol, *The Logic of Care* (London: Routledge, 2008), and Hannah Landecker, "Antibiotic Resistance," *New Biologies* 22, no. 4 (2016): 44. For the seat belt, see Bruno Latour,

"Where Are the Missing Masses?," in *Shaping Technology / Building Society: Studies in Sociotechnical Change*, ed. Wiebe Bijker and John Law (Cambridge, MA: MIT Press, 1992), 225–259. For AI and decision-making as a search process, see Nathan Ensmenger, "Is Chess the Drosophila of Artificial Intelligence? A Social History of an Algorithm," *Social Studies of Science* 42, no. 1 (2011): 16. For the critical role of the analyst, see Sheila Jasanoff, "Imagined and Invented Worlds," in *Dreamscapes of Modernity: Sociotechnical Imageries and the Fabrication of Power*, ed. Sheila Jasanoff and Sang-Hyun Kim (Chicago: University of Chicago Press, 2015), 339.

The body of literature on this topic across several fields of research is huge. For a study of the diverging ethics of actors in the framework of colonial history, see Suzanne Moon, *Technology and Ethical Idealism—A History of Development in the Netherlands East Indies* (Leiden: CNWS, 2007).

6. Immanuel Kant, *Kritik der reinen Vernunft*, ed. Benno Erdmann (Berlin: De Gruyter, 1990). For a standard English translation, see *Critique of Pure Reason*, ed. and trans. Paul Guyer and Allen W. Wood (Cambridge: Cambridge University Press, 1998).

7. Knowing and owning can certainly be independent from each other, as when actors take ownership of an object without knowing anything about it. This we distinguish as property and possession. But there is no way that owning does not always affect the possibilities of knowing. This is why we talk about kn/own/ability, i.e., a process of facilitation. We thank the anonymous reviewers for asking us to clarify this point.

8. Wiebe E. Bijker, "Technology, Social Construction of," in *International Encyclopedia of the Social & Behavioral Sciences*, ed. James D. Wright 2nd ed. (Amsterdam: Elsevier, 2015), 135–140. For English scholarship of sociology of knowledge, see David Bloor, "Anti-Latour," *Studies in History and Philosophy of Science* 30, no. 1 (1999); and Steven Shapin and Simon Schaffer, *Leviathan and the Air-Pump: Hobbes, Boyle, and the Experimental Life* (Princeton, NJ: Princeton University Press, 2018).

9. See Manuel DeLanda, *Assemblage Theory* (Edinburgh: Edinburgh University Press, 2016), which takes recourse to French philosophers such as Deleuze and Guattari rather than to Henri Lefebvre's "ensembles" in his *Critique of Everyday Life* (London: Verso, 1991), 298. For an example of how work in STS activates the notion of assemblages, see Israel Rodriguez-Giralt, *Reassembling Activism: Activating Assemblages* (London: Routledge 2020); and Bruno Latour, *Reassembling the Social: An Introduction to Actor-Network-Theory* (Oxford: Oxford University Press, 2005).

10. See, e.g., David Wengrow, "Prehistories of Commodity Branding," *Current Anthropology* 49, no. 1 (2008): 7–34.

11. Sheila Jasanoff, "Making Order: Law and Science in Action," in *The Handbook of Science and Technology Studies*, ed. Edward J. Hackett et al., 3rd. ed. (Cambridge, MA: MIT Press, 2007), 761.

12. The reverse is also true—by declaring them illegitimate when they cannot be made legible to law and science.

13. Amy E. Wendling, *Karl Marx on Technology and Alienation* (London: Palgrave Macmillan, 2009).

14. In his 1844 manuscript, Karl Marx emphasized estrangement as a distancing from reality and the separation of work as a process of "objectification" (*Vergegenständlichung*). *Karl Marx, frühe*

Schriften, ed. Hans-Joachim Lieber (Stuttgart: Cotta, 1960). Ernst Bloch, Anne Halley, and Darko Suvin, "*Entfremdung, Verfremdung*: Alienation, Estrangement," *The Drama Review: TDR* 15, no. 1 (1970): 120–125, noted in the 1960s a growing misapprehension between the German and English renderings of Marx's ideas. See also Marcello Musto, "Revisiting Marx's Concept of Alienation," *Socialism and Democracy* 24, no. 3 (2010): 79–101. For Karl Polanyi's approach to Marxism, originally published in 1944, see Karl Polanyi, *The Great Transformation*, 2nd ed. (Boston: Beacon Press, 2001). See also Michael Polanyi, *The Tacit Dimension* (New York: Doubleday, 1966).

15. Robert K. Merton, *Social Theory and Social Structure* (Glencoe, IL: Free Press, 1957), 195.

16. Geoffrey R. Scott, "The Cultural Property Laws of Japan: Social, Political, and Legal Influences," *Washington International Law Journal* 12, no. 2 (2003): 315–402; Dagmar Schäfer and Florence Hsia, "History of Science, Technology, and Medicine: A Second Look at Joseph Needham," *Isis* 110, no. 1 (2019): 94–99.

17. Polanyi, *Tacit Dimension*.

18. Michael Polanyi, "Tacit Knowing: Its Bearing on Some Problems of Philosophy," *Reviews of Modern Physics* 34, no. 4 (1962): 601–616.

19. Harry Collins, "The Tea Set: Tacit Knowledge and Scientific Networks," *Science Studies* 4, no. 2 (1974): 165–185.

20. As represented by the California school or the Annales movement originating in France.

21. Pamela O. Long, "Invention, Authorship, 'Intellectual Property,' and the Origin of Patents: Notes toward a Conceptual History," *Technology and Culture* 32, no. 4 (1991): 846–884.

22. Pamela O. Long, *Openness, Secrecy, Authorship: Technical Arts and the Culture of Knowledge from Antiquity to the Renaissance* (Baltimore: Johns Hopkins University Press, 2004). For the social constructivist approach and its relation to the material and immaterial construction of knowledge, see Trevor J. Pinch and Wiebe E. Bijker, "The Social Construction of Facts and Artifacts: Or How the Sociology of Science and the Sociology of Technology Might Benefit Each Other," in *The Social Construction of Technological Systems*, ed. Wiebe E. Bijker, Thomas P. Hughes, and Trevor Pinch (Cambridge, MA: MIT Press, 1999), 12–44.

23. Two examples of this growing literature are Pamela H. Smith, *The Body of the Artisan: Art and Experience in the Scientific Revolution* (Chicago: University of Chicago Press, 2004), and Lissa Roberts, Simon Schaffer, and Peter Dear, *The Mindful Hand: Inquiry and Invention from the Late Renaissance to Early Industrialisation* (Amsterdam: Koninklijke Nederlandse Akademie van Wetenschappen, 2007).

24. Michel Foucault, *Discipline and Punish: The Birth of the Prison*, trans. Alan Sheridan (New York: Vintage Books, 1995), 26.

25. Philip Mirowski, "On Playing the Economics Trump Card in the Philosophy of Science: Why It Did Not Work for Michael Polanyi," in "Proceedings of the 1996 Biennial Meetings of the Philosophy of Science Association. Part II: Symposia Papers," ed. Lindley Darden, supplement, *Philosophy of Science* 64, no. S4 (1997): S127–S138.

26. Maxine Berg, "In Pursuit of Luxury: Global History and British Consumer Goods in the Eighteenth Century," *Past and Present* 182, no. 1 (2004): 85–142.

27. Alfred Chatterton, *Agricultural and Industrial Problems in India* (Madras: G. A. Natesan, 1904).

28. Abigail McGowan, *Crafting the Nation in Colonial India* (New York: Palgrave Macmillan, 2009).

29. For a detailed discussion on the kind of ethnographic accounts of knowledge generated by the colonial state, see McGowan.

30. McGowan.

31. Ananda Kentish Coomaraswamy, *The Indian Craftsman* (London: Probsthain, 1909); Stella Kramrisch, *Exploring India's Sacred Art: Selected Writings of Stella Kramrisch*, ed. Barbara Stoler Miller, 1st Indian ed. (Delhi: IGNCA and Motilal Banarsidass, 1994); Yanagi Sōetsu, *The Unknown Craftsman: A Japanese Insight into Beauty*, ed. Bernard Leach (Tokyo: Kodansha International, 1972).

32. The commodification of craft as a marketable good certainly has much older traces. See, e.g., Peter Betjemann, *Talking Shop: The Language of Craft in an Age of Consumption* (Charlottesville: University of Virginia Press, 2011), 71–74.

33. Andreas Philippopoulos-Mihalopoulos, *Niklas Luhmann: Law, Justice, Society* (Abingdon, UK: Routledge, 2010) 71. See also Mario Biagoli and Marius Buning, "Technologies of the Law / Law as a Technology," introduction to special issue, *History of Science* 57, no. 1 (2019): 3–17, as well as the entire special issue in which this article appears.

34. John Pickstone, *Ways of Knowing: A New History of Science, Technology, and Medicine* (Chicago: University of Chicago Press, 2001); Andrew Pickering, ed., *Science as Practice and Culture* (Chicago: University of Chicago Press, 1992).

35. Such dichotomies also connected to world orders or periphery and center; see, e.g., Hyungsub Choi, "The Social Construction of Imported Technologies," *Technology and Culture* 58, no. 4 (2017): 905–920. For an approach to ontological issues, see Annemarie Mol, *The Body Multiple: Ontology in Medical Practice* (Durham, NC: Duke University Press, 2008), or Tiago Saraiva, *Fascist Pigs* (Cambridge, MA: MIT Press, 2018).

36. Such was also addressed and implemented by reuniting terms in Ursula Klein, "Technoscience avant la lettre," *Perspectives on Science* 13, no. 2 (2005): 226–266.

37. Sheila Jasanoff, ed., *States of Knowledge: The Co-Production of Science and the Social Order* (London: Routledge, 2004); Jürgen Renn, *The Evolution of Knowledge: Rethinking Science for the Anthropocene* (Princeton, NJ: Princeton University Press, 2020).

38. See Isaac Mazonde and Thomas Pradip, eds., *Indigenous Knowledge System and Intellectual Property Rights in the Twenty-First Century: Perspectives from Southern Africa* (Dakar: CODESRIA, 2007), 95–103, for an example of how such "global" regimes play out locally.

39. This is evident in Aristotelian notions of *techne* and *episteme*, as well as Chinese notions of scholarly written traditions vis-à-vis farmers' bodies, or the Indian caste hierarchy.

40. Barker, *Archeological Excavation*, 111.

41. Only recently, the shifting emphasis on diversity has also stimulated neurologists to develop new models of distributed cognition and multimodal knowing, along with research programs to address the social and cultural nature of the brain and body, and to consider that their methods must account for all kinds of knowledge, textual and embodied. Edwin Hutchins, "Cognition, Distributed," in *International Encyclopedia of the Social & Behavioral Sciences*, ed. James Wright (Oxford: Pergamon, 2001), 2068–2072.

42. Lorraine Daston, ed., *Things That Talk: Object Lessons from Art and Science* (New York: Zone Books, 2004).

Bibliography

Barker, Philip. *Techniques of Archeological Excavation*. 3rd ed. London: Routledge, 2015.

Berg, Maxine. "In Pursuit of Luxury: Global History and British Consumer Goods in the Eighteenth Century." *Past and Present* 182, no. 1 (2004): 85–142.

Betjemann, Peter. *Talking Shop: The Language of Craft in an Age of Consumption*. Charlottesville: University of Virginia Press, 2011.

Bijker, Wiebe E. "Technology, Social Construction of." In *International Encyclopedia of the Social & Behavioral Sciences*, edited by James D. Wright, 135–140. 2nd ed. Amsterdam: Elsevier, 2015.

Bloch, Ernst, Anne Halley, and Darko Suvin. "*Entfremdung*, *Verfremdung*: Alienation, Estrangement." *The Drama Review: TDR* 15, no. 1 (1970): 120–125.

Bloor, David, Steven Shapin, and Simon Schaffer. *Leviathan and the Air-Pump: Hobbes, Boyle, and the Experimental Life*. Princeton, NJ: Princeton University Press, 2018.

Chatterton, Alfred. *Agricultural and Industrial Problems in India*. Madras: G. A. Natesan, 1904.

Choi, Hyungsub. "The Social Construction of Imported Technologies." *Technology and Culture* 58, no. 4 (2017): 905–920.

Collins, Harry. "The Tea Set: Tacit Knowledge and Scientific Networks." *Science Studies* 4, no. 2 (1974): 165–185.

Coomaraswamy, Ananda Kentish. *The Indian Craftsman*. London: Probsthain, 1909.

Daston, Lorraine, ed. *Things That Talk: Object Lessons from Art and Science*. New York: Zone Books, 2004.

DeLanda, Manuel. *Assemblage Theory*. Edinburgh: Edinburgh University Press, 2016.

Ensmenger, Nathan. "Is Chess the Drosophila of Artificial Intelligence? A Social History of an Algorithm." *Social Studies of Science* 42, no. 1 (2011): 5–30.

Foucault, Michel. *The Archaeology of Knowledge and the Discourse on Language*. Translated by A. M. Sheridan Smith. New York: Pantheon Books, 1972.

Foucault, Michel. *Discipline and Punish: The Birth of the Prison*. Translated by Alan Sheridan. New York: Vintage Books, 1995.

Hård, Mikael, and Andrew Jamison. *Hubris and Hybrid*. London: Routledge, 2005.

Hutchins, Edwin. "Cognition, Distributed." In *International Encyclopedia of the Social and Behavioral Sciences*, edited by James Wright, 2068–2072. Oxford: Pergamon, 2001.

Jasanoff, Sheila. "Imagined and Invented Worlds." In *Dreamscapes of Modernity: Sociotechnical Imageries and the Fabrication of Power*, edited by Sheila Jasanoff and Sang-Hyun Kim, 321–342. Chicago: University of Chicago Press, 2015.

Jasanoff, Sheila. "Making Order: Law and Science in Action." In *The Handbook of Science and Technology Studies*, 3rd ed., edited by Edward J. Hackett, Olga Amsterdamska, Michael E. Lynch, and Judy Wajcman, 761–786. Cambridge, MA: MIT Press, 2007.

Jasanoff, Sheila, ed. *States of Knowledge: The Co-production of Science and the Social Order*. London: Routledge, 2004.

Kant, Immanuel. *Kritik der reinen Vernunft*. Edited by Benno Erdmann. Berlin: De Gruyter, 1990.

Klein, Ursula. "Technoscience avant la lettre." *Perspectives on Science* 13, no. 2 (2005): 226–266.

Knappett, Carl. *Thinking through Material Culture*. Philadelphia: University of Pennsylvania Press, 2005.

Kramrisch, Stella. *Exploring India's Sacred Art: Selected Writings of Stella Kramrisch*. Edited by Barbara Stoler Miller. 1st Indian ed. Delhi: IGNCA and Motilal Banarsidass, 1994.

Landecker, Hannah. "Antibiotic Resistance." *New Biologies* 22, no. 4 (2016): 19–52.

Latour, Bruno. *Reassembling the Social: An Introduction to Actor-Network-Theory*. Oxford: Oxford University Press, 2005.

Latour, Bruno. "Where Are the Missing Masses?" In *Shaping Technology / Building Society: Studies in Sociotechnical Change*, edited by Wiebe Bijker and John Law, 225–259. Cambridge, MA: MIT Press, 1992.

Law, John, ed. "Power, Action, and Belief: A New Sociology of Knowledge." Special issue, *Sociological Review* 32, no. 1 (1984).

LeCain, Timothy J. *The Matter of History: How Things Create the Past*. Cambridge: Cambridge University Press, 2017.

Lefebvre, Henri. *Critique of Everyday Life*. London: Verso, 1991.

Lemonnier, Pierre. *Mundane Objects: Materiality and Non-verbal Communication*. Walnut Creek, CA: Left Coast Press, 2012.

Lieber, Hans-Joachim, ed. *Karl Marx, frühe Schriften*. Stuttgart: Cotta, 1960.

Long, Pamela O. "Invention, Authorship, 'Intellectual Property,' and the Origin of Patents: Notes toward a Conceptual History." *Technology and Culture* 32, no. 4 (1991): 846–884.

Long, Pamela O. *Openness, Secrecy, Authorship: Technical Arts and the Culture of Knowledge from Antiquity to the Renaissance*. Baltimore: Johns Hopkins University Press, 2004.

Mathur, Saloni. *India by Design: Colonial History and Cultural Display*. Berkeley: University of California Press, 2007.

Mauss, Marcel. *The Gift: The Form and Reason for Exchange in Archaic Societies*. London: Routledge, 2002.

Mazonde, Isaac, and Thomas Pradip, eds. *Indigenous Knowledge System and Intellectual Property Rights in the Twenty-First Century: Perspectives from Southern Africa*. Dakar: CODESRIA, 2007.

McGowan, Abigail. *Crafting the Nation in Colonial India*. New York: Palgrave Macmillan, 2009.

Merton, Robert K. *Social Theory and Social Structure*. Glencoe, IL: Free Press, 1957.

Mirowski, Philip. "On Playing the Economics Trump Card in the Philosophy of Science: Why It Did Not Work for Michael Polanyi." In "Proceedings of the 1996 Biennial Meetings of the Philosophy of Science Association. Part II: Symposia Papers," edited by Lindley Darden. Supplement, *Philosophy of Science* 64, no. S4 (1997): S127–S138.

Mol, Annemarie. *The Body Multiple: Ontology in Medical Practice*. Durham, NC: Duke University Press, 2008.

Mol, Annemarie. *The Logic of Care*. London: Routledge, 2008.

Moon, Suzanne. *Technology and Ethical Idealism—A History of Development in the Netherlands East Indies*. Leiden: CNWS, 2007.

Musto, Marcello. "Revisiting Marx's Concept of Alienation." *Socialism and Democracy* 24, no. 3 (2010): 79–101.

Parr, Joy. *Sensing Changes: Technologies, Environments, and the Everyday, 1953–2000*. Seattle: University of Washington Press, 2010.

Philippopoulos-Mihalopoulos, Andreas. *Niklas Luhmann: Law, Justice, Society*. Abingdon, UK: Routledge, 2010.

Pickering, Andrew, ed. *Science as Practice and Culture*. Chicago: University of Chicago Press, 1992.

Pickstone, John. *Ways of Knowing: A New History of Science, Technology, and Medicine*. Chicago: University of Chicago Press, 2001.

Pinch, Trevor J., and Wiebe E. Bijker. "The Social Construction of Facts and Artifacts: Or How the Sociology of Science and the Sociology of Technology Might Benefit Each Other." In *The Social Construction of Technological Systems*, edited by Wiebe E. Bijker, Thomas P. Hughes, and Trevor Pinch, 12–44. Cambridge, MA: MIT Press, 1999.

Polanyi, Karl. *The Great Transformation*. 2nd ed. Boston: Beacon Press, 2001.

Polanyi, Michael. *The Tacit Dimension*. New York: Doubleday, 1966.

Polanyi, Michael. "Tacit Knowing: Its Bearing on Some Problems of Philosophy." *Reviews of Modern Physics* 34, no. 4 (1962): 601–616.

Renn, Jürgen. *The Evolution of Knowledge: Rethinking Science for the Anthropocene.* Princeton, NJ: Princeton University Press, 2020.

Roberts, Lissa, Simon Schaffer, and Peter Dear. *The Mindful Hand: Inquiry and Invention from the Late Renaissance to Early Industrialisation.* Amsterdam: Koninklijke Nederlandse Akademie van Wetenschappen, 2007.

Rodriguez-Giralt, Israel. *Reassembling Activism: Activating Assemblages.* London: Routledge, 2020.

Saraiva, Tiago. *Fascist Pigs.* Cambridge, MA: MIT Press, 2018.

Schäfer, Dagmar, and Florence Hsia. "History of Science, Technology, and Medicine: A Second Look at Joseph Needham." *Isis* 110, no. 1 (2019): 94–99.

Scott, Geoffrey R. "The Cultural Property Laws of Japan: Social, Political, and Legal Influences." *Washington International Law Journal* 12, no. 2 (2003): 315–402.

Smith, Pamela H. *The Body of the Artisan: Art and Experience in the Scientific Revolution.* Chicago: University of Chicago Press, 2004.

Sōetsu, Yanagi. *The Unknown Craftsman: A Japanese Insight into Beauty.* Edited by Bernard Leach. Tokyo: Kodansha International, 1972.

Wendling, Amy E. *Karl Marx on Technology and Alienation.* London: Palgrave Macmillan, 2009.

Wengrow, David "Prehistories of Commodity Branding." *Current Anthropology* 49, no. 1 (2008): 7–34.

1 MUTUAL CONDITIONING

2

INTELLECTUAL PROPERTY WITH CHINESE CHARACTERISTICS

Cynthia Brokaw

Examination of knowledge creation, possession, and dissemination in different cultures provides one avenue for the investigation of the various meanings of the ownership of knowledge. Such an examination can be particularly fruitful as a means of "provincializing" the understanding of intellectual property (IP) as a concept of universal utility. To this end, I focus here on the meanings of the ownership of knowledge as they arose in later imperial China—that is, the period from the eleventh through most of the nineteenth century—and then turn briefly to consider how these meanings were reformulated, rather dramatically, in the early years of the People's Republic of China (PRC), a second moment that illuminates a related, but different, alternative framework for the ownership of knowledge. As this chronological leap suggests, I am interested less in tracing the development or history of conceptions of the ownership of knowledge than in what we can learn from analysis of strikingly different—if not entirely disjunctive—expressions of what "intellectual property" and the ownership of knowledge meant at two points in Chinese history: first, when Western conceptions were simply not relevant (the imperial period as I treat it here); and second, when Western legal protections of intellectual property were seen as a site of opposition (the early decades of the PRC).

Rather than cleaving to the term *intellectual property*, my discussion of Chinese regimes of knowledge expression and the ownership of inventions proceeds from a looser, more open-ended inquiry. Currently, normalized categories of intellectual property (which, as Marius Buning explains in chapter 3, in fact originated in very particular nineteenth-century European and American ideas of knowledge ownership) include copyright—the protection of text ownership; patents—the protection of the ownership of inventions; and trademarks—the protection of exclusive ownership of commercial brands. My focus is almost entirely on the ownership of the expressions of ideas and knowledge, with some attention to the ownership of craft inventions and patents (particularly in the discussion of the PRC). Although the development of publishers' trademarks and the commercial branding of print in China are topics that sorely need investigation, because trademarks communicate a different kind of knowledge ownership, I leave exploration of the role they played in Chinese concepts of intellectual property to future research.

This project, although it focuses on the past, developed out of an interest in searching in the present for alternatives to the now globalized regime of intellectual property that generally reflects the hegemony of Western notions of the ownership of knowledge. The global IP regime, in its conceptualization of intellectual property and the legal restrictions it places on knowledge sharing, faces increasing criticism and calls for reform or abolition. Attacks come from all sides. Critics of copyright argue that delaying the entry of works into the public domain—now, according to the United States Copyright Term Extension Act of 1998, for up to ninety-five years—is incommensurate with two of copyright's ostensible goals: the wide and rapid dissemination of information and the innovative use of new works. Inventors complain that instead of encouraging innovation—again, one of its stated goals—the patent system in fact inhibits the development of new inventions.[1] The concept of creation embedded in IP law is seen as problematic on two counts. Associated in various instantiations with a romantic notion of artistic and literary authorship and of technological invention as the work of individual genius, globally circulating IP regimes universalize legal constructions that developed from distinctively European origins.[2] By failing to acknowledge that creation and invention are social acts,[3] the results of the contributions of many men and women, such understandings of IP privilege the knowledge-ownership claims of the individual over the open dissemination of knowledge. In global capitalism, this privileging of individual claims has been extended to serve the profit goals of private corporations. Geopolitics have normalized such understandings of intellectual property and IP law and thereby enabled multinational corporations in developed countries to inhibit the economic advancement of poorer countries by invoking copyright and patent protections and to exploit indigenous peoples in many countries by violating their "traditional" knowledge and resource rights.[4]

I turn to China to look at other ways of thinking about the ownership of knowledge. The case of China is particularly useful because it offers an opportunity to examine distinctive indigenous concepts of both knowledge production and knowledge control ("intellectual property with Chinese characteristics")[5] as they evolved through the late imperial period. Moreover, it offers the opportunity to observe, after China's struggles to adapt to globally circulating IP regimes in the early twentieth century, the decision taken in the early decades of the People's Republic to repudiate externally normative notions of intellectual property and its associated legal protections.

It is necessary first to confront the tendency of many Western (and some Chinese) scholars to ask why China did not develop intellectual property—or even any general notion of intellectual property—throughout its long premodern history. The implicit assumption of this question—that somehow the Chinese *should* have developed IP law—has led to some misleading characterizations of Chinese conceptions of authorship, creation, and invention. I look critically at these characterizations, first, by examining Chinese notions of authorship and invention and, second, by looking at specific examples of Chinese authors—and, to a limited degree, inventors—who *did* emphatically assert their ownership of the knowledge they produced. From this I move to a discussion of "intellectual property with Chinese characteristics." Finally, I conclude with a brief reflection on the recent efforts of the People's Republic to abandon entirely the notion of intellectual property and individual ownership of knowledge.

INTELLECTUAL PROPERTY IN IMPERIAL CHINA

Answering the Wrong Question: Why Didn't China Develop Intellectual Property Law?

The worldwide normalization of intellectual property law is nowhere more evident than in the ongoing conflict between the United States and the People's Republic of China over the widespread and frequent violation of IP law in China. The long and tangled history of the concept of intellectual property and IP law in Europe and the US—a history that encompasses changing and conflicting understandings of authorship, natural rights, and state-society relations; long-term negotiations between states, authors, and publishers; and complex legal debates about the meaning of "intellectual property"[6]—is overlooked in the apparent assumption that there is something natural and inherently rational in the notion of intellectual property narrowly conceived as an object of legal protection.

In looking to China, then, difference becomes coded as lack. Legal and intellectual historians have tried to identify what it is in Chinese history or culture that prevented, or at least inhibited, the development of a notion of intellectual property. Among Western scholars, the consensus seems to be that long-held attitudes toward the ownership of knowledge, authorship, and creation precluded this development. William Alford highlights the constant "need to interact with the past" in the Chinese scholarly and literary tradition. The writings of the ancient sages had established the truth; it was then the duty of all those who followed to recover and explain that truth, not to attempt the vain creation of new truths:

> The content of expectations concerning the appropriateness of individuals and groups exercising control over the expression of particular ideas derived, in turn, from the critical role that the shared past played in the Confucian understanding of both individual moral and collective social development. Simply stated, the need to interact with the past sharply curtailed the extent to which it was proper for anyone other than persons acting in a fiducial capacity to restrict access to its expressions.[7]

Theories of literature, calligraphy, and painting all centered on the artist's relationship to the past. As a result, the replication of writing or art from the past—what in the West might be considered at best, imitation, and at worst, plagiarism—never had the "dark connotations . . . it does in the West."[8]

Philip Ivanhoe, in contrast, invokes underlying Chinese "attitudes and beliefs about the nature of truth" and the conviction that knowledge was "for" society, and therefore any measure that would limit its dissemination was immoral. He explains:

> The idea is that the most important kind of knowledge is knowledge about the Way. Anyone who discovers some truth about the Way is uncovering some facet of a much greater pattern, something that by its very nature cannot be owned by any one person, for it belongs to everyone. Moreover, because the Way nurtures and benefits all under Heaven, knowledge about it exists for this greater good. On such a view, to regard any discovery or invention as one's own property manifests a profound ignorance of the nature of such truth and tends to interfere with the role that such knowledge is to play in the greater scheme of things.[9]

Ivanhoe (and he is not alone) is here arguing, in essence, that the Chinese did not conceive of knowledge—at least knowledge of the Way—as something that could be owned at all; existing for the greater good of all, it could never be legitimately possessed by any individual or group within human society.

Chinese Concepts of Human Creation: Authorship and Invention

Authorship There is something to be said for both of these arguments. It is certainly the case that Chinese scholarly authors (and artists) were often deeply engaged in interacting with the past. In this sense, what constituted a major realm of textual knowledge (although, it must be emphasized, not the only realm of textual knowledge) was not conceived in terms of innovation or invention. One could argue that the long and complex history of Confucian thought, for example, was driven by repeated efforts to engage with ancient writings attributed to Confucius. And this kind of knowledge—that is, the knowledge embedded in the Confucian Classics—was perceived (to varying degrees in different periods) as knowledge about the Way ideally accessible to all. Through much of imperial history, schooling began with memorizing a portion of these texts, and the civil service examination system assumed mastery of all of them.

But a closer look at Chinese notions of textual creation and invention might allow for a somewhat more complicated and nuanced understanding of how writing and textual knowledge were understood.[10]

As Michael Puett has argued, when discursive writing emerged in early China, it was understood as a product of moral and political degeneracy. The earliest sages, the creators of culture (e.g., agriculture, characters, fire), were rulers, not authors; as creators of perfect order, they had no need for writing. With the decline of that perfect order, however, writing became the means of preserving some memory of that golden age and, more practically, the principles, beliefs, and practices that had sustained it. New sages arose from among the scholarly scribal class of the Zhou—Confucius was one of them—and they were the first authors, the first men who, deprived of the opportunity to rule, created (*zuo* 作) the works that recorded the Sagely Way for their own and future generations. Others might collect these works or write about them; they were simply "transmitters" (*shuzhe* 述者) or "discussers" (*lunzhe* 論者) of the works of the sages."[11]

As the number of "sage" authors proliferated and the collections of "sagely" writings grew—and as it became clear that there was little agreement among them about the nature of the Way—there developed, roughly around the turn to the Common Era, some reservations about the necessary sageliness of these works and their authors, many of whom remained anonymous. As Stephen Owen explains: "The Sage maker (*zuozhe* 作者) formulates how things both should be and historically were. By the [Eastern] Han, this grander sense of sagely 'making' had diminished ... to a weaker and broader sense of 'writing' or 'composition.'"[12] To be sure, some of the early texts maintained their status as the creations of sages—or had that status bestowed on them as they were canonized in later Chinese history, the fate of the Five Classics attributed to Confucius. These are among the "past" works with which later scholars were expected to engage. But from the first century CE, anyone could be an author.

What did this debased kind of authorship mean? How did one become an author in an age far removed from the perfect order of the sages? A good author, an author worth reading, was, first, one who had mastered the scholarly literary tradition through wide reading and extensive memorization. He (they are almost all male) then might create a text that was an artful pastiche of passages drawn from this knowledge. The author was establishing his scholarly credentials—and, of course, assuming that his equally learned readers would know the sources of these passages. The artistry and brilliance of this act of writerly creation came from the skillful selection of passages to reproduce and their meaningful juxtaposition in the new text.

Charles Stone nicely explains the process in his discussion of the masterpiece of the historian and official Sima Guang 司馬光 (1019–1086), the massive *Comprehensive Mirror for Aid in Government* (*Zizhi tongjian* 資治通鑑). As Stone notes, this work is "almost entirely comprised of unattributed verbatim quotations from other works," yet its author "is not only not considered a plagiarist, he is considered one of the foremost historians of his age."[13] Part of what makes his work great is the wide-ranging verbatim quotation; it revealed not only the author's enormous erudition (and excellent memory), but also his commitment to accuracy. Rewording his sources—as would be demanded of a modern

Western historian—would have meant introducing inaccuracies into writings that were supposed to "speak for themselves."[14]

The other part of what makes *Comprehensive Mirror* great is Sima Guang's brilliant arrangement of his quotations, the means he uses to signal his understanding of events. By juxtaposing quotations that seem to qualify or contradict one another, he requires the dedicated reader—who is expected to be able to identify the sources of the quotations and thus the purpose of the juxtaposition—to work out his interpretation. Robert LaFleur explains,

> There is a dissonance of voices in the *Zizhi tongjian* that is not resolved by the author's commentaries. The reader is expected to play an integral role in shaping the meaning of the text, in working through the multiple assertions and meanings found in the quoted materials which make up the work. The *Zizhi tongjian* is hard to read, in short, beyond the most basic relation of events in time. It assumes a broad classical education as well as a reader capable of putting the text back together in his own vision.[15]

One could argue that Sima Guang represents a special case, as historians are by definition reliant on the use of previous writings. But a similar approach can be found in other works as well. *Plum in the Golden Vase* (*Jin Ping Mei* 金瓶梅), one of the great vernacular novels of the early modern period—and a work of unquestioned originality—opens with chapters lifted almost verbatim from another famous novel of the period, *Water Margin* (*Shuihu zhuan* 水滸傳). And much of the novel, as Shang Wei has pointed out, combines phrases and verses from the popular daily-life encyclopedias of the day.[16] In this case, the unattributed quotations from texts that elite readers dismissed as trash points up the anonymous author's attack on nouveau riche merchant culture. Another roughly contemporary work, Tang Xianzu's 湯顯祖 (1550–1616) drama *Peony Pavilion* (*Mudan ting* 牡丹亭), includes many arias that string together separate lines from the great Tang poets and other well-known works, often for comic effect.[17]

The ways in which writing was taught suggests an understanding of authorship that included, if it did not consist entirely of, the notion that a text was created through the artful combination of passages and phrases borrowed from other works. Students were of course expected to be able to quote the Classics, and since every educated person knew the Classics, there was no need—indeed, it would be gauche and insulting—to provide a reference to a classical citation in one's writings. But the writer was also expected to be able to borrow from a wide range of other ancient texts, including poetry, histories, and philosophical essays. One of the basic children's textbooks for writing, *Treasury of Allusions for Young Students* (*Youxue gushi qionglin* 幼學故事瓊林), is an encyclopedia of phrases and allusions drawn from major works of Chinese literature and history. By providing the sources of the phrases and allusions, it introduced children to the literary and historical canon; the phrases and allusions themselves were to be memorized so that the student could easily and quickly select excerpts to be combined into an essay, omitting, of course, the unnecessary source references. Popular literary encyclopedias often also offered lists of set phrases for the use of poorly educated writers hoping to achieve some elegance in composition. An essay produced by this method very often formed, as one scholar has put it, "a beautiful patchwork."[18] Similarly, the missionary John Livingston Nevius (1829–1893), writing in the late nineteenth century, described the typical examination essay as "a kind of literary mosaic, composed of ethical axioms, historical references, obscure allusions, and hints, poetical, biographical, and historical with which [the students'] memories are stored; while they almost unconsciously fall into the style and forms of expression with which their minds have become familiar in the course of their *memoriter* studies."[19]

This rough characterization of concepts of authorship in imperial China might appear at first glance to support both Alford's and Ivanhoe's identifications of the reasons why China failed to develop a notion of intellectual property and IP law. The author was supposed to engage with the past quite literally by borrowing vigorously and repeatedly from it. Past knowledge was apparently seen as unowned (or as the property of all), so attribution of quotations from past writings was seen as unnecessary.

But not *all* Chinese writings were pastiches of passages from previous texts. Indeed, such works were generally not considered of any literary value unless, as was the case with Sima Guang's great history, the novel *Plum in the Golden Vase*, and the drama *Peony Pavilion*, they were crafting the presentation of quotations in original, significant, and interesting ways. For the most part, writings like the "mosaic" examination essays described by Nevius and the stilted compositions produced by users of the popular literary encyclopedias would be dismissed as literarily worthless. Readers (and writers) distinguished between imitation and invention that might—or might not—draw on the literary past. Gu Yanwu 顧炎武 (1613–1682), the great Qing polymath, argued in his *Record of Daily Knowledge* (*Rizhilu* 日知錄) that writing should transmit new and essential knowledge— "what the ancients had not achieved and what future generations cannot do without." His *Record* was intended to fulfill just this purpose.[20]

Theories of literature and art emphasized the central importance of works of art as expressions of the individual self. What distinguished "good" painting and writing from the mediocre or bad was the power with which they conveyed the emotions and character of the individual artist or writer. This distinction was the basis for the disdain literati felt for professional artists and writers, who, it was assumed, worked to support rather than to express themselves. What made Li Bai 李白 (701–762), Du Fu 杜甫 (712–770), and Su Shi 蘇軾 (1037–1101) great poets was not their facility in recycling lines from earlier writings but their ability to create their own distinctive voices. Critics of poetry often attacked writers for their lack of originality and failure to develop a distinctive voice. Xu Wei 徐渭 (1521–1593) complained of contemporary poets:

> If a person imitates a bird's voice, even if his voice is that of a bird, his nature is still that of a man. If a bird imitates a person, even if it does sound like a person, its nature is still that of a bird. How are those who write poems today different from this? They do not write from what they know and simply steal things people once said, saying that such and such a piece is in such and such a style. . . . They cannot avoid seeming like birds imitating the speech of men.[21]

Imitation, in both the visual arts and writing, was a tool of teaching, but even those literati who promoted the occasional "return to the ancients" (*fugu* 復古) movements were, for the most part, advocating not imitation but imaginative recreation of ancient styles.

Invention   Less attention has been devoted to premodern Chinese understandings of invention. As is the case in many cultures, technologies fundamental to human life and society were believed to be the creation of mythical figures. Cang Jie 倉頡 (ca. 2650 BCE), mythical official historian to the equally mythical Yellow Emperor, was credited with inventing Chinese characters, and his contemporaries Fuxi 伏羲 and his sister (and wife) Nüwa 女媧, with introducing hunting, the domestication of animals, and cooking. Less essential but no less impressive inventions were attributed, often accurately, to real historical players, including Cai Lun 蔡倫 (48?–121), the court eunuch reputed to have invented paper; Zhang Heng 張衡 (78–139), the inventor of the seismoscope; and Su Song 蘇頌 (1020–1101), the creator of a hydromechanical astronomical clocktower. These named inventors tended to be members of the official (or, in the case of Cai Lun, court) elite; their inventions can be seen as manifestations of the kind of ownerless "knowledge of the Way" that Ivanhoe describes as serving the public good. There is no evidence that any of them took any measures to profitably "own" their creations.

Of course, a great many technologies were anonymous inventions, most likely the product of the tinkerings and mundane experiments by a host of unnamed peasants, artisans, industrial workers, and entrepreneurs. The treadle pump, blast furnace, woodblock printing, ceramic kiln, repeating crossbow (to give just a few random examples) were each the product of an inventor or, more likely, a succession of inventors, who relied on these tools and technologies for their livelihoods. Their humble social status would preclude their inclusion in the written record.

On those occasions when court officials or scholars chose to write about crafts and manufactures, they were, indeed, far more attentive to processes than to individual inventors. The records of Qing imperial workshops, for example, provide often quite detailed accounts of techniques of production, as, for example, Jin Jian's 金簡 (Kim Kan;

d. 1795) *Program for Printing with Movable Types at the Imperial Printing Office* (*Wuying dian juzhenban chengshi* 武英殿聚珍版程式, 1777), which describes the creation of 253,000 movable types for the production of a collectanea of rare books commissioned by the Qianlong 乾隆 emperor (r. 1735–1796)—but little information about individual inventors and workers.[22] Song Yingxing 宋應星 (1587–1666?), whose *Works of Heaven and the Inception of Things* (*Tiangong kaiwu* 天工開物, 1637) not only described but also theorized craft and industrial production, did not even credit human invention as the source of manufacturing processes. Rather, he interpreted craft technologies as manifestations of the movements of *qi* 氣, the stuff of the cosmos. As Dagmar Schäfer explains, for Song Yingxing, "Man's creative activity, crafts, and technological efforts . . . enacted universal patterns in the same way as natural phenomena."[23] The craftsman or the craftsman-inventor was merely an unknowing tool of these universal patterns, and craft work simply "a performance of general principles," a display of "the works of heaven" that could be known only by the scholar.[24] Song Yingxing's work, as Schäfer makes clear, was in many ways unrepresentative of literati understandings of craft processes, but he shared with his class an assumption that, however deeply peasants, artisans, and industrial workers may have been involved in material production, they were not recognized as individual creators or inventors, but as mere cogs in the wheel of the universe.

Assertions of Ownership of Intellectual Property

The ownership of knowledge in texts Chinese attitudes toward literary creation and technological invention were far too complex to support broad-brush claims of cultural orientations that inhibited the development of notions of intellectual property. Another, perhaps even more compelling reason to doubt such claims is the very considerable evidence that many Chinese *did* in fact believe that they owned the products of their intellectual labor (even if it might be considered "knowledge of the Way"), that, in short, they did cherish a notion of knowledge ownership or intellectual property—albeit one very different from that assumed in Western IP law.

The actual responses of Chinese authors and publishers to unauthorized reproduction of their work reveals a complexity that is lost in grand generalizations about attitudes toward the past and communal ownership of knowledge about the Way. To be sure, some authors appeared to accept unauthorized copying of their work with at least a good show of Confucian indifference to the loss of profit or fame that it might mean. The poet Yuan Mei 袁枚 (1716–1797), for example, seems to have been unmoved by reports that his *Poetry Talks from the Garden of Leisure* (*Suiyuan shihua* 隨園詩話), along with some of his other writings, had been pirated.[25] And Alford notes, citing art historian Wen Fong, "a general attitude of tolerance, or indeed receptivity, shown on the part of the great Chinese painters towards the forging of their own works." The Ming painter Shen Zhou 沈周 (1427–1509), on learning that his works were being forged, is said to have remarked calmly, "If my poems and paintings, which are only small efforts to me, should prove to be of some aid to the forgers, what is there for me to [begrudge]?"[26]

Noteworthy here is that, in both these examples, the expectation was that Yuan and Shen would be angry that their work had been copied, suggesting that their indifference was quite unusual—otherwise, it would not have been worthy of comment. But more direct evidence of the existence of some notion of intellectual property can be provided in the many cases of other writers and artists who did not take unauthorized reproduction of their work quite so lightly. Unsurprisingly, although cases exist prior to the tenth century, it is in the Song (960–1279), in the midst of China's first publishing boom, that we begin to encounter frequent complaints about what we now call the pirating of texts.

Until the late twentieth century, when the terms for "to steal printing blocks" "to steal a text edition" (*dao ban* 盜版), and "to steal printing" or "to pirate" (*daoyin* 盜印) were introduced,[27] the two-character phrase "fanke" (翻刻, literally "to recut [printing blocks]") was most commonly used to refer to the unauthorized reproduction of texts.[28] Unauthorized recutting of texts might take several different forms. A publisher might cut and print a draft manuscript without the author's permission; it was, in fact, this form of theft that landed the great Song poet Su Shi in so much trouble in 1079, when his poems, which had been circulating widely in unauthorized editions, were banned by the government.[29] A publisher might recut a work with a different title or author's name to make it appear a new work; thus in 1042 Di Zhaoying 翟昭應, when serving as district magistrate of Renhe County, Hangzhou, had the *Commentary on Statutes Governing Punishment* (*Xingtong lüshu* 刑統律疏) recut, printed, and sold as *Correct Text of Statutes and Substatutes* (*Jinke zhengyi* 金科正義). Another way of presenting a text as a new work was to abridge, expand, or rearrange its contents (again, without authorization); this was the fate of Zhu Mu's 祝穆 (?–1255) *Survey of Topography* (*Fangyu shenglan* 方輿勝覽, 1238), which was altered and then published under the title *Abbreviated Survey of Regions* (*Jielüe yudi jisheng* 節略輿地紀勝). Last is the simple unauthorized reprinting of an already published work, perhaps the most common form of "pirating," practiced very often on bestselling titles.[30]

The poet Su Shi did not share Yuan Mei's or Shen Zhou's casual acceptance of the unauthorized reproduction of his work. Worried about the damage that circulation of inferior editions might do to his reputation (and the threat to his life that the circulation of *any* edition posed after his work had been proscribed), he told a friend that he would happily destroy all the blocks used to print his poems if that act would prevent commercial publishers from recutting—and in the process often altering—his writings.[31] As Susan Cherniack has pointed out, as woodblock printing came to be accepted as a convenient means of textual reproduction in the Song, scholarly and literary reputations came to rest more and more on a writer's "performance in print." As a result, authors had a strong incentive to "control and monitor" the publication of their works.[32] When Sima Guang discovered that a pirate had stolen his notes for organizing the material for the *Comprehensive Mirror*, rearranged and changed the text—in the process, introducing many errors—and then published it under the title *Ditong* 帝統 (Imperial rule), his response was to put out his own carefully edited version under the title *Linian tu* 歷年圖 (Chronological chart). It quickly outsold the unauthorized version.[33]

Nor was the great Song Confucian thinker Zhu Xi 朱熹 (1130–1200) willing to shrug off the threat of unauthorized reproduction of his writings. As the leader of a controversial intellectual movement, he had a natural interest in ensuring the accurate transmission of his teachings. When he learned that an academy director in Wuzhou had published copies of his *Questions about the Four Books* (*Sishu huowen* 四書或問) without authorization, he simply bought up the whole stock of the printed texts. He explained this unusual move to a friend: he wanted to prevent the circulation of a variant text that, as it might contain serious errors, would mislead his readers and undermine his scholarly standing, but at the same time he did not wish to arouse the anger of the director by demanding that the blocks be destroyed.[34] In other cases he was not so considerate; in 1177, on hearing that another publisher was recutting the same work, he petitioned the local government to have the blocks seized and destroyed.

Zhu Xi, who seems to have been a favored target of pirate publishers, also had financial reasons for controlling publication of his works, as he relied in part on income from publishing to make a living. On learning that a publisher was planning to recut[35] his *Essential Meaning of the Analects* (*Lunyu jingyi* 論語精義) and *Essential Meaning of the Mencius* (*Mengzi jingyi* 孟子精義) without permission, he sought help from his friend (and an official) Lü Zuqian 呂祖謙 (1137–1181). In this case, Zhu Xi's primary concern was financial. "If Shen [the pirate] can be stopped before he has invested much money, neither he nor I will be hurt," he apologetically explained to Lü Zuqian. "Because I am poor and have to find means to eat, I am forced to this extreme. I hope you will excuse me."[36]

Several centuries later, the highly successful author Li Yu 李漁 (1611–1680), who also relied on the income from his publication and sale of his own writings (in Li Yu's case these were dramas and short stories, not commentaries on the Four Books), felt the need to move his publishing operation from Hangzhou to Nanjing (both major cities in the Jiangnan region, the cultural center of China) because the market in Hangzhou was flooded with pirated editions of his work. But once he had settled in Nanjing, he learned that in yet another major Jiangnan city, Suzhou, book merchants "with greedy hearts" were also recutting his writings. He complained to a friend, "I wipe them out in the east, then [have to] expel them in the west; I attack them in the south, then [have to] exterminate them in the north. When will I be able to stop fighting?"[37]

When authors and author-publishers did take the step of petitioning a local magistrate to prohibit unauthorized publication, they might couch their suit in terms not unfamiliar to a modern author demanding copyright protection. Take the case of Zhu Mu, author of the aforementioned *Survey of Topography*. A decree (*bangwen* 榜文) issued by the circuit intendant of Liangzhe and printed in the first edition of this work stated that it (and one other title by Zhu) "was the product of his own compilation (*size bianji* 私自編輯) and several years of labor." The family, the publisher of the work, had "invested heavily in cutting the blocks." If "publishers greedy for profit" recut (*fankai* 翻開) the work, they "will severely damage both Zhu's intellectual effort and the family's investment." Almost three decades later, well after Zhu Mu's death, his son obtained another decree to protect his father's writings, this time specifically against recutting by the notorious Masha commercial publishers of northern Fujian. Again, he claimed that these works deserved protection because they "were the result of a lifetime of hard intellectual labor." In contrast, "commercial publishers greedy for profit," "not capable themselves of writing works that express their own ideas and opinions, resort to recutting the work of others and, it is feared, changing the titles or abridging the text, misleading literati (*shidafu* 士大夫) readers and causing real harm."[38] Here the author was very much an individual expressing his own distinctive ideas and creating intellectual value, as well as a publisher investing heavily in the production of a text and expecting to recover this investment in profits from sales.

All these cases—and many more could be produced[39]—star prominent scholar-publishers and literary author-publishers. But commercial publishers—though very often the guilty parties in any dispute over pirating—were also sensitive to the problem of unauthorized reproduction. The phrase "recutting certain to be investigated" (*fanke bijiu* 翻刻必究) was often printed on the cover pages (*fengmian* 封面) of commercial publications in an effort to discourage pirates. Some publishers offered more forceful and colorful threats: "He who recuts these blocks will have a thief for a son and a whore for a daughter" (*fancibanzhe nan dao nü chang* 翻此板者男盜女娼).[40] Publishers and authors, then, shared notions of ownership.

The ownership of craft knowledge   As indicated above, recognition of invention in craft and industrial processes was usually associated either with an official's technological innovation for the public good (absent of any desire for individual profit) or with the unconscious expression by craft workers of the changing patterns of the cosmos.

Both views might suggest that technological invention or innovation is irrelevant to any consideration of the discourse of knowledge ownership in imperial China.

Nonetheless, there is evidence, both implicit and explicit, of claims to the ownership of craft knowledge. The *Zhuangzi* 莊子, composed in the late Warring States period (476–221 BCE), provides clear evidence that certain kinds of craft knowledge were very early on seen as commercially owned. The putative author Zhuang Zhou 莊周 (ca. 369–ca. 286) tells the story of a family whose ancestor had made a salve to prevent chapped hands; ever after, the family had made a living by bleaching silk in water, using the salve to protect their hands—and of course carefully guarding the secret of its manufacture. A traveler hears about the salve and offers to buy the secret for one hundred measures of gold. The family, dazzled by the riches the sale will bring them, agrees. Zhuangzi is making a point about skillful use (the traveler finds a better use for the salve and becomes wealthy),[41] but along the way makes clear that certain unique craft or manufacturing processes were considered the property of the individual or family that had developed or inherited them.

Indeed, craft knowledge was often kept secret or transmitted within families, who might impose inheritance and marriage restrictions on their members in order to prevent rival families from learning about special techniques.[42] In order to preserve family or master-apprentice traditions, craft techniques were very rarely recorded, much less published. To be sure, in many cases this absence of written manuals reflected the bodily nature of craft transmission. As Jacob Eyferth explains in his study of papermaking in Sichuan, since the procedures of manufacture were embodied and best learned through the repeated imitation of a master, there was no need for texts describing the processes—until the modernizing state attempted to "rationalize" the craft in the early twentieth century.[43]

Yet, when it was useful or necessary to record manufacturing processes, as, for example, in the making of medicines, secrecy was usually the rule. Pharmaceutical manuals or *fangshu* 方書 were often handwritten and closely guarded against theft, for a physician might prosper or fail on the strength of a claim to possess knowledge of a wonderfully efficacious "secret" prescription. It is noteworthy that prefaces to published prescription guides almost always praise the compiler for his willingness to make public pharmaceutical prescriptions that most physicians would, it was assumed, want to keep secret.[44]

Intellectual Property with Chinese Characteristics

It seems, then, that at least from the time in the early Song dynasty when the widespread use of printing ensured greater access to texts, many Chinese assumed that they owned the products of their intellectual labor (although, as we shall see, those products took a distinctive form). This belief included, to varying degrees depending on the individual, an expectation of fair profit from the dissemination of their works and/or a sense of responsibility (whether to their own reputations or to the public good) to ensure their accurate reproduction.

This is not to say that premodern Chinese notions of authorship and ownership of knowledge can be seen as equivalent to modern understandings of intellectual property as embedded in IP law. They were animated by other considerations, conventions, and concerns. Major differences in the ownership of knowledge and in the understanding of state-society relations and in the definition of the form that intellectual property took complicate any effort to assert (as several Chinese scholars do) that China developed a Western-style notion of intellectual property and "copyright" (*banquan* 版權) several hundred years before the West.[45]

The Imperial State, Society, and the Ownership of Knowledge

The assumptions that underlie modern IP law emerged in Western European society over the course of the seventeenth, eighteenth, and nineteenth centuries, products of a liberal ideology that celebrated a cluster of interlocking concepts: natural rights, individualism, the social contract, laissez-faire free trade, political representation, public interest, and "the marketplace of ideas."[46] Very different notions of governance and the relationship between the state, society, and economy underpinned the Chinese political and legal system. The Confucian ideal of the sage ruler who governed by moral example and ritual practice was of course never attained in practice, but the ideal supported paternalistic governance, whereby the ruler and officials, devoted to ensuring the welfare of the people, were responsible for deciding how best to achieve this goal; the people themselves were ignorant of what was best for them. To be sure, the ruler might want to learn the sentiments of his people, to understand how his policies were working, but the people were not to be "represented" in any formal institution that would allow them to hold the ruler to anything like a social contract. As Legalist thinkers prescribed, laws were the creation of the ruler; necessary for the maintenance of order and the ruler's authority, they were certainly not designed to protect anyone's individual rights. Individuals, all subjects of the emperor, were powerful primarily as they belonged to larger social units—the family, but also the household, the lineage, the gentry community, the examination cohort, and so on. The notion that each individual had "natural rights" to be protected by the state did not exist.[47]

Though Confucian and Legalist political theorists were at loggerheads on most points, they agreed on at least one: that the ruler had a responsibility to regulate and control what the people thought. However, each camp had a different motive for this view.

Legalists saw ideological control as essential to the maintenance of the ruler's power. Confucians understood it as one of the moral requirements of a virtuous ruler: it was his duty to ensure that his subjects thought "good" and correct thoughts. As William Alford puts it, "Central to [the ruler's fiduciary] responsibility was the need to determine which knowledge warranted dissemination and which ought to be circumscribed in the best interests of the commonwealth."[48] In a sense, then, the state claimed—*assumed* might be a better word—stewardship or control, if not ownership, of knowledge.

This very crude and necessarily incomplete characterization of Chinese political ideology as it relates to the issue of individual literary, artistic, or technological production helps to explain why, as Alford has more or less correctly pointed out, "virtually all known examples by the state to provide protection for what we now term intellectual property in China prior to the twentieth century seem to have been directed overwhelmingly toward sustaining imperial power. These official efforts were only tangentially, if at all, concerned with the creation or maintenance of property interests of persons or entities other than the state or with the promotion of authorship or inventiveness."[49] There is plenty of evidence that the imperial state was deeply interested in asserting its exclusive authority to control and thus "own" at least certain types of knowledge. One could trace this interest all the way back to Qin Shihuangdi's 秦始皇帝 (r. 221–210 BCE) destruction of works he deemed harmful to his state or, more positively, to the identification by Han Wudi 漢武帝 (r. 141–87 BCE) of five works—the *Classic of Songs, Book of Documents*, a trilogy of ritual texts, the *Classic of Changes*, and the *Spring and Autumn Annals*—as "Classics," works of sagely wisdom to be mastered by aspiring government officials.

But, not surprisingly, government efforts to control the publication and circulation of knowledge intensified around the same time that woodblock printing, invented no later than the early eighth century, made the reproduction and dissemination of texts easier. In 835 Emperor Tang Wenzong 文宗 (r. 827–840), alarmed at reports that privately produced calendars (*sizhi liriban* 私置歷日板) were for sale in far-flung markets in Jiannan, Liangchuan, and Huainan circuits (roughly, modern Sichuan and Jiangsu provinces), issued an edict, "probably the oldest publication ordinance in history,"[50] prohibiting the private publication of calendars.[51] Producing an accurate calendar, the work that annually established the rhythms of agriculture, was one of the most important responsibilities of imperial governance and thus the prerogative solely of the emperor. The Tang state also prohibited the reproduction, in manuscript or print, of astronomical charts (required for the creation of the calendar), government statutes, the dynastic histories, and works of prognostication.[52]

The establishment of the Song by rulers devoted to the project of centralizing state power—at the same time that woodblock printing finally emerged as a major means of textual reproduction—marked a new stage in the government's (implicit) claim to exclusive, authoritative use or determination of specific forms of knowledge. The state expanded the scope of the textual knowledge it assumed the authority to control and stepped up efforts to prevent, or at least regulate, extra-official publication of a wider range of works. As before (and throughout the rest of imperial history), the publication of calendars and astronomical charts remained, in law if not in fact, the exclusive preserve of the state. Prohibitions against their private publication were included in the *Criminal Code of the Song, Revised in the Jianlong Era* (960–963; *Song Jianlong chongxiangding xingtong* 宋建隆重詳定刑統) in 962 and repeated in the *Administrative Laws of the Qingyuan Era* [1195–1200], *Classified* (*Qingyuan tiaofa shilei* 慶元條法事類) in 1202, while separate imperial edicts reiterated the prohibition in 1071 and 1080.

Under the Song a whole range of government documents, legal writings, and historical works were added to the category of texts under the exclusive domain of state authority. The private publication and dissemination of imperial edicts, civil and military dispatches, memorials, records of the court, and other state documents were outlawed. Certainly, one of the goals was to protect state prerogative, but the prohibition was also designed to prevent enemies of the Song from learning state secrets. Indeed, an edict of 1090 outlining the harsh punishments to be imposed on anyone caught violating the proscription seems to have been issued in response to a warning by imperial envoy Su Che 蘇轍 (1039–1112), who had been disturbed by the number of Song government works he had seen during a visit to the neighboring Liao court.[53]

Such state prerogative or "ownership" can be reframed as an expression of the indivisibility of the state and the public good. The state recognized this indivisibility in taking on the responsibility to see to the production of "correct" and standardized editions of important works like the Classics. Just before the Song was founded, two ministers of state during the Five Dynasties period, Feng Dao 馮道 (882–954) and Li Yu 李愚 (?–935), published the first state-sponsored print edition of the "orthodox" texts of the Nine Classics (the number had grown from five to nine since the Han). And the later expansion of the civil service examination system in the Song—one of the centralizing measures taken by the early Song emperors—made it even more important for the state to establish correct editions of the Classics and other examination texts, as well as to prohibit the circulation of unauthorized and thus possibly inaccurate versions. The Directorate of Education (*Guozijian* 國子監) was responsible for producing standard editions of the Classics, which came to be known as directorate editions, or *jianben* 監本. The state also assumed the authority to compile and publish collections of examination essays by successful candidates (*chengwen* 成文) to serve as exemplars for students.

Doubtless the Song rulers, in making efforts to assert control over—and thus, implicitly, ownership of—the production of examination texts, were aiming, as Alford insists, to maintain themselves in power. But they were also attempting to fulfill a responsibility to ensure the accuracy and standardization of the texts to be tested. Some action of this sort was clearly necessary, as the widespread circulation of illegal (and cheap and error-ridden) commercial editions of the Classics had, as early as the late Northern Song, become a headache for the state. Particularly notorious were the *Mashaben* 麻沙本 (editions from Masha) published in a village of that name in northern Fujian. An oft-repeated anecdote recounts how a teacher in the Hangzhou prefectural school made a fool of himself by basing a question to his students on a Masha edition of the *Classic of Changes*. "The teacher asked, '[The hexagram] *qian* 乾 corresponds to "metal" [*jin* 金] and *kun* 坤 also corresponds to "metal," how is that?' Upon checking their Directorate of Education edition of the *Classic of Changes*, the students responded, 'Sir, you must be using a Masha edition, because the Directorate edition says that *kun* corresponds to "receptacle" [*fu* 釜].'"[54] Officials memorialized the throne repeatedly over the course of the eleventh, twelfth, and early thirteenth centuries about the harm the proliferation of such texts did to the integrity of the examination system. A series of imperial edicts, some specifically targeting the Fujian publishers, regularly (but largely ineffectually) proscribed private publication of examination literature without authorization and threatened violators with beatings and the destruction of their woodblocks.[55]

In asserting its ownership of examination literature, then, the Song government was acting in large part out of a concern for the integrity of the examination system. But the state was also interested in ensuring access to these and other important texts, for they did allow private publishers and local governments to reproduce them, as long as the reprints were of the "correct" editions authorized by the Directorate of Education.[56] The 986 edition of the Han-era dictionary *Explaining Patterns and Analyzing Characters* (*Shuowen jiezi* 說文解字), for example, includes a statement to the effect that individuals could use the government's woodblocks to reprint the text as long as they paid for the costs of the paper and the printing. (Since cutting the woodblocks was by far the largest portion of a book's production costs, this was not a bad deal.) The Nine Classics, in the orthodox edition compiled during the Five Dynasties period and endorsed by the Song state, could also be reprinted, as long as the directorate authorized the reprint and a fee was paid.[57]

Publishers could also cut their own set of blocks to produce facsimiles of government editions of certain medical treatises (the early Song state was committed to both centralizing medical knowledge and ensuring its broad dissemination in accessible texts), institutional compendia, Buddhist and Daoist texts, and dictionaries.[58] The government was, in effect, licensing private individuals to reproduce works in government editions from government woodblocks (or facsimiles thereof); the state could thereby both continue to control and implicitly own the knowledge being reproduced (and ensure some standardization and quality control) *and* encourage its broader dissemination.[59]

The Song state also instituted a system of publication registration that channeled new, not yet officially sanctioned textual knowledge into the realm of state control. We might think of this as an ownership-sharing system, in which the state, in order to maintain control over what was published and circulated, acknowledged some private ownership of knowledge. Imperial edicts in 1009 and again in 1090 ordered all nonofficial publishers to submit the manuscripts of texts they planned to print to the Directorate of Education in the capital for prepublication approval; the latter edict also stated what kinds of works would not be approved. Again in 1159 the government prohibited the publication of any work that had not received approval from the directorate and, in addition, required that a copy of each approved work, printed on special yellow paper, be deposited at the directorate. Between 1195 and 1201 a series of edicts codified these regulations and asserted the state's authority to censor improper (that is, heterodox and "frivolous and licentious") content and even inappropriate writing styles.[60]

Only a few extant Song imprints provide evidence of how this system of registration worked. Approved works apparently received warrants of publication (*gongju* 公據 or *wendie* 文牒),[61] statements certifying that the publisher had submitted the work for approval and often giving quite detailed information about the production of the book, including the number of characters in the text, the number of blocks used to print the work, the cost of an impression for a single copy, and the retail price of the book. At a certain point, too, it seems that local officials could stand in for the Directorate of Education. For instance, a colophon to the *Xiaochu ji* 小畜集, the collected works of the poet and official Wang Yucheng 王禹偁 (954–1001; *js* 983), certifies that the book, containing 163,848 characters, had conformed to the requirement, stated in the *Statutes of the Shaoxing Era* [1131–1162] (*Shaoxing ling* 紹興令), that all private publications be submitted to the authorities for inspection. The issuer of the warrant, however, was the director of the postal relay service in Huangzhou (modern Huanggangshi, Hubei), not the Directorate of Education; he found the *Xiaochu ji* "advantageous to learning" and thus granted permission for publication and distribution of the text. Such warrants might be printed as a colophon in the book, as in this example, or inserted on the title page.[62]

By the late twelfth century at the latest, the warrants might also include prohibitions against unauthorized reprinting or alteration of the works. The earliest extant example of such a warrant appears in an edition of Wang Cheng's 王稱 *Record of the*

Eastern Capital (*Dongdu shilüe* 東都事略), published in Meishan 眉山 (Sichuan) between 1190 and 1194 by the author's son; a colophon on the last page of the table of contents reads, "Published by the house of Secretary Cheng of Meishan. It is already reported to the authorities. Reprinting is not permitted."[63] Evidence that the Directorate of Education did in fact act against pirates comes from a warrant issued by that office commanding a local official in Ganzhou (Jiangxi) to stop the unauthorized reprinting of Duan Changwu's 段昌武 (fl. thirteenth century) *Collected Glosses on the* Mao Odes *by the Cinnamon Grove Studio* (*Conggui Maoshi jijie* 叢桂毛詩集解, preface dated 1248) by an unscrupulous relative of the original publisher. He was profiting handsomely from sales of the pirated text in neighboring Fujian Province.[64]

These strategies for managing the intellectual property of the state were devised, as Alford asserts, from a desire to "[sustain] imperial power" in a political and legal context quite alien from that which produced the early modern Western notion of IP and copyright. But it is worth noting that the phrase "sustain imperial power" does not do full justice to the varied concerns of the state in its attempt to control and, in a sense, possess knowledge. Its efforts to regulate the reproduction of texts grew out of its Confucian responsibility to ensure that people were exposed only to correct ideas, and that these ideas were determined by the paternalistic state. But, on the evidence of the licensing system developed in the Song and continued at least through the Ming, the state also wanted to promote the broad dissemination of correct ideas, even if it meant granting (in theory, tightly controlled) reproduction privileges of state editions to nonofficial players.

Print Technology, Tangible Property in Knowledge Production, and the Ownership of Knowledge

The Song state may have envisioned itself as the sole owner of important forms of knowledge; certainly, the rules and institutions that it established to assert control over text creation and reproduction suggest that this was the case. But there is little evidence that, over time, the Song state was routinely able to enforce this regime of control. As the Song empire crumbled over the course of the twelfth and thirteenth centuries, and as private and commercial publishing flourished, it became increasingly difficult for the state to effect its claims of knowledge ownership through control of print.

There is no sign that any later imperial state wavered in its assurance of its claims to authority over knowledge production, although, to be sure, actual enforcement of this authority was frequently inadequate or neglected. During the Ming (1368–1644) the government was still trying (in vain) to prevent Masha publishers from producing error-ridden editions of the Classics; publishers were allowed to publish facsimile reprints of the government editions, but prohibited (ineffectually) from producing their own versions. Throughout the rest of imperial history there were intermittent efforts at post-publication censorship of varying intensity and success. The brutal literary inquisitions of Ming Taizu 明太祖 (r. 1368–1398) and the early emperors of the Qing (1644–1911: Kangxi 康熙, r. 1661–1722; Yongzheng 雍正, r. 1722–1735; Qianlong, r. 1735–1796) managed to obliterate thousands of works these rulers considered threats to their power. Occasionally, zealous officials would initiate campaigns to destroy texts perceived as pornographic and harmful to the public good, as in 1868, when the governor of Jiangsu Province Ding Richang 丁日昌 (1813–1882) prohibited a long list of historical romances, "scholar-beauty" love stories, and jokebooks that he deemed "licentious."[65] But no imperial state succeeded in establishing institutions that were effective in systematically enforcing its claim to control and "own" knowledge.

In the face of state failure to institute a thorough system of control, there was, particularly from the Song on, a widespread implicit acceptance of some de facto private ownership of textual knowledge in the form of the capacity to produce texts not approved by the state. And here it is necessary to consider one other major difference between Chinese governance of knowledge production and modern IP regimes. As He Zhaohui has pointed out, because of the nature of the dominant premodern print technology, intellectual property was understood to be tangible,[66] not intangible (the conclusion reached after much legal debate in the West[67]). Although the Chinese had long used a variety of methods of movable type printing—employing earthenware, wooden, or metal fonts—woodblock printing or xylography remained, from the seventh to the late nineteenth century, the major mode of text reproduction. Ownership of the woodblocks for a text, which represented by far the largest share of the capital investment in book production, defined the publisher of that text. Typically, the publisher of a woodblock text was identified on the cover page (or in a colophon elsewhere in the book or on the first page of the main text) with the phrase, such-and-such a shop/studio "holds/owns the blocks" (*cang ban* 藏版／板).[68] Tellingly, on those rare occasions when a pirate was prosecuted successfully, he was required to destroy the blocks of the work he had stolen.

The customary regulations governing relationships among publishers also indicate that intellectual property was *material* property in the premodern Chinese book trade. If a publisher simply rented blocks from another publisher, he was not to change the name of the publishing house (*tangming* 堂名) that identified the owner of the blocks on the cover page. But if he purchased cut blocks from another publisher, he became the owner of that piece of intellectual property and could erase evidence of the previous owner by having that man's shop name scooped out of the block and replaced by a piece of wood cut with his own.[69] And as the new owner of the work, he could also of course add new paratexts to the original text, rearrange or reformat it as he saw fit, or combine it with other texts in a collection or *congshu* 叢書.[70]

That intellectual property was considered to be material property might help to explain the continuing importance, into the early twentieth century, of what is called private (*sike* 私刻), family (*jiake* 家刻), or literati publishing, in contrast to government (*guanke* 官刻) and commercial (*fangke* 坊刻) publishing, in Chinese book history. Here, an individual, often representing a family, would hire block-cutters and printers to prepare woodblocks for the printing of, for example, a text or texts in the family library; a collection of an ancestor's or family member's poetry; a rare scholarly work, history, or medical text; or a literary creation of the individual's own composition.[71] As long as the individual publisher (or his family) owned the blocks, he owned the text (whether he or any other family member had written it or not), could control its content, and profited from its sale.[72] But as soon as he sold the woodblocks, the tangible property of knowledge reproduction, he lost control of a text's contents as well as the profits from its sale.

Authors in this context would have a particular incentive to publish their own writings. Indeed, it is noteworthy that in almost all the premodern cases we have of publishers pursuing pirates, the publishers are also the authors of the works pirated, authors who were presumably keen to preserve the integrity of their writing, but also dependent on income from publishing to support themselves. In order to profit both in terms of reputation and sales, they had to own their knowledge—their intellectual property—in the material form of woodblocks.

Extra-Official Efforts to Claim Ownership of Textual Knowledge

Cutting woodblocks for the unauthorized printing of a text was thus perceived as not intellectual but rather economic theft. As the examples above demonstrate, both officials and author-publishers did understand pirating to be illegal—not, to be sure, because it was a violation of an author's or a publisher's "rights," but because it was a form of property theft.[73] Western observers of publishing in nineteenth-century China testified to this categorization as property theft of what in the West would be considered a violation of copyright:

> Even though the Chinese criminal code contains no sections referring to publishing law, a violation of the latter is nevertheless punishable in the Middle Kingdom. Those who violate an author's right may be ultimately punished with a hundred strokes of the bamboo or a three-year exile, if they printed and sold works without authorization. However, if the offense is only such that the work was printed but had not yet been sold, then the person who has transgressed is punished with fifty lashes of the bamboo and the confiscation of

the books and the woodblocks by which such books were printed. The transgressing publisher is sentenced under the section of the criminal code that refers to serious thefts and is subject to the same penalties as if he had robbed commodities.[74]

If unauthorized publication of a text was a form of property theft, then there was no need for the imperial state to develop separate laws or legal mechanisms specifically designed to facilitate routine prosecution of piracy.

In this context, how did author-publishers and commercial publishers protect their tangible intellectual property?

Customary regulations and publishing communities As already suggested, commercial publishers—at least those operating within a circumscribed publishing and bookselling network—developed some unwritten rules or conventions to establish "ownership" of a text and to restrict unauthorized reprinting. For example, the Ma and Zou lineage-based publishing industries, established in the late seventeenth century in two contiguous villages in remote and mountainous western Fujian, formulated a set of rules designed to regulate competition, while at the same time allowing for both shared benefits from the publication of tried-and-true bestsellers and the profits an individual publisher might reap from the introduction of new texts. These rules required the publishing households to post, at the end of each year, the cover pages of the texts they planned to publish in the new year (*suiyi shua xin* 歲一刷新, "each year publishing anew"), so that multiple claims to the same title could be negotiated ahead of time.[75] Once negotiations were completed, a publisher who held the blocks "owned" them (*cangban suoyou* 藏版所有) and thus the title at least for that year. Rented blocks remained the property of the renter; purchased blocks became the property of the purchaser. At the same time, enormously popular works like the literacy primer *The Three Character Classic* (*Sanzi jing* 三字經) were considered the property of any publisher who wished to cut the blocks for a new edition.[76]

It is likely that the efficacy of such regulations depended heavily on the limited size and closed nature of the publishing community and the support of the most powerful local publishers. Although the Zou and Ma publishers were competitors, they were interacting (and intermarrying) within a circumscribed local community. It is not clear how effective these rules would be in a more expansive business context. They were certainly not powerful enough to regulate the publishing industry as a whole; the Zou and Ma publishers never hesitated to pirate works of "outside" houses. There is little evidence that there ever developed within the commercial book trade the sorts of guilds or associations that, as was the case in other industries and businesses, might routinely and consistently enforce the informal rules governing intellectual property.[77]

In the absence of effective trade organizations or industry-wide regulations, both private and commercial publishers asserted their proprietary status by employing a variety of paratextual elements. The identification of the owner of the blocks, followed by the warning "reprinting will be investigated/prosecuted" (both discussed above), was the simplest and most common expression of ownership. Colophons (*paiji* 牌記), initially printed at the end of texts but eventually moved to the front, identified the publisher of a text and at times provided other information that supported the claim of ownership, such as the publisher's address, the date of publication, the relationship of the edition to previous editions, and claims about the edition's accuracy.[78] The commercial publishers of Jianyang outdid all others in "branding" their work with very distinctive colophon formats and quite elaborate colophon texts. Beginning in the late Yuan or early Ming, these publishers printed their colophons within a lotus-leaf design that functioned rather like a trademark (although a lotus design was used by other publishers as well).[79]

These men also used colophons to advertise their texts, assert proprietary claims to them (in the form of editorial labor and costs), and explicitly address the problem of piracy. For example, Liu Ziming 劉子明 (*zi* Shuangsong 雙松), includes a colophon on the cover page of his popular encyclopedia, *Marvelous Complete Book of Myriad Treasures from the Forest of Literati and for the Convenient Use of All under Heaven, Newly Published and Fully Supplemented* (*Xinban quanbu tianxia bianyong wenlin miaojin wanbao quanshu* 新板全補天下便用文林妙錦萬寶全書, 1612) that both makes claims about his own involvement in the production of the text and warns the prospective reader against pirated editions:

> This text was originally compiled by this house [Liu Ziming's Anzheng tang 安正堂] and circulated widely. Now, because the second set of blocks is worn, we have taken the trouble to hire cutters and printers and asked Master Liu Shuangsong to revise the old and add the new, to extract the pure and select the outstanding, so that all knowledgeable men in the world will praise it. Recently, profit-hungry villains have, under a false registration, fraudulently sold incomplete reprints that will not only do readers no benefit, but will also cause users to make errors. Therefore, in the third authentic edition, we have cut the title "true compendium of myriad treasures" (*zhen wanbao quanshu* 真萬寶全書) and certified the text with the seal of twin pines [Liu's courtesy name, Shuangsong, means "twin pines"]. If the buyer finds that name, then there is no error.[80]

Here the publisher is calling on customers to aid in the fight against "fraudulent" reprints. He also asserts his personal engagement, both editorial and financial, in the work through his assurance that he himself has revised and corrected the work and has spared no expense to reprint an accurate edition; it is certified with his seal.[81] Other publishers, too, would print their seals in a work as a display of ownership. They were probably copying the more laborious—but perhaps more effective—practice of stamping cover pages of books with identifying seals, either pictorial or reproductions of the publisher's name in a fancy calligraphic script, in order to authenticate them.[82]

These measures, not surprisingly, did not work particularly well in preventing piracy. Tellingly, the most effective method of prosecuting piracy was ad hoc appeals for aid from local officials. Zhu Mu's appeal to the Liangzhe circuit intendent to stop the unauthorized publication of his *Survey of Topography* under a different title bore fruit in an official command to the pirate publisher to desist printing the work and destroy the woodblocks he had had cut.[83] We have seen that Zhu Xi and Li Yu both appealed to friends with official status to prosecute pirates. Perhaps the cleverest appeal to official assistance was made by Yu Anqi 俞安期 (fl. 1596), the author of the *Encyclopedia of Tang Dynasty Anthologies* (*Tang leihan* 唐類函). Anticipating that his encyclopedia would be pirated, he reported that the work had been illicitly reprinted even before it appeared on the market. The magistrate duly posted notices denouncing the pirates—and no one dared to reprint the work when it did come out.[84]

These examples demonstrate that officials did perceive unauthorized publication as a transgression, a form of property theft, that necessitated some action on their part. Petitioned for protection by individual author-publishers, they might issue decrees prohibiting the recutting of specific titles; appealed to for help by literati friends, they might personally intervene and order the destruction of the offending blocks. But these ad hoc actions were not effective in routinely preventing the pirating of texts. The sustained success of the Masha publishers, the most notorious pirates of the later imperial period, suggests that, despite some individual triumphs, authors and publishers without privileged access to powerful officials had little recourse when faced with pirating of their works.

For, as Ye Dehui has noted in his survey of this strategy,[85] such measures were not available to all; one had to have high status and excellent connections to be able to call on the assistance of local officials. And those who did not have connections or ranked rather low on the hierarchy of knowledge producers were reduced to printing ineffectual threats, like Cao Shiheng's 曹士珩 warning at the opening of his *qigong* manual *Dao yuan yiqi* 道元一氣 (late Ming)—not the sort of text likely to secure the respect of officials unless the author was *very* well connected: "If there are ignoramuses who furtively copy my book for profit, I will surely report them to the government."[86] Equally excluded from official favor, commercial publishers of popular fiction in the Qing occasionally attempted to identify their works as publications of the local government office by placing the phrase *benya cang ban* 本衙藏版, "the local yamen owns the blocks" on the cover pages, hoping that pirates, fearful of government investigation, would not dare to reprint them.[87]

And of course, even if appeals to local officials were effective, in the end the geographical and administrative scope of their efficacy was limited to the locality.[88] The case of Li Yu illustrates this point nicely: he was able to rely on his friend, the local official Sun Picheng 孫丕承, to protect his *cang ban*, ownership of blocks, in Suzhou; but for the same problem in Hangzhou, he had to call on his son-in-law to approach local officials there for assistance. Pirates were clearly aware of the local limits of official action, as the case of the unauthorized reprinting of Duan Changwu's *Collected Glosses on the* Mao Odes *by the Cinnamon Grove Studio*, mentioned above, indicates; here the pirate cleverly saw to it that his unauthorized reprints were sold outside the site of production, in a neighboring province. Successful efforts to protect either private or commercial "ownership of knowledge" / "ownership of woodblocks" depended on ad hoc and geographically limited efforts by local officials in response to appeals from literati members of their own status group and network of connections.

THE PEOPLE'S REPUBLIC OF CHINA, 1949–1976: THE CHALLENGE TO INTELLECTUAL PROPERTY

From the late nineteenth century on, Western nations eager to expand their business in China began pressuring the Qing government to legislate IP law in accordance with multiple Western models (and Japanese variants). The array of bilateral and multilateral copyright treaties signed in this era indicate that the exhausted Qing state, outnumbered and overpowered, had little choice but to comply (Alford titles his chapter on this stage in China's intellectual property history "Learning the Law at Gunpoint"). In 1910, the government issued a provisional copyright act that gave "certain very limited exclusive rights to Chinese authors";[89] this law aligned with the Western understanding of IP in that it acknowledged "authors' rights" (*zhuzuoquan* 著作權)—that is, an author's right to their intangible intellectual property—rather than protecting the exclusive access to publication by the owners of tangible woodblock property.

However, as Fei-hsien Wang argues, the most vigorous and effective practical efforts to regulate "copyright" (*banquan* 版權) came from Shanghai publishers who still understood it not as a right to intangible property, but as a right held by those who possessed the material means of production—even though lithography and letterpress printing had by this time replaced xylography as the major technologies of text reproduction.[90] That they should think of copyright in this way is hardly surprising when we look closely at the compound adopted to translate "copyright." *Banquan* 版權, introduced from Japan in the late nineteenth century, means literally "the block (*ban*) right, authority, or power (*quan*)," or "the right to the blocks." Thus, incommensurable understandings of the relationship between intangible intellectual labor and tangible means of text production were made equivalent in translation, creating a source of much later confusion.

The unstable translation of foreign IP law and its implementation in legal reforms of the early twentieth century were interrupted, and ultimately radically rejected, in the early decades of the People's Republic of China. In line with the ideology of the new state, there was a move away from a liberal, capitalist endorsement of individual property rights toward an approach that first attempted to privilege the rapid and widespread dissemination of knowledge over individual ownership rights, and then embraced a full-bodied Marxist repudiation of the whole notion of individual ownership of intellectual property. In this second stage, inventions, scientific discoveries, and literary and scholarly writings were interpreted as the products not of an individual mind and effort, but of social forces (to which the individual inventor, scientist, or author may have had privileged access). As society (or "the public") was therefore the only legitimate beneficiary of intellectual labor, no one person had the right to own the products of his or her individual intellectual labor.

Unfortunately, little has been written about the actual implementation of the rules developed during these two stages—the first of adaptation, the second of repudiation. This discussion will thus necessarily be brief.

After the PRC was founded in 1949, the government repealed all the laws of the previous, republican, government, including regulations around copyright and publishing. Not surprisingly, given the overwhelming emphasis on the need for scientific and technological development and modernization of the economy, early efforts to develop new regulations of intellectual property focused on inventions and patent law. Initially, in 1950, a patent system remained in place, although inventors were encouraged to accept, in place of a patent (*zhuanli* 專利),[91] a certificate of authorship and a monetary award or bonus. In 1954 patents were abolished, but certificates of authorship were still issued and material incentives provided for invention. In 1963, however, the Regulations on Awards for Inventions ("Faming jiangli tiaoli" 发明奖励条例) declared that all inventions were the property of the state. Inventors were to be rewarded with honorary certificates and a one-off payment matching the value of their invention. In addition, inventors might receive other noncash "rewards" such as privileged medical treatment or travel subsidies.[92]

The regulations did, then, continue to acknowledge the usefulness of material rewards as incentives for invention, while imposing greater state control over invention.[93] Under both ideological pressure to embrace the socialist understanding of invention as a social product that naturally "belonged" to society and economic pressure to encourage technological development, it seems that the state was attempting to develop regulations that would ensure (through the organs of the state) immediate and widespread dissemination of useful inventions and, at the same time, offer the material incentives that in theory drove invention and innovation. But in the 1963 regulations, there is clearly an interest in reducing material incentives and retreating from any notion of intellectual property "rights" held by inventors.

Regulations on publishing and copyright followed a roughly similar path. The 1950 Resolution on the Improvement and Development of Publishing Work ("Guanyu gaijin he fazhan chuban gongzuo de jueyi" 关于改进和发展出版工作的决议) called on publishers to respect copyrights and prohibited the unlawful reproduction and alteration of texts.[94] Royalties were to be determined by the publisher in consultation with authors, and "based on the principle of giving consideration to the three interests of authors, readers, and publishers"; and were to be calculated on the basis of the nature and quality of the work, the number of characters, and the number of copies printed.[95] Subsequently, in 1952, the General Publishing Office (GPO) urged publishers and authors to sign contracts clearly stating royalty payments and copyright regulations.[96]

Through the 1950s to 1958, these arrangements, although doing little to stop copyright violations, "resulted in relatively large payments to authors," because they allowed continuing royalties to be paid to authors on a per-copy basis. Even an effort by the GPO to establish a uniform remuneration standard for royalty calculations did little to reduce "sizable payments to authors." In 1958, however, in the wake of the Anti-Rightist movement, both the per-character and per-copy royalties were cut by half. Three years later, per-copy royalties were eliminated altogether, as were royalties for reprints. Instead, it was decreed that authors would be given one-off "basic payments" (*gaofei* 稿费) for their work; these payments were to be calculated on the basis of the nature of the work (with scientific writings more highly valued than others) and the quality and quantity of the writing (the latter measured by the number of characters). Publishers were to respect "the rights both of authors and of [other] publishers," and thus unauthorized reproduction or emendation of texts and plagiarism was prohibited.[97]

But the state provided no legal mechanism for the prosecution of copyright violation. This decision of the state not to provide legal protection for authors and publishers has been seen as an indication of the government's "distaste for legislating extensive rights for an elite group" (in particular, its determination to prevent the development of a literary or cultural elite such as that which dominated late imperial Chinese society) and its related goal of minimizing "the differences between income groups."[98]

The Cultural Revolution of course overturned the balance between providing incentives for invention and ensuring public access to new knowledge that the regulations governing invention and publication issued between 1950 and 1963 had, to varying degrees, tried to strike. Material incentives were abandoned, and the Marxist conviction that all technological advances and artistic and literary creations were the products of society, not an individual, was ardently pressed. As the New China News Agency (Xinhua tongxun she 新华通讯社) announced in October 1966, the year the Cultural Revolution began, "In China's major inventions, it is impossible in many cases to establish who are the inventors, because the combined effort of so many people and so many units are involved, and no one claims credit."[99] One effect of this view was the proliferation in the late 1960s and 1970s of works authored by committees whose individual members were not identified.

The government thus repudiated the notion that an individual might lay claim to any intellectual property "right." Dietrich Loeber explains the impact this move had on authorship and copyright:

> An author generally receives no compensation for works which he has produced. If his writing activities are part of his employment responsibilities, he receives only his salary. If an author engages in literary or artistic activity outside of his profession, then this is regarded as honorary service to the people. Consequently, there is no room for the commercial transfer of rights to exploit in an author's work. As a general rule, an author transfers the right to use his work to state enterprises (e.g., publishers, film studios, theater troupes) without receiving any compensation for doing so. Thus, the use of copyright works is a state monopoly.[100]

Yet, as Loeber notes, the transfer from author to state is not a *legal* requirement, but a political one: "In China, an author's right to use his works without payment is not alienated against his will by compulsory legal means. Rather, the author is persuaded by political means to transfer the right of use voluntarily to state organizations."[101]

During the Cultural Revolution, then, the Chinese state reclaimed—but with much fuller force than had been possible in the Song—its "ownership" of knowledge and knowledge production by assuming the authority to decide which inventions and creations were for the public good and to oversee their immediate dissemination. As Alford notes, the state "freely reproduced or tolerated the reproduction of [works deemed worthy of publication] without obtaining the permission of the author or original publisher, providing any remuneration, or, in some instances, even acknowledging authorship."[102]

The insistence on both the communal sources of invention/creation and the moral imperative that invention/creation (as defined by the state) should immediately serve society has, I imagine, never received more vigorous promotion than it did during the Cultural Revolution. Here, the repudiation of liberal Western notions of intellectual property as a possession of the individual and the "rationality of the market" was complete.

Inventors and creators were social creatures, not individual geniuses standing above society. They therefore had no "rights" in (and thus no claim to profit privately from) their inventions or creations, which, as products of society, had to be disseminated throughout society, at no cost, for the common good. The state, not the market, determined the suitability of inventions and creations and managed their dissemination.

CONCLUSION

At certain periods in both premodern and modern China, there were moments when knowledge, as evaluated and channeled through the state, was seen as "ownable" not by any one individual but by the public at large, the people, as represented by state. In the 1960s, this point was made explicit in part as a reaction against the Western notion of IP; in the Song, we have to extrapolate it from what little we know of the institutions and regulations the imperial state established to ensure that the Classics and other important texts—in state-approved editions—were disseminated to the literate population. We might argue that the noticeable absence of interest by later imperial states in developing and routinely enforcing laws for the prosecution of piracy offers a kind of tacit acceptance of that stance.

But these cases also suggest that the only way in which knowledge legitimately becomes the "possession" of the public is through the vigorous intervention of the state. When the Song state allowed private individuals and publishers to license Directorate of Education blocks to print the Classics and orthodox medical texts, it was hoping to ensure the widespread dissemination of knowledge for the public good. And in granting warrants to publish preapproved texts by members of the educated elite—and, in the process, offering protection against unauthorized recutting—it was attempting to regulate the flow of new knowledge to the reading public.

When the modern Chinese state legislated the immediate entry of inventions (and by extension, writing, computer software, works of art, and other creative works) into the public domain, it set up an award system that provided authors and inventors with proofs of authorship or invention and one-off monetary awards of standardized, predetermined amounts, while obviating the need to protect the private profit of publishers by making all publishing operations public. The contributions of authors and inventors were acknowledged (and they were encouraged to continue creating and inventing), and their works were open to all on publication and circulation.

But in both cases, the government was also empowered not only to decide what was published and disseminated, but also to favor certain forms of new knowledge over others—that is, to decide what knowledge is. Insofar as control makes a claim of ownership, then the Song state and the PRC state during the Cultural Revolution were equally claiming ownership of knowledge. It might be noted that both made this claim on the basis of an assumption that the state was indivisible from the public good. In the case of the Song and later dynasties, the claim was one of screening knowledge, blocking the production of "heterodox" and "licentious" works that would be harmful to public morals, while promoting the publication and dissemination of "orthodox" works, primarily for the consumption of the educated elite. In the case of the PRC, the state was claiming to represent the revolutionary masses in determining what useful knowledge was and in ensuring its immediate and widespread dissemination.

ACKNOWLEDGMENTS

I thank Bryna Goodman, the participants in the two MPIWG conferences on the ownership of knowledge, and the two reviewers for their very helpful corrections and comments on this chapter.

Notes

1. Peter Drahos, "Intellectual Property Law and Basic Science: Extinguishing Prometheus?," *Law in Context* 10 (1992): 56–79. See also Michael A. Heller and Rebecca S. Eisenberg, "Can Patents Deter Innovation? The Anticommons in Biomedical Research," *Science* 280, no. 5364 (1998): 698–701.

2. Peter Baldwin provides a far more nuanced account of the early modern and modern Western debates over the relative importance of an author's individual rights and service for the public good, which ended in the early twentieth century with the enshrinement of authors' rights in modern IP law. Peter Baldwin, *The Copyright Wars: Three Centuries of Trans-Atlantic Battle* (Princeton, NJ: Princeton University Press, 2014), 16.

3. Tao-tai Hsia and Kathryn A. Haun, "Laws of the People's Republic of China on Industrial and Intellectual Property," *Law and Contemporary Problems* 5, no. 3 (1973): 275–277.

4. Peter Drahos, *Intellectual Property, Indigenous People and Their Knowledge* (Cambridge: Cambridge University Press, 2014).

5. The phrase is from William P. Alford, *To Steal a Book Is an Elegant Offense: Intellectual Property Law in Chinese Civilization* (Stanford, CA: Stanford University Press, 1995), 56.

6. Very generally speaking, IP rests on several developments distinctive to European history: a liberal idea that society advances through the pursuit of individual interests, a related romantic notion of the author/inventor as an individual who has transcended the prior knowledge of society, and the emergence of the "modern political subject" and the idea of inventors' "rights." These assumptions (along with, not surprisingly, a stronger notion of plagiarism) developed at a

particular moment in Western society and have earned the force they have today only because of the geopolitical balance of power.

7. Alford, *Elegant Offense*, 25. The "persons acting in a fiducial capacity" would of course be members of the highly educated elite, ideally scholar-officials who had achieved their official status in part through mastery of the writings of the sages.

8. Wen Fong, "The Problem of Forgeries in Chinese Painting," *Artibus Asiae* 25, no. 2–3 (1962): 100, quoted in Alford, *Elegant Offense*, 28.

9. Philip J. Ivanhoe, "Intellectual Property and Traditional Chinese Culture," in *Law and Social Justice*, ed. Joseph Keim Campbell, Michael O'Rourke, and David Shier (Cambridge, MA: MIT Press, 2005), 134–135. Ivanhoe also argues that, since "the vast majority of intellectuals were in one way or another working for the state," their writings would be defined as "works for hire" and not regarded as their personal property, even in the West today.

10. For a fuller critique of Alford's argument than can be offered here, see Jonathan Ocko's excellent review of *To Steal a Book Is an Elegant Offense*: "Copying, Culture, and Control: Chinese Intellectual Property Law in Historical Context," *Yale Journal of Law & the Humanities* 8, no. 2, (1996): 559–578.

11. See Michael Puett, "The Temptations of Sagehood, or: The Rise and Decline of Sagely Writing in Early China," in *Books in Numbers: Seventy-Fifth Anniversary of the Harvard-Yenching Library*, ed. Wilt L. Idema (Cambridge, MA: Harvard-Yenching Library, Harvard University, 2007), 23–48, for a more detailed discussion of the relationships between these different concepts of authorship and the ways in which authors manipulated these concepts to claim sagely authorship for themselves. Confucius of course famously identified himself as a transmitter, explicitly rejecting the notion that he created anything. Puett argues, first, that by the Han (206 BCE–220 CE) he had come to be identified as a creator; and second, that his disclaimer came to be employed by authors who were asserting themselves as creators, but recognized that they could not do so explicitly.

12. Stephen Owen, *Readings in Chinese Literary Thought* (Cambridge, MA: Council on East Asian Studies, 1992), 77, quoted in Puett, "Temptations of Sagehood," 42.

13. Charles R. Stone, "What Plagiarism Was Not: Some Preliminary Observations on Classical Chinese Attitudes toward What the West Calls Intellectual Property," *Marquette Law Review* 92, no. 1 (2008): 208. Sima Guang and his assistants cited 322 different sources; and in this very long work—a recent edition is ten thousand pages—there are only a little over one hundred instances in which Sima Guang uses his own words to write *lun* or commentaries/discussions. Robert André LaFleur, "Literary Borrowing and Historical Compilation in Medieval China," in *Perspectives on Plagiarism and Intellectual Property in a Postmodern World*, ed. Lise Buranen and Alice M. Roy (Albany: State University of New York Press, 1999), 141.

14. Stone, "What Plagiarism Was Not," 208.

15. LaFleur, "Literary Borrowing," 149; see also Stone, "What Plagiarism Was Not," 209.

16. Shang Wei, "The Making of the Everyday World: *Jin Ping Mei cihua* and Encyclopedias for Daily Use," in *Dynastic Crisis and Cultural Innovation: From the Late Ming to the Late Qing and Beyond*, ed. David Der-wei Wang and Shang Wei (Cambridge, MA: Harvard University Asia Center, 2005), 67.

17. See the translation by Cyril Birch: Tang Xianzu, *The Peony Pavilion*, 2nd ed. (Bloomington: Indiana University Press, 2002).

18. L. M. Dryden, "A Distant Mirror or Through the Looking Glass? Plagiarism and Intellectual Property in Japanese Education," in Buranen and Roy, *Perspectives on Plagiarism*, 80.

19. John Livingston Nevius, *China and the Chinese* (New York: Harper & Brothers, 1869), 65.

20. Gu Yanwu 顧炎武, "Zhu shu zhi nan" 著書之難, in *Rizhilu jishi* 日知錄集釋, ed. Huang Rucheng 黃汝成 (Shanghai: Shanghai Guji Chubanshe, 2006), quoted in Ke Shao, "The Justice of Balance: Understanding Intellectual Property from Chinese Historical and Philosophical Perspectives" (PhD diss., University of London, 2007), 50.

21. Xu Wei, "Ye Zisu shixu" 葉子肅詩序, in *Xu Wei Ji* 徐渭集 (Beijing: Zhonghua Shuju, 1983), 2:519; translation from Sophie Volpp, *Worldly Stage: Theatricality in Seventeenth-Century China* (Cambridge, MA: Harvard University Asia Center, 2011), 123. Writers of the late Ming seem to have been particularly contemptuous of literary imitation. Jiao Hong 焦竑 (1541–1620) delivered this pungent attack on contemporary authors: "Since they want to write great works but have no ideas of their own, what recourse do they have but to borrow phrases from Zuo Qiuming and Sima Qian [famous historians and literary stylists of ancient China]—like begging and stealing piss and shit? If one were to rub out all the archaisms and cliched expressions in their writings, one would end up with nothing more than a blank sheet of paper." Jiao Hong, "Yu youren lun shu" 與友人論書, in *Danyuan ji* 澹園集 (Beijing: Zhonghua Shuju, 1999), 1.93; reference and translation from Rivi Handler-Spitz, *Symptoms of an Unruly Age: Li Zhi and the Culture of Early Modernity* (Seattle: University of Washington Press, 2017), 22–23. See this work for similar complaints from late Ming authors like Wang Shizhen 王世貞 (1526–1590) and Jiang Yingke 江盈科 (1553–1605), 23–24; and Shao, "Justice of Balance," 50.

22. See Christine Moll-Murata, *State and Crafts in the Qing Dynasty (1644–1911)* (Amsterdam: Amsterdam University Press, 2018), 223.

23. Dagmar Schäfer, *The Crafting of the 10,000 Things: Knowledge and Technology in Seventeenth-Century China* (Chicago: University of Chicago Press, 2011), 88.

24. Schäfer, 235.

25. He Zhaohui 何朝晖, "Individual Protection, Local Coordination, and Material Attachment: The Copyright of Woodblock Imprints in Imperial China" (paper presented at the "Authorship, Copyright, and Editions: The Circulation of Works in Late Imperial China" workshop, Harvard University, May 5, 2007), 10.

26. Fong, "Problem of Forgeries," 100, quoted in Alford, *Elegant Offense*, 30.

27. These terms first appeared in the 1980 *Hui'an shuhua* 晦庵書話 of Tang Tao 唐弢 (1913–1992).

28. This term reflects the nature of the dominant publishing technology in the later imperial period: woodblock publishing or xylography. To reproduce a text without authorization, it was usually necessary to cut a new set of blocks, either using a print copy to create a facsimile or simply cutting what would be considered a new edition of the work.

29. Li Mingshan 李明山, *Zhongguo gudai banquan shi* 中国古代版权史 (Beijing: Shehui Kexue Wenxian Chubanshe, 2012), 130.

30. Li, 130. Li also provides the case of Liao Xingzhi 廖行之 (1137–1189), who published a collection of his own writings, *Xingzhai ji* 省齋集, under the name of the more famous Zhou Bida 周必大 (1126–1204), presumably in the hope that it would sell better, as an example of copyright violation. This strategy might better be seen, however, as a form of trademark violation rather than unauthorized reproduction.

31. Ronald Egan, "To Count Grains of Sand on the Ocean Floor: Changing Perceptions of Books and Learning in the Song Dynasty," in *Knowledge and Print Production in an Age of Print: China, 900–1400*, ed. Lucille Chia and Hilde De Weerdt (Leiden: Brill, 2011), 55; and Shao, "Justice of Balance," 77.

32. Susan Cherniack, "Book Culture and Textual Transmission in Sung China," *Harvard Journal of Asiatic Studies* 54, no. 1 (1994): 65.

33. Sima Guang 司馬光, *Sima wenzheng gong chuanjia ji* 司馬文正公傳家集 (Taipei: Taiwan Shangwu Yinshuguan, 1965), 71.19ab, quoted in Li Mingshan, *Zhongguo gudai banquan shi*, 143–144.

34. Zhu Xi 朱熹, "Yu Yang jiaoshou shu" 與楊教授書, in *Hui'an xiansheng Zhu wengong wenji* 晦庵先生朱文公文集 (Shanghai: Shangwu Yinshuguan, 1919), *juan* 26.5b–7a.

35. Zhu uses the term *fankai* 番 (for 翻) 開, which I take to mean "redo" or "recut."

36. Zhu Xi 朱熹, "Da Lü Bogong" 答呂伯恭, *Hui'an xiansheng Zhu wengong wenji*, *juan* 33. Translation, slightly modified, from Wing-tsit Chan in *Chu Hsi: New Studies* (Honolulu: University of Hawai'i Press, 1989), 78.

37. Li Yu 李漁, "Yu Zhao Shengbo wenxue" 與趙聲伯文學, in *Li Yu suibi quanji* 李漁隨筆全集 (Chengdu: Bashu Shushe, 2003), 444, quoted in He Zhaohui 何朝暉, "Shilun Zhongguo gudai diaoban yinshua banquan xingtai de jiben tezheng" 试论中国古代雕版印刷版权形态的基本特征, *Tushu yu qingbao*, no. 3 (2008): 114.

38. Zhou Lin 周林 and Li Mingshan 李明山, eds., *Zhongguo banquan shi yanjiu wenxian* 中国版权史研究文献 (Beijing: Zhongguo Fangzheng Chubanshe, 1999), 3; see also p. 4 for another example.

39. For multiple other examples, see Ye Dehui 叶德辉, *Shulin qinghua* 书林清话 (1921; repr., Shenyang: Liaoning Jiaoyu Chubanshe, 1998), 30–34; and Shao, "Justice of Balance," 80–87.

40. This curse appeared in *Court Cases of [Magistrate] Peng (Peng gong'an* 彭公案), published by the Benli tang 本立堂 in Beijing in 1892. I am grateful to Sören Edgren for this information, presented in the course of the "Authorship, Copyright, and Editions: The Circulation of Works in Late Imperial China" workshop, Harvard University, May 5, 2007.

41. Zhuangzi, *Basic Writings*, trans. Burton Watson (New York: Columbia University Press, 2003), 29.

42. Shao ("Balance of Justice," 89) provides the example of a lineage in Yongqing County, Hebei, whose family instructions dictated that their method of making special measuring instruments—from which they profited handsomely—never be passed down to the women in the family for fear that they might, on marriage, transmit the method to village rivals.

43. Jacob Eyferth, "Artisans into Peasants," in *Eating Rice from Bamboo Shoots: The Social History of a Community of Handicraft Papermakers in Rural Sichuan, 1920–2000* (Cambridge, MA: Harvard University Asia Center, 2009), 92–115.

44. See Cynthia Brokaw, *Commerce in Culture: The Sibao Book Trade in the Qing and Republican Periods* (Cambridge, MA: Harvard University Asia Center, 2007), 443.

45. Li, *Zhongguo gudai banquan shi*, 132. See also Zou Shencheng 邹身城, "Baohu banquan shiyu heshi heguo?" 保护版权始于何时何国?, *Faxue yanjiu* 法学研究 63, no. 2 (1984): 63; and Chengsi Zheng and Michael D. Pendleton, *Copyright Law in China* (North Ryde, Australia: CCH International, 1991), 11.

46. See Baldwin, *Copyright Wars*, for a detailed history of the debates over the meaning of copyright and authors' rights in Europe and the United States.

47. For this reason, *copyright* is not a useful designation for the Chinese notion of intellectual property before the twentieth century.

48. Alford, *Elegant Offense*, 23.

49. Alford, 16.

50. Hok-lam Chan, *Control of Publishing in China, Past and Present* (Canberra: Australian National University, 1983), 2.

51. Liu Xu 劉昫, "Wenzong ji" 文宗記, in *Jiu Tangshu* 舊唐書 (Beijing: Zhonghua Shuju, 1975), quoted in Li Mingshan, *Zhongguo gudai banquan shi*, 112.

52. Chan, *Control of Publishing*, 2. Throughout imperial history, the state prohibited the private publication of calendars and almanacs. The 1579 *Calendar of the Seventh Year of the Wanli Reign of the Great Ming Dynasty* (*Da Ming Wanli qinian sui yimao datong li* 大明萬曆七年歲已卯大統曆), for example, offers a reward of 50 ounces of silver to anyone who reported such private publication. See Cao Zhi 曹之, *Zhongguo guji banbenxue* 中国古籍版本学 (Wuhan: Wuhan Daxue Chubanshe, 1992), 449.

53. Hilde De Weerdt, "What Did Su Che See in the North? Publishing Laws, State Security, and Political Culture in Song China," *T'oung Pao: International Journal of Chinese Studies* 92, no. 4–5 (2006): 466–494.

54. Translation, slightly modified, from Lucille Chia, *Printing for Profit: The Commercial Publishers of Jianyang, Fujian (11th–17th Centuries)* (Cambridge, MA: Harvard University Asia Center, 2002), 116.

55. Chia, 121–123.

56. Denis Twitchett, *Printing and Publishing in Medieval China* (New York: Frederic C. Beil, 1983), 61.

57. Li, *Zhongguo gudai banquan shi*, 132, 141.

58. Chan, *Control of Publishing*, 19. The Ming state employed a similar policy; it permitted the reprinting of standardized texts of the Classics, but prohibited publishers from producing their own editions. See Ye, *Shulin qinghua*, 148–149, for a directive addressed specifically to the commercial publishers of Jianyang, who were accused of producing error-ridden editions of the Classics.

59. Twitchett, *Printing and Publishing*, 61.

60. Twitchett, 61.

61. Li, *Zhongguo gudai banquan shi*, 139.

62. Twitchett, *Printing and Publishing*, 65.

63. Zhou and Li, *Zhongguo banquan shi yanjiu wenxian*, 2–3. Translation from Twitchett, *Printing and Publishing*, 63.

64. Zhou and Li, *Zhongguo banquan shi yanjiu wenxian*, 4.

65. Brokaw, *Commerce in Culture*, 497.

66. He, "Shilun Zhongguo gudai diaoban yinshua," 116–117. He uses the Chinese term *tizhai yilai* 體載依賴 (depending on the material substance) to designate "material ownership" of blocks. I am indebted to Mario Biagioli for emphasizing this point during the initial workshop, "Ownership of Knowledge," at the Max Planck Institute for the History of Science, Berlin, November 2016.

67. Brad Sherman and Lionel Bently, *The Making of Modern Intellectual Property Law: The British Experience, 1760–1911* (Cambridge: Cambridge University Press, 1999), 9–59.

68. *Cang* literally means to "store up," "hold," "keep," or "hide away." But here it designates ownership. A person might commission a printshop to print a work on demand from blocks he provided; in this case, the printshop would physically hold the blocks in its shop, but the person who had paid for the blocks to be cut and provided them to the printshop was the publisher, the owner of the blocks, the one who *cang ban*. Variants on *cang ban* existed. *Zi* 梓, "catalpa," sometimes replaces *cang ban*, but as catalpa is one of the woods most commonly used to make woodblocks, this is just another way of identifying ownership through the possession of cut blocks.

69. These rules about block purchase, although never to my knowledge codified or given legal status before the twentieth century, seem to have been widely acknowledged. See Chia, *Printing for Profit*, 165, 168, 170, and He, "Shilun Zhongguo gudai diaoban yinshua," 116, both describe a lively market in woodblocks in late Ming Jiangnan and Jianyang (northern Fujian).

70. As one might imagine, there are many variants of this process, and the practice has caused much confusion for bibliophiles. Some publishers were very thorough, replacing *all* references to an earlier publisher, making it very difficult to know who had published the original edition; they might even add other "evidence" to suggest that they were the original publishers of a work. Thus, after Shen Shangjie 沈尚傑 purchased the blocks of Qian Zeng's 錢曾 *Reading Select Works* (*Dushu minqiu ji* 讀書敏求記), first published in 1726 by Zhao Mengsheng 趙孟升, he removed all traces of Zhao's publishing house, Songxue Zhai 松雪齋, and, for good measure, added a preface claiming that he was publishing the first edition of the work, in 1745. But others, either through laziness or limited funds, replaced the publishing house's name only on the cover page, leaving the original house name on the first leaf of the main text or in the lower margins at the folds of all the folio pages. See He, "Shilun Zhongguo gudai diaoban yinshua," 117. One nineteenth-century edition of the collected medical works of the popularizing physician and official Chen Nianzu 陳念祖 (1753–1823), *Sixteen Medical Works Composed at Leisure* (*Gongyu shiliuzhong yixue*

quanshu 公餘十六種醫學全書) (n.p.: Shancheng Tang, n.d.), for example, is identified as a reprint by the Shancheng Tang on its cover page. But at least two other houses are listed as publishers in the works that make up the collection. For other examples and the range of complications that could arise from these practices, see He, "Shilun Zhongguo gudai diaoban yinshua banquan xingtai de jiben tezheng," 116–117.

71. The cost of block-cutting required by far the largest outlay of a publisher's capital, as paper and printing and binding costs were relatively low; this is perhaps one of the reasons why ownership of the blocks determined who owned the intellectual property. An individual or family who wished to publish a work might commission a "character-cutting shop" (*kezidian* 刻字店) to cut the woodblocks; the person or entity that paid for the cutting labor was considered the owner of the blocks and, thus, the publisher.

72. For example, the Zhu lineage of Jianyang, Fujian, which claimed descent from Zhu Xi, owned a set of blocks for their famous member's literary collection, the *Collected Writings of Master Zhu* (*Zhuzi wenji* 朱子文集). Anyone who wished to print and sell copies of this work was required to pay for the privilege by giving one out of every ten copies printed to the corporate lineage. The lineage's ownership of the blocks, not their relationship to Zhu Xi, enabled them to control publication so effectively that no other editions of the work survive from Jianyang, one of the most prolific commercial publishing sites of the Song, Yuan, and Ming. Xie Shuishun 谢水顺 and Li Ting 李珽, *Fujian gudai keshu* 福建古代刻书 (Fuzhou: Fujian Renmin Chubanshe, 1997), 103.

73. Chengsi Zheng and Michael D. Pendleton, *Chinese Intellectual Property and Technology Transfer Law* (London: Sweet & Maxwell, 1987), 87.

74. "Verlagsrecht in China," *Der Ostasiatische Lloyd*, September 21, 1889. I am very grateful to Rudolf Wagner for informing me of this source, and to James Wang for the translation from German into English.

75. Yang Lan 楊瀾, *Linting huikao* 臨汀彙考 (preface dated 1878), 4, no. 8a.

76. Brokaw, *Commerce in Culture*, 177–185.

77. It should be noted, however, that there is some evidence that publishers might have formed associations of some sort, possibly even as early as the Song. Ye (*Shulin qinghua*, 30–34) uncovered references to a "book society" (*shuhui* 書會) in the Southern Song and to a "book guild" (*tushu hang* 圖書行) in late-Ming Beijing, but unfortunately no information about their precise nature or operation survives. In the very early Qing, the owners of the important publishing house Saoye Shanfang 掃葉山房 established the Chongde gongsuo 崇德公所 (Chongde Guild) in the major book center of Suzhou as a mutual aid society for publishers. In 1670, six years after its founding, the guild established the Chongde Academy (Chongde shuyuan 崇德書院) so that publishers would have a place to edit texts and discuss methods of collation. But the activities of both the guild and the academy are rather hazy. By the 1830s at the latest, the guild was functioning as an inspection station for prohibited books, apparently in cooperation with the local government. Booksellers from outside Suzhou had to have their products checked, and any prohibited titles were destroyed. See Joseph P. McDermott, "Rare Book Collections in Qing Dynasty

Suzhou: Owners, Dealers, and Uses," in *Jinshi Zhongguo de ruxue yu shuji: jiating, zongzu, wuzhi de wangluo* 近世中國的儒學與書籍: 家庭, 宗族, 物質的網絡, ed. Lü Miaw-fen (Taipei: Institute of Modern History, Academia Sinica, 2013), 242–246. It is possible that the guild also checked for pirated texts; unfortunately, the scanty sources on this operation do not allow for anything more than speculation. Early twentieth-century guilds founded to regulate the book business do not seem to have concerned themselves with piracy disputes. It was not until the establishment of "book-trade commercial associations" (*shuye shanghui* 書業商會) in 1905 that organizations of publishers began cooperating to control piracy and protect the foreign, imported "copyright." For these and later organizations, see Fei-Hsien Wang, *Pirates and Publishers: A Social History of Copyright in Modern China* (Princeton, NJ: Princeton University Press, 2019), 118–157, 158–210, 211–252. Interestingly, Japanese publishers of the early modern period, through guild organizations and collaboration with the state, were able to regulate the reproduction of texts much more successfully than their Chinese counterparts. See Peter Kornicki, *The Book in Japan: A Cultural History from the Beginnings to the Nineteenth Century* (Leiden: Brill, 1998), 181–182, 245.

78. For a very early example of this sort of colophon, see Sören Edgren, "Southern Song Printing at Hangzhou," *Museum of the Far Eastern Antiquities*, Bulletin 61 (1989): 32–33.

79. Chia, *Printing for Profit*, 217.

80. "Xinban quan[zeng]bu Tianxia bianyong wenlin miaojin wanbao quanshu 新板全[增]補天下便用文林妙錦萬寶全書," in *Mingdai tongsu riyong leishu jikan* 明代通俗日用類書集刊, vol. 10, ed. Zhongguo shehui kexue yuan Lishi yanjiusuo Wenhua shi 中國社會科學院歷史研究所文化室 (Chongqing: Xinan Shifan Daxue Chubanshe, 2011), 243. This strategy was practiced by literati-publishers as well. For example, when Gu Jian 顧梘 learned that his (legitimate) reprint of Lu Guimeng's 陸龜蒙 (?–881) *Collectanea from Lize* (*Lize congshu* 笠澤叢書) had been pirated, he inserted this announcement in a later edition: "I published this book with great care, but the number of copies was limited, and it was not distributed widely. Recently a merchant in Weiyang 維揚 copied the book without my permission, for the sake of profit. The characters are badly cut and the work lacks the [elegant] look of my book. I fear that knowledgeable gentlemen might mistake the reprint as the original from the Biyun caotang 碧筠草堂 and so make this announcement here." He, "Shilun Zhongguo gudai diaoban yinshua," 115.

81. Yu Xiangdou 余象斗, another late-Ming Jianyang publisher, established an even more deeply personal proprietary claim by having his portrait reproduced in several of his publications. See Chia, *Printing for Profit*, 217–220.

82. See Wang, *Pirates and Publishers*, 57–58, for information on the continuation of this practice in the late nineteenth and early twentieth centuries.

83. Zhou and Li, *Zhongguo banquan shi yanjiu wenxian*, 3.

84. He, "Individual Protection," 4.

85. Ye, *Shulin qinghua*, 30–34.

86. He, "Shilun Zhongguo gudai diaoban yinshua," 114.

87. In the course of discussion at the workshop on Chinese descriptive bibliography (Harvard-Yenching Library, May 16–19, 2017), Sören Edgren has pointed out that it is equally, if not more, likely that these publishers were trying to present an appearance of official approval in the hopes that censors might not discover that these works were "licentious."

88. He, "Shilun Zhongguo gudai diaoban yinshua," 115–116.

89. Norwood Allman, *Handbook on the Protection of Trademarks, Patents, Copyrights and Trade-names in China* (Shanghai: Kelly & Walsh, 1924), 178–179, quoted in Alford, *Elegant Offense*, 42.

90. See Wang, *Pirates and Publishers*, 40–55, for an account of the translation of "copyright" into the Japanese *hanken* (*banquan* in Chinese pronunciation) in the 1870s and its introduction to China later in the century. Wang equates ownership of the blocks with ownership of the means of production. She also provides finely researched accounts of the efforts of publishers, who had little support from the government, to protect their material intellectual property rights against piracy in Shanghai and Beijing in the early twentieth century.

91. *Zhuanli*, the Chinese compound used to translate "patent," first appears in the *Discourses of the States* (*Guoyu* 國語) with the meaning "exclusive [control of] profit" or "monopoly"—a translation that suggests nothing of the sense of "made public," "made open," that is included in the English word *patent* and that, not surprisingly, made the concept difficult for a socialist state to accept. Cheng and Pendleton, *Chinese Intellectual Property*, 51.

92. Barden N. Gale, "The Concept of Intellectual Property in the People's Republic of China: Inventors and Inventions," *China Quarterly* 74 (1978): 347–350.

93. Gale, 348–349.

94. Zhou and Li, *Zhongguo banquan shi yanjiu wenxian*, 267; see also Cheng and Pendleton, *Chinese Intellectual Property*, 88–90. Although the 1950 "Resolution" was granted the force of law with its promulgation that same year by the General Publishing Office of the Central People's Government, it did not dictate any procedures for enforcement of copyright protections. When, again that same year, the Dalian Bookstore reproduced five thousand copies of *The International Situation after the Korean War* without permission from the original publisher, the World Knowledge Press in Beijing, the General Publishing Office could only issue a report scolding the Dalian Bookstore for its "extremely improper" act and demand self-criticism and compensation. The purely administrative remedies available to victims of pirating were largely ineffectual. See Mark Sidel, "The Legal Protection of Copyright and the Rights of Authors in the People's Republic of China, 1949–1984: Prelude to the Chinese Copyright Law," *Columbia Journal of Art and the Law* 9 (1984): 482.

95. It is noteworthy that this document perpetuates the terminological confusion mentioned above by using two different words to express "copyright": *zhuzuoquan*, "author's rights," and *chubanquan* 出版權, "publishing rights." See Sidel, "Legal Protection of Copyright," 480. For another example of a similarly ineffectual effort to restrict piracy in the early PRC, see Fei-Hsien Wang's description of the conflict between the Shanghai Booksellers' Guild and Chunming Bookstore in "A Crime of Being Self-Interested: Literary Piracy in Early Communist China," *Twentieth-Century China* 43, no. 4 (2018): 275–294.

96. For sample contracts, see Zhou and Li, *Zhongguo banquan shi yanjiu wenxian*, 267–286.

97. Alford, *Elegant Offense*, 59–60. See also Hsia and Haun, "Industrial and Intellectual Property," 288–290.

98. Hsia and Haun, "Industrial and Intellectual Property," 290; and Dietrich Loeber, "Copyright Law and Publishing in the People's Republic of China," *UCLA Law Review* 24 (1977): 907.

99. "China's Unique Road for Developing Science and Technology," *New China News Agency*, October 17, 1966, in *Survey of China Mainland Press* 3805 (1966): 24, quoted in Gale, "Concept of Intellectual Property," 351.

100. Loeber, "Copyright Law and Publishing," 910.

101. Loeber, 911.

102. Alford, *Elegant Offense*, 64–65.

Bibliography

Alford, William P. *To Steal a Book Is an Elegant Offense: Intellectual Property Law in Chinese Civilization*. Stanford, CA: Stanford University Press, 1995.

Allman, Norwood. *Handbook on the Protection of Trademarks, Patents, Copyrights and Trade-Names in China*. Shanghai: Kelly & Walsh, 1924.

Baldwin, Peter. *The Copyright Wars: Three Centuries of Trans-Atlantic Battle*. Princeton, NJ: Princeton University Press, 2014.

Brokaw, Cynthia. *Commerce in Culture: The Sibao Book Trade in the Qing and Republican Periods*. Cambridge, MA: Harvard University Asia Center, 2007.

Buranen, Lise, and Alice M. Roy, eds. *Perspectives on Plagiarism and Intellectual Property in a Postmodern World*. Albany: State University of New York Press, 1999.

Cao Zhi 曹之. *Zhongguo guji banbenxue* 中国古籍版本学. Wuhan: Wuhan Daxue Chubanshe, 1992.

Chan, Hok-lam. *Control of Publishing in China, Past and Present*. Canberra: Australian National University, 1983.

Chan, Wing-tsit. *Chu Hsi: New Studies*. Honolulu: University of Hawai'i Press, 1989.

Chen Nianzu 陳念祖. *Gongyu shiliuzhong yixue quanshu* 公餘十六種醫學全書 [Sixteen medical titles composed at leisure]. N.p.: Shancheng Tang, n.d. Held in the Jinan Municipal Library (Shandong).

Cherniack, Susan. "Book Culture and Textual Transmission in Sung China." *Harvard Journal of Asiatic Studies* 54, no. 1 (1994): 5–125.

Chia, Lucille. *Printing for Profit: The Commercial Publishers of Jianyang, Fujian (11th–17th Centuries)*. Cambridge, MA: Harvard University Asia Center, 2002.

"China's Unique Road for Developing Science and Technology," *New China News Agency*, October 17, 1966. In *Survey of China Mainland Press* 3805 (1966): 24.

De Weerdt, Hilde. "What Did Su Che See in the North? Publishing Laws, State Security, and Political Culture in Song China." *T'oung Pao: International Journal of Chinese Studies* 92, no. 4–5 (2006): 466–494.

Drahos, Peter. *Intellectual Property, Indigenous People and Their Knowledge*. Cambridge: Cambridge University Press, 2014.

Drahos, Peter. "Intellectual Property Law and Basic Science: Extinguishing Prometheus?" *Law in Context* 10, no. 2 (1992): 56–79.

Dryden, L. M. "A Distant Mirror or Through the Looking Glass? Plagiarism and Intellectual Property in Japanese Education." In Buranen and Roy, *Perspectives on Plagiarism*, 75–85.

Edgren, Sören. "Southern Song Printing at Hangzhou." *Museum of the Far Eastern Antiquities*, Bulletin 61 (1989): 1–212.

Egan, Ronald. "To Count Grains of Sand on the Ocean Floor: Changing Perceptions of Books and Learning in the Song Dynasty." In *Knowledge and Print Production in an Age of Print: China, 900–1400*, edited by Lucille Chia and Hilde De Weerdt, 33–62. Leiden: Brill, 2011.

Eyferth, Jacob. *Eating Rice from Bamboo Shoots: The Social History of a Community of Handicraft Papermakers in Rural Sichuan, 1920–2000*. Cambridge, MA: Harvard University Asia Center, 2009.

Fong, Wen. "The Problem of Forgeries in Chinese Painting." *Artibus Asiae* 25, no. 2–3 (1962): 95–140.

Gale, Barden N. "The Concept of Intellectual Property in the People's Republic of China: Inventors and Inventions." *China Quarterly* 74 (1978): 347–350.

Gu Yanwu 顧炎武. "Zhu shu zhi nan" 著書之難. In *Rizhilu jishi* 日知錄集釋, edited by Huang Rucheng 黃汝成, 1083–1084. Shanghai: Shanghai Guji Chubanshe, 2006.

Handler-Spitz, Rivi. *Symptoms of an Unruly Age: Li Zhi and the Culture of Early Modernity*. Seattle: University of Washington Press, 2017.

Heller, Michael A. and Rebecca S. Eisenberg. "Can Patents Deter Innovation? The Anticommons in Biomedical Research." *Science* 280, no. 5364 (1998): 698–701.

He Zhaohui 何朝晖. "Individual Protection, Local Coordination, and Material Attachment: The Copyright of Woodblock Imprints in Imperial China." Paper presented at the "Authorship, Copyright, and Editions: The Circulation of Works in Late Imperial China" workshop, Harvard University, May 5, 2007.

He Zhaohui 何朝晖. "Shilun Zhongguo gudai diaoban yinshua banquan xingtai de jiben tezheng" 试论中国古代雕版印刷版权形态的基本特征. *Tushu yu qingbao* 图书与情报, no. 3 (2008): 113–118, 125.

Hsia, Tao-tai, and Kathryn A. Haun. "Laws of the People's Republic of China on Industrial and Intellectual Property." *Law and Contemporary Problems* 5, no. 3 (1973): 274–291.

Ivanhoe, Philip J. "Intellectual Property and Traditional Chinese Culture." In *Law and Social Justice*, edited by Joseph Keim Campbell, Michael O'Rourke, and David Shier, 125–142. Cambridge, MA: MIT Press, 2005.

Jiao Hong 焦竑. *Danyuan ji* 澹園集. Beijing: Zhonghua Shuju, 1999.

Kornicki, Peter. *The Book in Japan: A Cultural History from the Beginnings to the Nineteenth Century*. Leiden: Brill, 1998.

LaFleur, Robert André. "Literary Borrowing and Historical Compilation in Medieval China." In Buranen and Roy, *Perspectives on Plagiarism*, 141–150.

Li Mingshan 李明山. *Zhongguo gudai banquan shi* 中国古代版权史. Beijing: Shehui Kexue Wenxian Chubanshe, 2012.

Liu Xu 劉昫. "Wenzong ji" 文宗記. In *Jiu Tangshu* 舊唐書. Beijing: Zhonghua Shuju, 1975.

Li Yu 李漁. "Yu Zhao Shengbo wenxue" 與趙聲伯文學. In *Li Yu suibi quanji* 李漁隨筆全集. Chengdu: Bashu Shushe, 2003.

Loeber, Dietrich. "Copyright Law and Publishing in the People's Republic of China." *UCLA Law Review* 24 (1977): 907–913.

McDermott, Joseph P. "Rare Book Collections in Qing Dynasty Suzhou: Owners, Dealers, and Uses." In *Jinshi Zhongguo de ruxue yu shuji: Jiating, zongzu, wuzhi de wangluo* 近世中國的儒學與書籍：家庭, 宗族, 物質的網絡, edited by Lü Miaw-fen 呂妙芬, 199–249. Taipei: Institute of Modern History, Academia Sinica, 2013.

Moll-Murata, Christine. *State and Crafts in the Qing Dynasty (1644–1911)*. Amsterdam: Amsterdam University Press, 2018.

Nevius, John Livingston. *China and the Chinese*. New York: Harper & Brothers, 1869.

Ocko, Jonathan. "Copying, Culture, and Control: Chinese Intellectual Property Law in Historical Context." *Yale Journal of Law & the Humanities* 8, no. 2 (1996): 559–578.

Owen, Stephen. *Readings in Chinese Literary Thought*. Cambridge, MA: Council on East Asian Studies, 1992.

Puett, Michael. "The Temptations of Sagehood, or: The Rise and Decline of Sagely Writing in Early China." In *Books in Numbers: Seventy-Fifth Anniversary of the Harvard-Yenching Library*, edited by Wilt L. Idema, 23–48. Cambridge, MA: Harvard-Yenching Library, Harvard University, 2007.

Schäfer, Dagmar. *The Crafting of the 10,000 Things: Knowledge and Technology in Seventeenth-Century China*. Chicago: University of Chicago Press, 2011.

Shang, Wei. "The Making of the Everyday World: *Jin Ping Mei cihua* and Encyclopedias for Daily Use." In *Dynastic Crisis and Cultural Innovation: From the Late Ming to the Late Qing and Beyond*, edited by David Der-wei Wang and Shang Wei, 63–92. Cambridge, MA: Harvard University Asia Center, 2005.

Shao, Ke. "The Justice of Balance: Understanding Intellectual Property from Chinese Historical and Philosophical Perspectives." PhD diss., University of London, 2007.

Sherman, Brad, and Lionel Bently. *The Making of Modern Intellectual Property Law: The British Experience, 1760–1911.* Cambridge: Cambridge University Press, 1999.

Sidel, Mark. "The Legal Protection of Copyright and the Rights of Authors in the People's Republic of China, 1949–1984: Prelude to the Chinese Copyright Law." *Columbia Journal of Art and the Law* 9 (1984): 477–508.

Sima Guang 司馬光. *Sima wenzheng gong chuanjia ji* 司馬文正公傳家集. Taipei: Taiwan Shangwu Yinshuguan, 1965.

Stone, Charles R. "What Plagiarism Was Not: Some Preliminary Observations on Classical Chinese Attitudes toward What the West Calls Intellectual Property." *Marquette Law Review* 92, no. 1 (2008): 198–227.

Tang, Xianzu. *The Peony Pavilion.* Translated by Cyril Birch. 2nd ed. Bloomington: Indiana University Press, 2002.

Twitchett, Denis. *Printing and Publishing in Medieval China.* New York: Frederic C. Beil, 1983.

Volpp, Sophie. *Worldly Stage: Theatricality in Seventeenth-Century China.* Cambridge, MA: Harvard University Asia Center, 2011.

Wang, Fei-Hsien. "A Crime of Being Self-Interested: Literary Piracy in Early Communist China." *Twentieth-Century China* 43, no. 4 (2018): 271–294.

Wang, Fei-Hsien. *Pirates and Publishers: A Social History of Copyright in Modern China.* Princeton, NJ: Princeton University Press, 2019.

Xie Shuishun 谢水顺 and Li Ting 李珽. *Fujian gudai keshu* 福建古代刻书. Fuzhou: Fujian Renmin Chubanshe, 1997.

Xinban quan[zeng]bu Tianxia bianyong wenlin miaojin wanbao quanshu 新板全[增]補天下便用文林妙錦萬寶全書. In *Mingdai tongsu riyong leishu jikan* 明代通俗日用類書集刊, edited by Zhongguo shehui kexue yuan Lishi yanjiusuo Wenhua shi 中國社會科學院歷史研究所文化室, vol. 10, 243–606. Chongqing: Xinan Shifan Daxue Chubanshe, 2011.

Xu Wei 徐渭. "Ye Zisu shixu" 葉子肅詩序. In *Xu Wei ji* 徐渭集. Vol. 2. Beijing: Zhonghua Shuju, 1983.

Yang Lan 楊瀾. *Linting huikao* 臨汀彙考. Preface dated 1878.

Ye Dehui 叶德辉. *Shulin qinghua* 书林清话. Shenyang: Liaoning Jiaoyu Chubanshe, 1998. First published 1921.

Zheng, Chengsi, and Michael D. Pendleton. *Chinese Intellectual Property and Technology Transfer Law.* London: Sweet & Maxwell, 1987.

Zheng, Chengsi, and Michael D. Pendleton. *Copyright Law in China.* North Ryde, Australia: CCH International, 1991.

Zhou Lin 周林 and Li Mingshan 李明山, eds. *Zhongguo banquan shi yanjiu wenxian* 中国版权史研究文献. Beijing: Zhongguo Fangzheng Chubanshe, 1999.

Zhuangzi. *Basic Writings*. Translated by Burton Watson. New York: Columbia University Press, 2003.

Zhu Xi 朱熹. "Hui'an xia⊐sheng Zhu wengong wenji" 晦庵先生朱文公文集. In *Sibu congkan chupian* 四部叢刊初篇. Shanghai: Shangwu Yinshuguan, 1919.

Zou Shencheng 邹身城. "Baohu banquan shiyu heshi heguo?" 保护版权始于何时何国? *Faxue yanjiu* 法学研究 63, no. 2 (1984): 63.

3

TEACHING INTELLECTUAL PROPERTY: CONSTRUCTING THE HISTORICAL NARRATIVE OF INTELLECTUAL PROPERTY IN UNIVERSITY TEXTBOOKS

Marius Buning

Since it is man, and man alone, who has required that his inventions be protected from unauthorized emulation by others, it is worth pausing to enquire as to why this is so.
—Jeremy Phillips and Alison Firth, *Introduction to Intellectual Property Law*, 24

Ideas about the present are based, in part, on our conception of the past. In that way, the past directly relates to history—that is, to the stories we tell about our predecessors and the kinds of things they did. The many books that have appeared on this topic each emphasize a different aspect, from Paul Ricoeur's ideas on historical time to Benedict Anderson's *Imagined Communities*, from Hayden White's *Metahistory* to Stephen Bann's *Clothing of Clio*.[1] In a nutshell, the literature argues that history consists of stories that we commonly hold to be true. Alternative histories are kept in check by professional scholars who relentlessly conduct a tense debate on how history should be told and which aspects from the infinitely rich past are worthy of mention. From this perspective, without denying that specific historical events actually took place, historical knowledge production is thus by definition a social construct.

Over the last 150 years or so, a lot of attention has been given to the social dimensions of the law as well. Exact definitions of "the law" remain in that context a matter of contestation. From Bruno Latour's actor-network-theory (ANT) to Eugen Ehrlich's "living law," from Niklas Luhmann's systematic approach to Marilyn Strathern's notion of "social control"—the literature is endless and filled with subtle differences of opinion on how the law operates.[2] To put it bluntly, and perhaps stating the blindingly obvious, one could say that the common denominator in the literature is a recognition that the law is whatever people recognize as being law. It follows that telling stories about the law forms an important part of what law is. This idea has received a particular boost over the last decennia in the form of the so-called law and literature movement, advocated by representatives such as Ronald Dworkin and James Boyd White.[3] Authors in this field focus on the linguistic aspects of the law, such as lawyers' pleas and court decisions, all the way to how law is linguistically experienced in society. The important insight coming from this literature is that narrative is crucial to successfully creating consistency within the law.

In this chapter I will try to reconstruct the historical narrative of what is commonly known as *intellectual property* (hereafter IP)—a collective term that runs to different forms of property that include intangible creations of the human intellect, such as patent law, copyrights, and trademark law. What I attempt to show is how the choice of historical categories and the specific use of rhetorical language in university textbooks on IP affects the way in which we think of IP. The framework of intellectual property law is a strong example of a system in which naming defines owning—that is to say, a system in which words function as a way to own knowledge. This chapter not only explores how this silences, excludes, or ignores other possible systems of knowledge and ownership; it also reconstructs how these words operate within the presentation of this system in textbooks to justify the system's own historical assumptions and theoretical preconditions.

The idea of studying the rhetoric of intellectual property is not entirely new. Jessica Reyman, for instance, has written on the topic and highlighted the implications of the rhetorical positioning of technology as being destructive to creative production.[4] Jessica Silbey has analyzed the mythical aspects of American IP law and concluded that "the origin stories of intellectual property are the mechanisms by which one area of law works to both embrace its founding and overcome its limitations to move forward."[5] Yet, neither Reyman nor Silbey have dealt with the question of how language and discourse is used in the specific context of legal textbooks. For that matter, remarkably little has been written about those legal textbooks.[6] With particular regard to IP law, Ronan Deazley has published on the making of Copinger's *Law of Copyright* (1870), Christopher Wadlow has written about *Terrell on the Law of Patents*, and Jose Bellido has contributed, with a number of excellent essays, to our understanding of how concepts and laws emerge within an educational setting.[7] None of these authors, however, have paid much attention to any rhetorical issues within the text. Such issues have instead been addressed in research on historical textbooks, where narratives and analogies play a central role.[8] Yet, these studies focus exclusively on historical works and primers, not on the question of how historical accounts are (being made) part of another discipline, such as the law.

Finally, however, there is an entirely different kind of literature in which the relationship between history and IP comes to the fore—namely, in works that focus on the justification and morality of the law. In this framework, attention has been given to the question of who has written the history of copyright and with what objective—it turns out that writing histories of IP is mainly a phenomenon of the last two centuries, which came about in parallel with the internationalization of legal concepts.[9] Less historical, yet more concretely related to semiotics, is the highly original work by Kelly Gates, Majid Yar, and Tarleton Gillespie on copyright outreach campaigns.[10] These outreach campaigns are particular in the sense that they are oriented toward the general public and often funded by specific organizations, with a clear idea of what they want to achieve. So far, no attention has been given to the question of how future legal professionals (such as lawyers, judges, and so on) are schooled in thinking about IP in a particular way. This chapter attempts to fill that gap by analyzing a concrete body of university textbooks, further defined in the next section.

THEORY AND SOURCES: THE SEMIOTICS OF IP

It would be impossible to deal exhaustively with the theme of "IP teaching" in the course of just one chapter. In addition to monographic textbooks, there are handbooks like encyclopedias, dictionaries, anthologies, and readers (as well as compendia of reading materials for particular courses) that such a study would have to consider. Moreover, it would be important to keep a geographical balance and to delve deep into various educational settings. I have chosen instead to single out a limited number of textbooks in the area of IP law, following a reading list used at the London School of Economics (LSE; see box 3.1) in 2018/2019.[11] While I will allude to all the books on the list, to achieve a more thorough analysis I have chosen to focus on one text in particular: Bainbridge's *Intellectual Property* (which celebrated its tenth edition in 2018). The reason for this is because the book by Bainbridge has been widely read for many decades (see also box 3.2), but also because narrative study is best served by the close reading of a text as a coherent entity; an integral analysis of all the materials on the LSE reading list would demand a separate monograph.

The use of Bainbridge's *Intellectual Property* differs from case to case. Thus, the course offered at the LSE is "available on the BA in Anthropology and Law and LLB in Laws . . . [and] . . . as an outside option to students on other programmes where regulations permit and to General Course students."[12] The aim of the optional course at the LSE for year 2 and year 3 students on a BA/LLB program is to provide students with an overview of the basic principles of IP, which they can then apply in their own specialization.[13] In the other courses listed in box 3.2, the use of Bainbridge's text might be different; the listed courses are not equal or do not teach the same content. Furthermore, the course at the LSE provides a *general* overview of IP, which is again different from other courses. For instance, the course at the City University of Hong Kong, IP Law:

Box 3.1

Sources of the Analysis

> The following literature is prescribed as essential reading in the LSE course Intellectual Property Law LL251:
>
> Bainbridge, David I. *Intellectual Property*. 9th ed. Harlow, UK: Pearson, 2012.
>
> Cornish, William, and David Llewelyn. *Intellectual Property: Patents, Copyright, Trademarks and Allied Rights*. 6th ed. London: Sweet & Maxwell, 2007. For this chapter, I have used the fourth edition (1999), written by William Cornish.
>
> **Background/further reading:**
>
> Aplin, Tanya, and Jennifer Davis. *Intellectual Property Law: Text, Cases, and Materials*. Oxford: Oxford University Press, 2009. For this chapter, I have used the third edition (2017).
>
> Bently, Lionel, and Brad Sherman. *Intellectual Property Law*. 4th ed. Oxford: Oxford University Press, 2014.
>
> MacQueen, Hector, Charlotte Waelde, Graeme Laurie, and Abbe Brown. *Contemporary Intellectual Property: Law and Policy*. 2nd ed. Oxford: Oxford University Press, 2010. For this chapter, I have used the third edition (2013), written by Charlotte Waelde, Graeme Laurie, Abbe Brown, Smita Kheria, and Jane Cornwell.

Theory, Patents and Trademarks LW 4642, does not deal with copyright, whereas in other courses—for example, Intellectual Property and Media Law LT5007, London Metropolitan University—the emphasis is on IP within the broader framework of media law, which creates a different focus. The consequence of this disciplinary fluidity is that the textbooks have to appeal to the greatest common denominator when it comes to areas of interest within IP. One could make the argument that the IP handbooks that shape the reading public are equally well shaped by this, with the result that important (but less accessible) topics within the field of IP law, such as plant and seed varieties, perhaps do not get the attention they deserve.

The specifically English context of the textbooks merits our further attention as well. Within law, both on doctrinal grounds and in practical terms (admission to the bar), the question of jurisdiction is central. So, it is logical that the practical textbooks zoom in on a law landscape with national lines of demarcation, unless the topic is treaties, supranational organizations, and international law. The textbooks aim to come to terms with a constantly changing legal system that is valid today. As the legal historian Frederic William Maitland (1850–1906) had already remarked toward the end of the nineteenth century, the logic of law is, as such, different from that of history. Whereas the discipline of history is guided by a "logic of evidence," and historians want to study

Box 3.2

Use of the Textbooks in University Courses

> *The selected textbooks are used, in various order and combinations, in—among many others—the following course syllabi:*
>
> Intellectual Property LA3026—University of London
>
> https://london.ac.uk/courses/intellectual-property-la3026
>
> Intellectual Property Law: Theory, Patents and Trademarks LW4642—City University of Hong Kong
>
> https://www.cityu.edu.hk/catalogue/ug/201516/course/LW4642.pdf
>
> Intellectual Property and Media Law LT5007 (2016/17)—London Metropolitan University
>
> https://intranet.londonmet.ac.uk/module-catalogue/record.cfm?mc=LT5007
>
> Intellectual Property Law LA4036 (2019/2020)—University of Limerick School of Law
>
> https://ulsites.ul.ie/law/sites/default/files/Law_Book%20of%20Modules%202019.2020%20B.pdf
>
> Media and Entertainment Law UJUTNG-30-3—University of the West of England
>
> https://info.uwe.ac.uk/modules/specification.asp?urn=2055146&file=Media_and_Entertainment_Law_UJUTNG-30-3.pdf
>
> Nature, Emergence and Development of IPR L4 RTDA2 C5—Guru Gobind Singh Indraprastha University
>
> http://www.ipu.ac.in/uslls/LawSyllabus/ipr070116.pdf

history in its own terms, the discipline of law is governed by a "logic of authority," whereby the past is seen only as a preamble to contemporary interpretation. According to this view, the most recent interpretation of the past with legal validity is considered as being the most "correct," whereas "from the historian's point of view it is almost of necessity a process of perversion and misunderstanding."[14] As we shall see, this distinction also plays out in the books on the LSE reading list.

Aside from the textbooks on that list, I have looked into some short IP law guidebooks for comparison. However, I did not examine how the material is taught in class or the role of the reader. A more exhaustive analysis of IP teaching material would certainly consider these elements, if only because IP courses are often upper-level courses where basic legal principles (such as the concept of justice) are no longer deliberated, on the assumption that they have been dealt with in general introductory courses. Moreover, it is, of course, very possible that lecturers each tell the history of IP differently in class. Still, the selected sources are some of the most important textbooks used today to explain IP to future generations of legal professionals. Thus, a better understanding of the ideological positioning in these textbooks by means of history will provide a better understanding of how specific narratives are framed alongside or against distinct ownership claims.

Based on the selected sources, I discuss the rhetorical framing of the genesis of IP. As classical theory on the topic tells us, this type of "history of origins" is a narrative with a plot that moves between two states, or more specifically, between "the transformation of equilibrium into disequilibrium and into a subsequent equilibrium."[15] In the course of the employment, our attention is focalized on certain aspects at the expense of others.[16] Telling and showing, naming and inscribing, routes our attention and thus leads us to "see" specific things whilst neglecting others. The positioning that takes place can be brought out by dismantling dichotomies, examining silences and disruptions, and identifying metaphors as well as the most alien elements in the narrative.

On the following pages, I shall follow a Greimasian approach centered on the "discursive," the "narrative," and the "thematic" levels.[17] The issues associated with these levels include the identification of places, objects, actors, opposites, and states of being ("discursive"), as well as the identification of the protagonists (subject/object, helper/opponent, sender/receiver) and the change that is being effectuated after a series of tests ("narrative"). On a deeper level, I shall question what the most abstract poles are in the story and what fundamental transformation of value is at stake ("thematic"). This will lead toward the construction of a semiotic square discussing unsaid elements in the history of IP. In conclusion, I shall discuss the importance of narrative analysis and the relevance of history of the making of IP's future.

Before we start, however, it is important to note that I have not considered the content in my analysis, but rather the form. The point is not to find the truth, the "real" history, but to show how the story is used to legitimize just one possible version of history. In this sense, my work unmistakably differs from that of, for instance, Kathy Bowrey, who argued that, whereas several people have written the history of copyright from a specific perspective, what "seems to be missing is a history of copyright that goes beyond a particular discipline's point of view."[18] I take the view that it is impossible to write a history that is value-free.

THE SCENERY

The first level of analysis must begin with what Gérard Genette has called the paratext.[19] Let us take the book by Bainbridge as our point of departure. On the back cover, the book is praised as one that "offers you unrivalled coverage of all aspects of the intellectual property syllabus, making it your essential guide through the intricacies of this dynamic subject."[20] The performative undertone is that the book is a key to success. This promise cashes in on the reader's hope of a successful professional career. The blurb points out that the textbook has been "trusted by generations of students and lecturers alike." As one endorsement (written by an anonymous author in the *Law Student Journal*) clarifies, it is clear that "those looking for an accessible and stimulating account of the nuts and bolts of intellectual property law need not, however, look any further."[21] The sense of confidence that is created serves not only to convince the potential buyer to purchase the book, but also to persuade the audience of the reliability of the information found inside the book. Positioning the credentials of the author in a clearly visible location contributes to the status of this book as well (in this case, "Emeritus Professor of IPL of Aston University and an honorary member of Hardwicke Building, Lincoln's Inn"). A statement on the back cover helps to distinguish the book from others on the market, declaring that it is "one of the best."[22] Similar claims are made in the other source material. For example, the cover of Waelde et al.'s *Contemporary Intellectual Property* announces that it "offers a unique perspective on intellectual property law, unrivalled amongst IP textbooks available today."[23] Bently and Sherman's *Intellectual Property Law* is presented as "the definitive textbook on the subject,"[24] and in this case, too, the authors' university positions are clearly mentioned to add institutional allure to the publication (Herchel Smith Professor of Intellectual Property at the University of Cambridge, and Professor of Law at the T.C. Beirne School of Law at the University of Queensland, respectively). Another text that was previously extolled, on its own back cover, as being the "definitive textbook on the subject," was written by the *former* Herchel Smith Professor of Intellectual Property at the University of Cambridge, W. R. Cornish, Q.C., LL.B., F.B.A. (*Intellectual Property: Patents, Copyrights, Trademarks and Allied Rights*, 4th ed., published by Sweet & Maxwell). Thus, the claim to authority is made even before the reader has opened the book. It helps that each of the selected books is published by renowned English publishers.

Moving on to the content, one is struck by the strong emphasis on the role England has played in the genesis of IP law. This one-sided focus can perhaps be explained by the need to limit the scope of the subject matter, with a view to the readership but also considering the jurisdictional reality of the present. The distinction between the logic of the law and the logic of history, as highlighted by Maitland, comes clearly to the fore here. Nonetheless, the closed or "circular" system of references remains remarkable. In a section on the "justification for patent rights," for instance, Bainbridge argues: "An inventor owns a property right in his invention. This is a natural right and accords with the views on property rights of philosophers such as Locke."[25] The text is punctuated by continuous references to English authorities, from Locke to the "great English philosopher" Jeremy Bentham (I will return to this example below). The neglect of non-English events and the contributions of non-English legal theory is sustained by the omission of existing literature on the topic written in a different context or in a different language. The references are to books written in English only, mainly on English topics.[26]

It seems that the role attributed to history in the coming about of the present almost inevitably leads to an idea of English uniqueness. As Bainbridge argues in his "brief historical perspective" on the patent system:

> As with the origins and development of other intellectual property rights, England has a prime place in world history and has set the mould for patent rights internationally. It is no coincidence that England was the country where the first major steps towards an industrial society were taken. Whether this was a direct result of the patent system is arguable, but it is without doubt that patents had an important role to play in the Industrial Revolution. Before this, the origins of patent law can be seen emerging in late medieval times.[27]

The sources of Bainbridge's information are Davenport's *The United Kingdom Patent System*, and Thorley et al.'s *Terrell on the Law of Patents*.[28] References to historical events throughout the book are to other scholarship in the field of law, not history. Later in the text, referencing the Statute of Monopolies, Bainbridge adds: "It seems that the world's first patents statute was passed in Venice in 1474: see Reid, B.C. (1998) *A Practical Guide to Patent Law* (3rd edn) Sweet & Maxwell, p 1."[29] This contribution from the non-English world is moved to a footnote and stated in terms much more uncertain ("It seems") than the decisive tone used in the rest of the text. Again, reference is made only to other legal scholarship. What emerges is the image of a closed system of references for scholars who are looking for the (ahistorical) antecedents of their own discipline instead of understanding the past in a broader social and intellectual context.

It should be stated here that the degree of historical sensitivity in the selected literature clearly varies from author to author. In Colston's *Principles of Intellectual Property Law* (1999) there is only one sentence, on the first page, acknowledging that "intellectual property law has a long history."[30] The idea is nowhere elaborated in the remainder of the book. Aplin and Davis deal with history more extensively, and rather prudently.[31] In a section on the history of patent law, for example, the authors include a reference to Bently and Sherman's *Making of Modern Intellectual Property Law* (208–209), warning against the tendency to trace patent law back to the 1624 Statute of Monopolies, as "it encourages us to gloss over the history of the patent system between 1624 to the present day and to treat patent law as predestined and timeless, as opposed to open and historically contingent."[32] Indeed, particularly in the case of Bently and Sherman, one cannot say that they are unaware of the way in which current IP is embedded in history. Bently has written extensively on the history of IP, and on the history of trademarks in particular.[33] Sherman has written elsewhere about how the writing of IP history originated in the second half of the nineteenth century, when a particular form of IP was confirmed in its existence on the international stage.[34] In a groundbreaking essay, Sherman has also pointed to the challenges of any essentialist approach to patent law based on a "consequentialist mode of thinking" about its history.[35] Indeed, the narratological analysis of several writings discussed in this chapter should not be seen as criticism of the authors concerned; it merely serves to gain a better understanding of *how* current legal practice is anchored in the past.

Let us briefly return to the "prime place in world history" that Bainbridge attributed to England in the example cited earlier. In the first instance, Bainbridge admits that the relationship between patents and the Industrial Revolution is "arguable," only to remove any reservation that such a connection does indeed exist in the same sentence ("without doubt"). After building consensus on the issue, the modern situation is set against the previous situation, in "medieval times." Speaking about the early history of patents, the text later goes on to argue:

> In this early form, there was no need for anything inventive; it had more to do with the practice of a trade and the granting of favors by the Crown.... Eventually, there was a strong need for an effective system that prevented unfair competition where, for example, one person had made some novel invention and wanted to stop others from simply copying it. A monopoly system developed in the reign of Elizabeth I and many letters patent were granted.[36]

The connotations and use of adjectives in this excerpt provide a clear sense of the direction in which the author wants to bring the reader. This can be highlighted by filling in the gaps and outlining tacit binary oppositions: "There was a strong [not weak] need [felt by whom?] for an effective system [which the patent system provided] that prevented [or imposed a ban on?] unfair [regular] competition." The specific wording reinforces the sentiment that the current patent system is the necessary outcome of history.

For Bainbridge, history is nothing but a stepping-stone to modern times. As such, this type of eschatological thinking fits well with the substantive structure found in most of the textbooks. Usually, any text on IP is divided into at least three sections: patents, copyright, and trademarks (supplemented by related or less-developed fields such as liability and design law). Each of these sections on IP's main components typically starts with "a brief historical introduction" before moving on to issues that play a more substantial role in the reality of the modern-day lawyer. Despite the sometimes limited attention to history, the justification for the very existence of IP is at all times clearly anchored in the past—as the next section of this chapter explores more fully.

FRAMING THE NARRATIVE: INTELLECTUAL GENEALOGIES

> Consequently, most IP law is statutory and the result of political and economic history.
> —Colston, *Principles of Intellectual Property Law*, 4

I have discussed some aspects of the claim to expertise and the varying degrees of attention that the different sources give to the history of IP. This variation is in part a matter of genre. In collections of jurisprudence and contemporary IP laws, for example, one rarely finds any reference to the distant past, since these works pay attention only to laws currently in use.[37] History is not mentioned in many of the dictionaries of law either. In this case, the neglect of the past can be explained by the size as well as the intended use of such books; in a dictionary that aims to cover all legal concepts in no more than around three hundred pages, a short definition of, for instance, *patent*, meets the requirements. Textbooks have a different function. They not only include the promise of a successful career but also give a sense of unity to a discipline. The attentiveness to history in that context is a tool to regulate the community; it suggests the existence of a temporal unity between the present and the past.

What is striking in the way this history is framed is the silent assumption that there is an inescapable route from past to present that coincides with the transition to modernity. The way the story is told very much resembles a Proppian fairy tale. Let us look, for illustration, at the way the history of patent law has been told so far. The subject (the patent notion) is depicted as going on a quest to become modern (the object). The necessity to act (*devoir faire*) is fueled by "a strong need for an effective system that prevented unfair competition" (the mandatory sender).[38] The patent notion establishes a contract, which is followed by three tests: the qualifying test, the decisive test, and the glorifying test. In the qualifying test, the subject "must acquire the necessary competence to perform the planned action or mission."[39] Patent law is hindered in its quest by the "odious monopolies" (the villain/opponent) issued by James I, and aided by parliamentary intervention in the form of the 1624 Statute of Monopolies (the helper) that provided exclusive rights to commercially exploit an invention for a duration of fourteen years.[40] The qualifying test thus enables the patent notion to progress, symbolizing "a first step towards the modern form of a right open to the world based upon legal principles and enduring for a specified period."[41] Later in the story, our hero encounters a decisive test ("the principal event or action for which the subject has been preparing, where the object of the quest is at stake")[42] in the form of the nineteenth-century reformulation of legal principles. There is a confrontation between the subject and the antisubject in the form of the conflict between the proponents of patents and the so-called patent abolitionist movement. As Bainbridge formulates it,

The Industrial Revolution brought a great many pressures upon the patent system, eventually leading to major reforms starting with the Patent Law Amendment Act 1852. During the preceding period there had been much debate about whether inventions should be afforded legal protection by the grant of patents, and indeed in Switzerland and the Netherlands patent law was dismantled, to be reintroduced later in the nineteenth century. The fact that this could happen and that the whole rationale for the granting of patents could be challenged in England now seems incredible.[43]

Withstanding the many pressures, the patent notion finally moves on to the "glorifying" test, which is "the stage in the story at which the outcome of the event is revealed."[44] Depending on the author's perspective, the final test was passed either in 1883 (the Patents, Designs, and Trade Marks Acts 1883 to 1888), or with the 1977 Patents Act. The performance of the subject is now recognized in accordance with the mandate instituted by the initial sender (the "strong need for an effective system that prevented unfair competition"). The sender-adjudicator (the author) can confidently conclude that "it is now unthinkable that the patent system would be abolished."[45] Patents slayed the dragon, and lived happily ever after.[46]

The narrative of the plot moves from chaos to a stable situation that coincides with the present. Historical events are singled out for their importance as markers of a legal discipline with its own right to existence, distinguishing IP from, for example, tort and property. The idea that this "long history" is significant for the present plays a role in all standard textbooks, and it reappears in every section on a different subdivision within IP law (I shall focus here on copyrights, patents, and trademarks). Only in the case of trademarks—something of a cuckoo in the nest in terms of theoretical reflection in IP law—is there any confusion over its true origins. On the one hand, it is claimed that "the use of trademarks has a long history, from the marks used by potters in Roman times, to the internationally known marks in use today, such as McDonald's 'golden arches,' the Nike 'swoosh' or the name Coca-Cola."[47] On the other hand, Lionel Bently in particular has advocated letting the history of modern trademark law begin in the nineteenth century, with

> a legal understanding of a trade mark as a sign which indicates trade origin; the establishment of a central registry in 1876; the conceptualization of the trade mark as an object of property; the recognition of a dual system of protection: one based on registration, the other based on use in the marketplace; and the development of international arrangements for the protection of marks in foreign territories.[48]

Of course, the debates in the nineteenth century had their roots in the past (the actors in those debates were certainly aware of a longer history). Nevertheless, one can speak of a radical break (or a "decisive test").

In the case of copyright and patents, Bently and his coauthor Sherman identify a similar break at a similar time. In another monograph on the topic, they argue their point extensively and emphasize that they are concerned with "the doctrine of intellectual property law, rather than in what, for example, economists or political philosophers may be able to tell us about intellectual property law."[49] They thus come to the "belief that during the middle period of the nineteenth century an important transformation took place in the law which granted property rights in mental labour."[50] Other proposals have been made, too. Mario Biagioli and Oren Bracha, for instance, have argued that a great shift in thinking about patents took place around the time of the French and American Revolutions, when former privileges became redefined as rights.[51] Once again, the point is not to evaluate the correctness of these various claims but to recognize that they emphasize a longer history that does not coincide with the time at which current laws became effective.

To explain this hankering for the past, ideas and theories about "situatedness" and "historical anchoring" might be worth exploring.[52] IP is socially legitimized *because* it is part of a longer tradition—and with that, the battle to claim the "true origins" of IP has begun. In the case of patent law, the Statute of Monopolies was hailed as the starting point of patent law when the British Empire was at its height, whereas in the interwar period, other scholars tried to put Italy and Germany on the map as important actors in the development of a patenting concept. In the case of copyright, the 1710 British Statute of Anne is usually considered to be the first copyright statute, even if Wikipedia tells us that "the earliest recorded historical case-law on the right to copy comes from ancient Ireland."[53] With the emergence of new economic powers on the global stage, attempts are being made to shift the focus elsewhere. Along these lines, one can read the expressions of disappointment on Wikipedia's Talk Page about the "History of Patent Law" article, with its "Eurocentric POV" and its failure to mention the importance of Muslim societies.[54] Reader Terry0051 replies that

> if there's evidence relevant to patent-relevant laws from Muslim sources from 600 AD to 1500 AD, then let's hear about it. But to assume (or even demand) that there be such a history, if (so far) there is no sign of any such facts, shows a POV of its own.[55]

What is all too easily forgotten in these debates is that the fundamental bone of contention is the question of definition: What exactly is IP, that it allows us to speak about it? Even if the term *intellectual property* does not appear to have existed before the end of the nineteenth century (and for most legal scholars, that settles the case), there is more going on than a simple conflict between disciplines, in which the lawyers look at the problem internally whereas a philosopher or a historian takes a broader view. By emphasizing mental labor and the reformulation of property as a right, alternative histories are silently suppressed.

This may not be an unknown phenomenon for those who occupy themselves with what is known as "traditional knowledge"—a field where conflicts between alternative definitions perhaps come most clearly to the fore. The views that have been formulated within this rich field of research are mostly aimed at solving the question of how to balance different understandings with one another. Vandana Shiva, for instance, criticized the IP "myth of stimulating creativity" by arguing that "science cannot be used to refer only to modern Western science. It should include the knowledge systems of diverse cultures in different periods of history."[56] Her work adds to a sizable body of literature that questions the need to take IP as a standard to which others should relate or conform, or even to revolt against. And this brings us to an important question: What exists outside of IP?

IN SEARCH OF THE PUBLIC DOMAIN

> Copyright and author's right are the two great legal traditions for protecting literary and artistic works. The copyright tradition is associated with the common law world—England, where the tradition began, the former British colonies, and the countries of the British Commonwealth. The tradition of author's right is rooted in the civil law system and prevails in the countries of the European continent and their former colonies in Latin America, Africa, and Asia.
>
> —Paul Goldstein, *International Copyright Principles, Law, and Practice*, 3

What is highlighted in the textbooks is that historically, there have been economic and legal reasons why "society" has implemented an ever-stronger program of intellectual property rights. The underlying dynamic is strongly focused on problem-solving, meaning that those who make the law find solutions for the problems they face. Attention is given in this context to issues (such as censorship and justice) or to particular actors (such as the Stationers, the Crown, John Locke, and so on), as well as to the various criticisms of the IP system (although this to a lesser degree). What is hardly ever mentioned, however, are the underlying processes of state formation that played a role in the reformulation of legal principles. The 1706 Acts of Union are never mentioned in relation to the 1710 Statute of Anne, for example. The impact of colonialism on the diffusion of IP principles is never problematized.

Another aspect that is silently passed over is the historic transformation of the public domain. Often confused with related notions such as the commons or the public sphere, the public domain refers to a distinct concept best identified as the space in which IP does not apply.[57] The public domain consists, in brief, of resources freely accessible to all to use without the need for special permission. Examples include works for which the copyright has expired and inventions made public without prior patent protection, as well as creations and discoveries that cannot be patented or copyrighted, such as products of nature, facts, government publications, and so on. The abundance of reflection on legal notions and the making of modern IP law in the textbooks is matched only by the paucity of attention given to the public domain. This is remarkable, to say the least, since the ultimate purpose of the IP system is to enlarge that public domain, thereby "promoting the progress of science and useful arts."[58]

It appears that the focalization on rights diverts attention away from the ultimate goal of those rights. In the different sections on justification in the university textbooks, we read about Locke, rewards, and a perpetual mantra that IP stimulates inventive labor. But we are rarely provided with any insight into what IP does in terms of expanding the public domain. In Bainbridge, for example, one finds two entries in the index, one short definition, and a number of passing references to the existence of the public domain, but with no further elaboration. There is no entry for "public domain" in the index of Bently and Sherman's *Intellectual Property Law*, except in relation to breach of confidence, where the notion has a different meaning than in other areas of IP (1147–1156). In the fourth editions of both Cornish's *Intellectual Property* and Phillips and Firth's *Introduction to Intellectual Property Law*, the public domain is not specifically defined or referenced in the index.[59] The major exception to this is Aplin and Davis's *Intellectual Property Law*, where the various ways to constitute a public domain are reviewed on pages 20 to 26 (out of 912). One would expect the public domain to be a more central subject in accounts aimed at understanding IP principles and justifications.

Whereas a full reflection on the public domain is missing, the various textbooks do pay attention to various (political and technological) challenges to IP law, ranging from alternative systems of IP protection all the way to complaints about market monopolization. Countering the panegyric on IP, over recent years, a growing number of authors have questioned the righteousness of the IP system, mostly by looking at the social benefits and effectiveness of the current system.[60] Proposed alternatives are plentiful; however, they usually lack a longer history, with the exception of traditional knowledge systems and the commons. These systems are not quite the opposite of IP, but they remain within the boundaries of exclusive use and ownership. If one really wanted to tell a different story, one would have to start paying attention to histories of the public domain or whatever is complementary to that.

It is useful at this point to invoke a semiotic square (see figure 3.1). A semiotic square is a map of logical possibilities; it can be made in many different ways.[61] I assume, however, that the complex contrary of IP is the public domain (hereafter PD).

At the top of the lower square, in between IP and PD, stand proposals such as Copyleft and Share-alike, where the author makes use of existing IP structures to enlarge the PD. At the bottom are alternative regimes of ownership that fall outside of the categories of IP and PD, such as sharing knowledge with an exclusive group of people (the commons

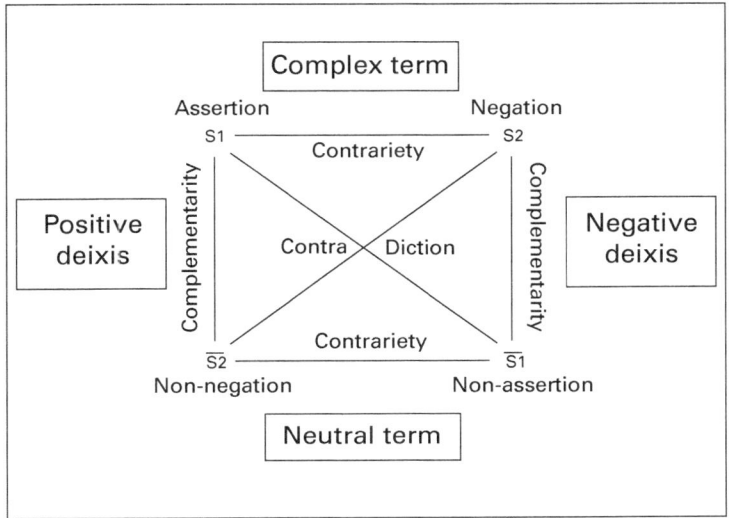

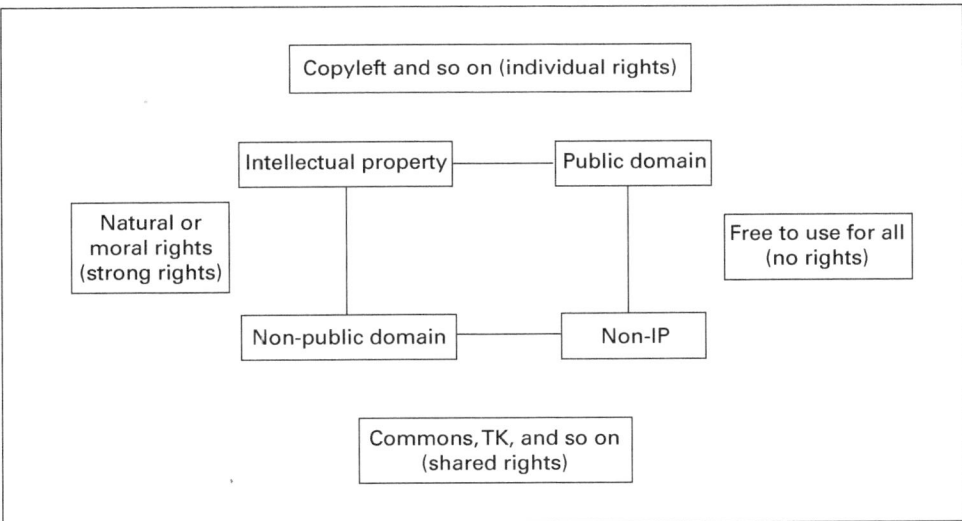

Figure 3.1
Semiotic square outlining the various relationships between IP and the PD. (Upper square adapted from Daniel Chandler, *Semiotics: The Basics*, 2nd ed. (London: Routledge, 2007), 107; lower square courtesy of the author).

and so on). Rules to ensure exclusive ownership apply in this case; however, they are not based on current IP regimes. On the left side of the square, we find programs that encourage strong rights, such as moral rights (not implemented everywhere in the same way) or proposals that are aimed at making IP indefinite (a logical step if one considers IP to be a natural right). These "strong rights" programs combine the notion of IP with other (non-PD) justifications for exclusive use. Finally, on the right side, we find a section that falls completely beyond the notion of "exclusive use": a combination of the PD and non-IP that is not based on exclusive use and that exists beyond the law. Whereas the first three alternatives to the current IP system are given some attention in the selected textbooks, the complete silence on the fourth possibility (the most abstract pole in the story) reveals the strategic boundaries around the narrative. Framing the historical narrative of intellectual property in terms of "legal solutions to problems" defines IP from the outset as a legal discourse, limiting the possibilities to consider IP beyond this framing. At the root of the historical narrative presented by these textbooks, however, we find a justification—the public domain—which the narrative systematically excludes in order to maintain this restriction.

THE BALANCE AND THE AFTERLIFE

In the final section of this chapter, I want to briefly focus on a metaphor that underpins the entire IP system, and patent law in particular.[62] At the basis of the patent system is the idea that inventors should obtain a reasonable temporary monopoly to commercially exploit their ideas in exchange for the proper disclosure of an invention; this is represented by the imagery of a balance between the interests of the inventor and society at large, an exchange in which both sides win.[63] In Bainbridge, for example, the view is formulated as follows:

> The conventional justification for a patent system is that inventors and investors are *rewarded* for their time, work and risk of capital by the grant of a limited, though *strong*, monopoly. This *benefits society* by *stimulating* investment and employment and because details of the invention are added to the store of available knowledge. Eventually, after a period of time, depending on how long the patent is renewed (subject to a maximum of 20 years), anyone will be free to put the invention to use. This *utilitarian* approach found favour with *great* English philosophers such as Jeremy Bentham, who argued that, because an invention involved a great deal of time, money and effort and also included a large element of risk, the exclusive use of the invention must be reserved for a period of time so that it could be exploited and thereafter used for the general increase of knowledge and wealth. He said that such exclusive use cannot ". . . otherwise be put upon any body but by the head of law: and hence the necessity and the use of the interposition of law to secure to an inventor the benefit of his invention."[64]

The idea of a contract that benefits both sides is presented as a self-evident fact that has even been confirmed by a "great" authority who functions as the focalizer in the story. Several assumptions are at work here, such as the notion that a "risk taker" should be "rewarded," that "strong" monopolies are good, and that all this is benefiting "society." One might wonder who "society" really is in this context, or who "gains" on the side opposite to the inventor—other inventor-entrepreneurs who can make use of the information provided, or the public at large, who will eventually benefit from its free use?

The idea that IP is based on a mutually beneficial contract, and that it has a long and successful history, is extremely powerful and has found its way from the textbooks into daily reality around the world, where it is highly influential in shaping future legislation. There are numerous examples of this, but I have selected one from the House Judiciary Subcommittee on Courts, Intellectual Property and the Internet, a subcommittee of the US House Judiciary Committee established in 2011. In its assessment of the effectiveness of current laws, the committee collects opinions from various actors in the field of IP law. One of those actors is the Honorable Sheila Jackson Lee, a representative in Congress from the state of Texas and a member of the committee, who declared in a prepared statement that

> the system [IP law] stands on principles of balance and fairness which allow for continued innovation while not infringing on the property rights of others. The roots of these laws go back many centuries, from the ancient Egyptians and people of the African Gold Coast, whose leader, Mansa Musa of ancient Ghana, traded books for gold, to the likes of political philosopher John Locke of Great Britain, who further wrote and expounded on the ideas and theory of property rights.[65]

The authority of the past is complemented in this example with the belief that Egyptians and other Africans had some notion of IP as well, and that the reach of the idea is therewith truly universal. Who would even begin to doubt such a system that "stands on principles of balance and fairness which allow for continued innovation"?

CONCLUSION: TRAJECTORIES OF OWNERSHIP

> The importance of intellectual property in the modern world goes far beyond the protection of the creations of the mind. It affects virtually all aspects of economic and cultural life. As a result, intellectual property education at the university level is of increasing relevance in educational programs.
> —WIPO, *WIPO Intellectual Property Handbook*, 422

This chapter has shown that historical introductions in legal textbooks on IP are marked by an ideology that is sustained by means of rhetorical techniques as well as strategic narratives. Scholarship itself is thus a powerful tool in producing social order, making explicit the politics of knowledge and ownership ideals. What stands out in the narratological analysis of the selected materials is a preference for ending embedded plots as well as the inclination to situate the beginning of IP in parallel to the beginning of modernity. Depending on the author's perspective, the "real" history of patents thus began either in the late Renaissance, with the scientific revolution and the discovery of the New World, or in the nineteenth century, with the making of empire and the internationalization of European patent laws. Earlier systems "provided no more than a germ of a functioning patent system."[66] Yet, they are invoked time and time again to create the impression that society was looking for a solution until an effective system finally came along. The emphasis on historical continuity is balanced by the omission of any alternative historical options and a silence with regard to alternative regimes of ownership over knowledge products in history.

As the German sociologist George Simmel has noted in his reflections on "historical time," authors are bound to make choices in the construction of a historical narrative.[67] I have tried to reveal some of these choices, whilst being fully aware that I, too, cannot escape from rhetoric. This contribution has no pretension to be complete, and one could quite rightly complain that it runs somewhat randomly through the enormously rich material, offering only limited insight into the distinctive rhetorical facets of how IP is taught in various contexts. Nevertheless, I hope that this short intervention may serve as a "germ" that inspires readers to think differently about IP, in terms of its genesis, its current implementation, as well as its future. If we want to create a change in the way IP is employed today, it is of little use to regard existing law as being ontologically different—as something that exists "out there" that has to be changed. What is needed is a new history, and a new plot.

ACKNOWLEDGMENTS

Funded by the European Union. Views and opinions expressed are however those of the author(s) only and do not necessarily reflect those of the European Union or the European Research Council. Neither the European Union nor the granting authority can be held responsible for them.

Notes

1. For an excellent overview of the history and narrative debate, see Geoffrey Roberts, *The History and Narrative Reader* (London: Routledge, 2010).

2. It is impossible to do justice to the vast body of literature that has been published on the topic of "law and society." The few contributions mentioned here are Eugen Ehrlich, *Fundamental Principles of the Sociology of Law* (New Brunswick, NJ: Transaction, 2001); Bruno Latour, *The Making of Law: An Ethnography of the Conseil d'Etat* (Cambridge: Polity, 2010); Niklas Luhmann and Fatima Kastner, *Law as a Social System* (Oxford: Oxford University Press, 2004); and Marilyn Strathern, "Discovering 'Social Control,'" *Journal of Law and Society* 12, no. 2 (1985): 111–134.

3. For a valuable introduction to some of the issues at stake, see Kieran Dolin, *Law and Literature* (Cambridge: Cambridge University Press, 2018); Timothy Endicott, "Law and Language," *Stanford Encyclopedia of Philosophy* Archive, ed. Edward N. Zalta, Metaphysics Research Lab, Stanford University, last updated April 15, 2016, https://plato.stanford.edu/archives/sum2016/entries/law-language/; and Bernard Jackson, "A Journey into Legal Semiotics," *Actes Sémiotiques* 120 (2017): 1–43.

4. Jessica Reyman, *The Rhetoric of Intellectual Property: Copyright Law and the Regulation of Digital Culture* (London: Routledge, 2012); Dan Burk and Jessica Reyman, "Patents as Genre: A Prospectus," *Law & Literature* 26, no. 2 (2014): 163–190.

5. Jessica Silbey, "The Mythical Beginnings of Intellectual Property," *George Mason Law Review* 15, no. 2 (2008): 379.

6. For an overview, see Richard Danner, "Foreword: Oh, the Treatise!," *Michigan Law Review* 111, no. 6 (2013): 821–834.

7. Ronan Deazley, "Commentary on Copinger's *Law of Copyright* (1870)," in *Primary Sources on Copyright (1450–1900)*, ed. Lionel Bently and Martin Kretschmer, http://www.copyrighthistory.org; Christopher Wadlow, "New Life and Vigour at Terrell?," *Journal of Intellectual Property Law & Practice* 6, no. 11 (2011): 833–836; Jose Bellido, "The Editorial Quest for International Copyright (1886–1896)," *Book History* 17, no. 1 (2014): 380–405; Jose Bellido, "The Constitution of Intellectual Property as an Academic Subject," *Legal Studies* 37, no. 3 (2017): 369–90.

8. See Maria Repoussi and Nicole Tutiaux-Guillon, "New Trends in History Textbook Research: Issues and Methodologies toward a School Historiography," *Journal of Educational Media, Memory, and Society* 2, no. 1 (2010): 154–170.

9. Kathy Bowrey, "Who's Writing Copyright's History?," *European Intellectual Property Review*, 18, no. 6 (1996): 322–329; Brad Sherman, "Remembering and Forgetting: The Birth of Modern Copyright Law," in *Comparing Legal Cultures*, ed. David Nelken (Aldershot, UK: Dartmouth, 1997), 237–266. Traces of the latter work can also be found in Lionel Bently and Brad Sherman, *The Making of Modern Intellectual Property Law: The British Experience, 1760–1911* (Cambridge: Cambridge University Press, 1999), 205–220.

10. Kelly Gates, "Will Work for Copyrights: The Cultural Policy of Anti-piracy Campaigns," *Social Semiotics* 16, no. 1 (2006): 57–73; Majid Yar, "The Rhetorics and Myths of Anti-piracy Campaigns: Criminalization, Moral Pedagogy and Capitalist Property Relations in the Classroom," *New Media & Society* 10, no. 4 (2008): 605–623; Tarleton Gillespie, "Characterizing Copyright in

the Classroom: The Cultural Work of Antipiracy Campaigns," *Communication, Culture & Critique* 2, no. 3 (2009): 274–318. See also the relevant chapters in Reyman, *Rhetoric of Intellectual Property*.

11. This course at the LSE was chosen at random and merely provides a guideline around which to shape this chapter. Nonetheless, it is worth noting that the LSE was the first university in Britain to offer a course on IP. After initial attempts to do so in the 1960s, in 1967 Bill Cornish drafted a syllabus for a postgraduate course titled "Industrial and Intellectual Property." It was only during the 1970s that IP was introduced as an undergraduate course (the first example dates from 1975, offered at the University of Southampton). For further details on the constitution of intellectual property as an academic subject in Britain, see Bellido, "Intellectual Property as an Academic Subject."

12. LSE, *Course Guides and Programme Regulations 2018/2019*, 240.

13. The course content specifies the learning goals as follows: "The curriculum of LL251 reflects the fact that it will be examined by means of an 8000-word essay. Instead of expecting students to acquire a more detailed knowledge of the mechanics of each of the principal branches of intellectual property law (copyright, patents, and trade marks) the course is structured around a strong theme that runs persistently through all parts of IP law, which will also be the basis of the dissertation topic that will be assigned at the start of the year. The objective will be to develop the skills required to engage critically with the mechanics of each branch." LSE, *Course Guides*, 240.

14. Frederic William Maitland, *Why the History of English Law Is Not Written* (London: C. J. Clay & Sons, 1888), 14.

15. Barbara Czarniawska-Joerges, *Narratives in Social Science Research* (London: Sage, 2004), 109. The classic theory I am referring to here is that of Todorov, who argued that "an 'ideal' narrative begins with a stable situation which is disturbed by some power of force. There results a state of disequilibrium; by the action of a force directed in the opposite direction, the equilibrium is re-established; the second equilibrium is similar to the first, but the two are never identical." Tzvetan Todorov, *The Poetics of Prose* (Ithaca, NY: Cornell University Press, 1977), 111.

16. For a succinct overview of the debate on focalization, see Burkhard Niederhoff, "Focalization," in *The Living Handbook of Narratology*, ed. Peter Hühn et al. (Hamburg: Hamburg University Press, 2009–2013), last modified September 24, 2013, https://www.lhn.uni-hamburg.de/node/18.html.

17. A. J. Greimas, "Narrative Grammar: Units and Levels," *MLN* 86, no. 6 (1971): 793–806.

18. Bowrey, "Who's Writing Copyright's History?," 322.

19. Gérard Genette, *Paratexts: Thresholds of Interpretation*, trans. Jane E. Lewin (Cambridge: Cambridge University Press, 1997).

20. David Bainbridge, *Intellectual Property*, 9th ed. (Harlow, UK: Pearson, 2012), back cover.

21. Bainbridge, back cover.

22. Bainbridge, back cover.

23. Charlotte Waelde et al., *Contemporary Intellectual Property: Law and Policy*, 3rd ed. (Oxford: Oxford University Press, 2013), back cover.

24. Lionel Bently and Brad Sherman, *Intellectual Property Law*, 4th ed. (Oxford: Oxford University Press, 2014), back cover.

25. Bainbridge, *Intellectual Property*, 393.

26. On the insularity of English legal studies, see also Peter Goodrich, "Critical Legal Studies in England: Prospective Histories," *Oxford Journal of Legal Studies* 12, no. 2 (1992): 195–236.

27. Bainbridge, *Intellectual Property*, 392.

28. Neil Davenport, *The United Kingdom Patent System: A Brief History* (Hampshire, UK: Kenneth Mason, 1979); Simon Thorley et al., *Terrell on the Law of Patents*, 16th ed. (London: Sweet & Maxwell, 2005), both quoted in Bainbridge, *Intellectual Property*, 392n11.

29. Bainbridge, *Intellectual Property*, 392n14.

30. Catherine Colston, *Principles of Intellectual Property Law* (London: Cavendish, 1999), 1.

31. Tanya Aplin and Jennifer Davis, *Intellectual Property Law: Text, Cases, and Materials*, 3rd ed. (Oxford: Oxford University Press, 2017).

32. Aplin and Davis, 535. I return later on to the meaning of the Statute of Monopolies, which provided a clause that gave inventors a fourteen-year term to exploit their invention.

33. Among the numerous publications, I single out: Lionel Bently, "The Making of Modern Trade Mark Law: The Construction of the Legal Concept of the Trade Mark (1860–1880)," in *Trade Marks and Brands: An Interdisciplinary Critique*, ed. Lionel Bently, Jane C. Ginsburg, and Jennifer Davis (Cambridge: Cambridge University Press, 2011), 3–41; Lionel Bently, Ronan Deazley, and Martin Kretschmer, eds., *Privilege and Property: Essays on the History of Copyright* (Cambridge: Open Book, 2010). Lionel Bently is one of the main editors of the project copyrighthistory.org, which provides a digital archive of primary sources on copyright from the invention of the printing press (c. 1450) to the Berne Convention (1886) and beyond.

34. Sherman, "Remembering and Forgetting."

35. Sherman, "Towards a History of Patent Law," 15.

36. Bainbridge, *Intellectual Property*, 392.

37. E.g., Roger Schechter, *Selected Intellectual Property and Unfair Competition Statutes, Regulations, and Treaties* (St. Paul, MN: West Academic, 2017); and Andrew Christie and Stephen Gare, *Blackstone's Statutes on Intellectual Property*, 13th ed. (Oxford: Oxford University Press, 2016).

38. Bainbridge, *Intellectual Property*, 392.

39. Bronwen Martin and Felizitas Ringham, *Dictionary of Semiotics* (London: Cassell, 2000), 11.

40. There are variations of this story, yet in principle they all distinguish between "monopolies in inventions, which were favourably viewed by Parliament and the public, and monopolies over things which were already invented, including a number of costumer staple products, which were viewed with great resentment by frustrated traders and distressed citizens." Phillips and Firth, *Introduction to Intellectual Property Law*, 34.

41. William Cornish, "Copyright I," in *The Oxford History of the Laws of England*, ed. William Cornish et al. (Oxford: Oxford University Press, 2010), 13:879.

42. Martin and Ringham, *Dictionary of Semiotics*, 11.

43. Bainbridge, *Intellectual Property*, 395.

44. Martin and Ringham, *Dictionary of Semiotics*, 12.

45. Bainbridge, *Intellectual Property*, 396.

46. One could think of different variations of the plot for the different sections of IP. In the case of copyright, for example, the different tests include the 1556 charter to the Stationers Company (the qualifying test), the 1710 Statute of Anne (the decisive test), and what Phillips and Firth call the "great consolidation of copyright in 1911" (the glorifying test). Phillips and Firth, *Introduction to Intellectual Property Law*, 128.

47. Colston, *Principles of Intellectual Property Law*, 343.

48. Bently, "Modern Trade Mark Law," 3–4 (notes omitted). This idea finds its way into the textbooks. For instance, Bainbridge argues that "although the application of distinguishing marks to goods has a long history, the law relating to trade marks is relatively young, going back to the early part of the nineteenth century." Bainbridge, *Intellectual Property*, 690.

49. Bently and Sherman, *Modern Intellectual Property Law*, 2.

50. Bently and Sherman, 2. Additional arguments concern the system of registration and the organization of the law. This idea is rehearsed in Bently and Sherman, *Intellectual Property Law*, 377.

51. Mario Biagioli, "Patent Specification and Political Representation: How Patents Became Rights," in *Making and Unmaking Intellectual Property*, ed. Mario Biagioli, Peter Jaszi, and Martha Woodmansee (Chicago: University of Chicago Press, 2011), 25–40; Oren Bracha, "Geniuses and Owners: The Construction of Inventors and the Emergence of American Intellectual Property," in *Transformations in American Legal History: Essays in Honor of Professor Morton J. Horwitz*, ed. Daniel W. Hamilton (Cambridge, MA: Harvard Law School, 2010), 1:369–390.

52. Cf. Ineke Sluiter, "Anchoring Innovation: A Classical Research Agenda," *European Review* 25, no. 1 (2017): 20–38; David Simpson, *Situatedness, or, Why We Keep Saying Where We're Coming From* (Durham, NC: Duke University Press, 2002).

53. Wikipedia, s.v. "History of Copyright," last modified June 12, 2020, https://en.wikipedia.org/wiki/History_of_copyright_law. The reference is to the Royal Irish Academy: "The Cathach/The Psalter of St. Columba," Library Cathach, archived from the original on July 2, 2014, https://web

.archive.org/web/20140702153948/http://www.ria.ie/Library/Special-Collections/Manuscripts/Cathach.aspx.

54. See Wikipedia, s.v. "Talk: History of Patent Law," last modified November 5, 2017, https://en.wikipedia.org/wiki/Talk:History_of_patent_law. The argument is that "whenever an article or a book claims to give a broad coverage on some topic and then talks about ancient Greece, just before jumping to Renaissance Europe (or vice versa) is Eurocentric POV.—The preceding unsigned comment was added by 74.103.17.98 (talk) 21:29, 8 May 2007 (UTC)."

55. Wikipedia, s.v. "Talk: History of Patent Law."

56. Vandana Shiva, *Protect or Plunder: Understanding Intellectual Property Rights* (London: Zed Books, 2001), 21.

57. On the complicated definition(s) of a public domain, see also Robert Merges and Amy Landers, *Intellectual Property and the Public Domain* (Cheltenham, UK: Edward Elgar, 2017).

58. This expression comes from Article I, Section 8, Clause 8 of the United States Constitution. The terminology changes from jurisdiction to jurisdiction; however, the underlying principle that IP should enlarge the public domain remains the same. As Walterscheid expresses it: "Indeed, it is precisely the unregulated and uncontrolled nature of knowledge in the public domain that renders it valuable for society. Patents and copyrights are deemed to be for the public good precisely because they are intended to enlarge the intellectual commons of knowledge available to all." Edward C. Walterscheid, *The Nature of the Intellectual Property Clause: A Study in Historical Perspective* (Buffalo, NY: W. S. Hein, 2002), 268.

59. See William Cornish, *Intellectual Property: Patents, Copyrights, Trademarks and Allied Rights*, 4th ed. (London: Sweet & Maxwell, 1999); Jeremy Phillips and Alison Firth, *Introduction to Intellectual Property Law*, 4th ed. (London: Butterworths, 2001).

60. Among the most vocal have been Michele Boldrin and David Levine, *Against Intellectual Monopoly* (Cambridge: Cambridge University Press, 2010).

61. See Algirdas Julien Greimas, *On Meaning: Selected Writings in Semiotic Theory* (Minneapolis: University of Minnesota Press, 1987). One could also think of a square where IP is contrasted to secrecy, for example.

62. This section has been shortened for reasons of space. I decided to single out the most central metaphor and not to look at other types of analogy, such as metonyms and allegories. One could add other metaphors as well: in copyright, for instance, the analogy is more about "the birth and caring of a baby," whereas in trademarks it is more about "personal identity and development." Still, one can find the idea of "a balance" in those domains of the law as well.

63. For a more elaborate reflection on the metaphorical use of "the balance" in IP law, see Mario Biagioli, "Weighing Intellectual Property: Can We Balance the Social Costs and Benefits of Patenting?," *History of Science* 57, no. 1 (2018): 140–163. For the philosophical underpinnings of this essay, which deals with the importance of the balance in the iconography of justice, see also Mario Biagioli, "Justice Out of Balance," *Critical Inquiry* 45, no. 2 (2019): 280–306.

64. Bainbridge, *Intellectual Property*, 384. Notes omitted; author's emphasis.

65. *Innovation in America, Hearings Before the Subcommittee on Courts, Intellectual Property, and the Internet of the Committee on the Judiciary, House of Representatives, One Hundred Thirteenth Congress, First Session, July 25 and August 1, 2013. Part I and II* (Washington, DC: U.S. Government Printing Office, 2014), 203.

66. Cornish, *Intellectual Property*, 123.

67. Georg Simmel, "Das Problem der historischen Zeit (1916)," in *Brücke und Tür: Essays des Philosophen zur Geschichte, Religion, Kunst, und Gesellschaft*, ed. Michael Landmann and Margarete Susman, 43–58 (Stuttgart: K. F. Koehler, 1957).

Bibliography

Aplin, Tanya, and Jennifer Davis. *Intellectual Property Law: Text, Cases, and Materials*. 3rd ed. Oxford: Oxford University Press, 2017.

Bainbridge, David I. *Intellectual Property*. 9th ed. Harlow, UK: Pearson, 2012.

Bellido, Jose. "The Constitution of Intellectual Property as an Academic Subject." *Legal Studies* 37, no. 3 (2017): 369–390.

Bellido, Jose. "The Editorial Quest for International Copyright (1886–1896)." *Book History* 17, no. 1 (2014): 380–405.

Bently, Lionel. "The Making of Modern Trade Mark Law: The Construction of the Legal Concept of the Trade Mark (1860–1880)." In *Trade Marks and Brands: An Interdisciplinary Critique*, edited by Lionel Bently, Jane C. Ginsburg, and Jennifer Davis, 3–41. Cambridge: Cambridge University Press, 2011.

Bently, Lionel, Ronan Deazley, and Martin Kretschmer, eds. *Privilege and Property: Essays on the History of Copyright*. Cambridge: Open Book, 2010.

Bently, Lionel, and Brad Sherman. *Intellectual Property Law*. 4th ed. Oxford: Oxford University Press, 2014.

Bently, Lionel, and Brad Sherman. *The Making of Modern Intellectual Property Law: The British Experience, 1760–1911*. Cambridge: Cambridge University Press, 1999.

Biagioli, Mario. "Justice Out of Balance." *Critical Inquiry* 45, no. 2 (2019): 280–306.

Biagioli, Mario. "Patent Specification and Political Representation: How Patents Became Rights." In *Making and Unmaking Intellectual Property*, edited by Mario Biagioli, Peter Jaszi, and Martha Woodmansee, 25–40. Chicago: University of Chicago Press, 2011.

Biagioli, Mario. "Weighing Intellectual Property: Can We Balance the Social Costs and Benefits of Patenting?" *History of Science* 57, no. 1 (2018): 140–163.

Boldrin, Michele, and David K. Levine. *Against Intellectual Monopoly*. Cambridge: Cambridge University Press, 2010.

Bowrey, Kathy. "Who's Writing Copyright's History?" *European Intellectual Property Review* 18, no. 6 (1996): 322–329.

Bracha, Oren. "Geniuses and Owners: The Construction of Inventors and the Emergence of American Intellectual Property." In *Transformations in American Legal History: Essays in Honor of Professor Morton J. Horwitz*. Vol. 1, edited by Daniel W Hamilton, 369–390. Cambridge, MA: Harvard Law School, 2010.

Burk, Dan L., and Jessica Reyman. "Patents as Genre: A Prospectus." *Law & Literature* 26, no. 2 (2014): 163–190.

Chandler, Daniel. *Semiotics: The Basics*. 2nd ed. London: Routledge, 2007.

Christie, Andrew, and Stephen Gare. *Blackstone's Statutes on Intellectual Property*. 13th ed. Oxford: Oxford University Press, 2016.

Colston, Catherine. *Principles of Intellectual Property Law*. London: Cavendish, 1999.

Cornish, William. "Copyright I." In *1820–1914, Fields of Development*, edited by William Cornish, J. Stuart Anderson, Ray Cocks, Michael Lobban, Patrick Polden, and Keith Smith, 879–930. Vol. 13 of *The Oxford History of the Laws of England*. Oxford: Oxford University Press, 2010.

Cornish, William. *Intellectual Property: Patents, Copyrights, Trademarks and Allied Rights*. 4th ed. London: Sweet & Maxwell, 1999.

Czarniawska-Joerges, Barbara. *Narratives in Social Science Research*. London: Sage, 2004.

Danner, Richard. "Foreword: Oh, the Treatise!" *Michigan Law Review* 111, no. 6 (2013): 821–834.

Deazley, Ronan. "Commentary on Copinger's *Law of Copyright* (1870)." In *Primary Sources on Copyright (1450–1900)*, edited by Lionel Bently and Martin Kretschmer. http://www.copyrighthistory.org.

Dolin, Kieran. *Law and Literature*. Cambridge: Cambridge University Press, 2018.

Ehrlich, Eugen. *Fundamental Principles of the Sociology of Law*. New Brunswick, NJ: Transaction, 2001.

Endicott, Timothy. "Law and Language." *Stanford Encyclopedia of Philosophy* Archive, edited by Edward N. Zalta. Metaphysics Research Lab, Stanford University, last updated April 15, 2016. https://plato.stanford.edu/archives/sum2016/entries/law-language/.

Gates, Kelly. "Will Work for Copyrights: The Cultural Policy of Anti-piracy Campaigns." *Social Semiotics* 16, no. 1 (2006): 57–73.

Genette, Gérard. *Paratexts: Thresholds of Interpretation*. Translated by Jane E. Lewin. Cambridge: Cambridge University Press, 1997.

Gillespie, Tarleton. "Characterizing Copyright in the Classroom: The Cultural Work of Antipiracy Campaigns." *Communication, Culture & Critique* 2, no. 3 (2009): 274–318.

Goldstein, Paul. *International Copyright Principles, Law, and Practice*. Oxford: Oxford University Press, 2001.

Goodrich, Peter. "Critical Legal Studies in England: Prospective Histories." *Oxford Journal of Legal Studies* 12, no. 2 (1992): 195–236.

Greimas, Algirdas Julien. "Narrative Grammar: Units and Levels." *MLN* 86, no. 6 (1971): 793–806.

Greimas, Algirdas Julien. *On Meaning: Selected Writings in Semiotic Theory*. Minneapolis: University of Minnesota Press, 1987.

Innovation in America. Hearings Before the Subcommittee on Courts, Intellectual Property, and the Internet of the Committee on the Judiciary, House of Representatives, One Hundred Thirteenth Congress, First Session, July 25 and August 1, 2013. Part I and II. Washington, DC: U.S. Government Printing Office, 2014. http://purl.fdlp.gov/GPO/gpo45855.

Jackson, Bernard S. "A Journey into Legal Semiotics." *Actes Sémiotiques* 120 (2017): 1–43.

Latour, Bruno. *The Making of Law: An Ethnography of the Conseil d'Etat*. Cambridge: Polity, 2010.

London School of Economics and Political Science (LSE). *Course Guides and Programme Regulations 2018/2019*. Accessed November 4, 2020. https://info.lse.ac.uk/staff/divisions/academic-registrars-division/Teaching-Quality-Assurance-and-Review-Office/Assets/Documents/Calendar/CourseGuidesProgrammeRegs18-19.pdf.

Luhmann, Niklas, and Fatima Kastner. *Law as a Social System*. Oxford: Oxford University Press, 2004.

Maitland, Frederic William. *Why the History of English Law Is Not Written*. London: C. J. Clay & Sons, 1888. https://en.wikisource.org/wiki/Why_the_History_of_English_Law_is_Not_Written.

Martin, Bronwen, and Felizitas Ringham. *Dictionary of Semiotics*. London: Cassell, 2000.

Merges, Robert P., and Amy L. Landers. *Intellectual Property and the Public Domain*. Cheltenham: Edward Elgar, 2017.

Niederhoff, Burkhard. "Focalization." In *The Living Handbook of Narratology*, edited by Peter Hühn, Jan Christoph Meister, John Pier, and Wolf Schmid (Hamburg: Hamburg University Press, 2009–2013). Last modified September 24, 2013. https://www.lhn.uni-hamburg.de/node/18.html.

Phillips, Jeremy, and Alison Firth. *Introduction to Intellectual Property Law*. 4th ed. London: Butterworths, 2001.

Repoussi, Maria, and Nicole Tutiaux-Guillon. "New Trends in History Textbook Research: Issues and Methodologies toward a School Historiography." *Journal of Educational Media, Memory, and Society* 2, no. 1 (2010): 154–170.

Reyman, Jessica. *The Rhetoric of Intellectual Property: Copyright Law and the Regulation of Digital Culture*. London: Routledge, 2012.

Roberts, Geoffrey. *The History and Narrative Reader*. London: Routledge, 2010.

Schechter, Roger. *Selected Intellectual Property and Unfair Competition Statutes, Regulations, and Treaties*. St. Paul, MN: West Academic, 2017.

Sherman, Brad. "Remembering and Forgetting: The Birth of Modern Copyright Law." In *Comparing Legal Cultures*, edited by David Nelken, 237–266. Aldershot, UK: Dartmouth, 1997.

Sherman, Brad. "Towards a History of Patent Law." In *Intellectual Property in Common Law and Civil Law*, edited by Toshiko Takenaka, 3–16. Cheltenham, UK: Edward Elgar, 2013.

Shiva, Vandana. *Protect or Plunder: Understanding Intellectual Property Rights*. London: Zed Books, 2001.

Silbey, Jessica M. "The Mythical Beginnings of Intellectual Property." *George Mason Law Review* 15, no. 2 (2008): 319–379.

Simmel, Georg. "Das Problem der historischen Zeit (1916)." In *Brücke und Tür: Essays des Philosophen zur Geschichte, Religion, Kunst, und Gesellschaft*, edited by Michael Landmann and Margarete Susman, 43–58. Stuttgart: K. F. Koehler, 1957.

Simpson, David. *Situatedness, or, Why We Keep Saying Where We're Coming From*. Durham, NC: Duke University Press, 2002.

Sluiter, Ineke. "Anchoring Innovation: A Classical Research Agenda." *European Review* 25, no. 1 (2017): 20–38.

Strathern, Marilyn. "Discovering 'Social Control.'" *Journal of Law and Society* 12, no. 2 (1985): 111–134.

Todorov, Tzvetan. *The Poetics of Prose*. Ithaca, NY: Cornell University Press, 1977.

Wadlow, Christopher. "New Life and Vigour at Terrell?" *Journal of Intellectual Property Law & Practice* 6, no. 11 (2011): 833–836.

Waelde, Charlotte, Graeme Laurie, Abbe Brown, Smita Kheria, and Jane Cornwell. *Contemporary Intellectual Property: Law and Policy*. 3rd ed. Oxford: Oxford University Press, 2013.

Walterscheid, Edward C. *The Nature of the Intellectual Property Clause: A Study in Historical Perspective*. Buffalo, NY: W. S. Hein, 2002.

World Intellectual Property Organization (WIPO). *WIPO Intellectual Property Handbook*. 2nd ed. Geneva: World Intellectual Property Organization, 2008.

Yar, Majid. "The Rhetorics and Myths of Anti-piracy Campaigns: Criminalization, Moral Pedagogy and Capitalist Property Relations in the Classroom." *New Media & Society* 10, no. 4 (2008): 605–623.

II THE THREE PRACTICES: PERFORMANCE, USE, NAMING

4

RAGA AND THE PROBLEM OF OWNERSHIP: KNOWLEDGE AND CULTURE IN CARNATIC MUSIC

Annapurna Mamidipudi and Viren Murthy

Capitalist society is usually thought of as being based on ownership—specifically, private ownership of the means of production. However, the problem of ownership pervades modern capitalist society in a manner that goes beyond owning the means of production and influences our relationship to intellectual production as well. Intellectual production, or knowledge, in this regime is usually subsumed under the concept of copyright, authorship, and patents, which delineates who has the rights over the sale, purchase, and symbolic capital related to the product of a specific person's intellectual labor. From this perspective, ownership is largely about exclusivity with respect to the use and exchange of objects.[1] In order that knowledge be owned in such a mode, it must be made detachable from the thing that is known, and mobilized[2] independent of the knower. This detachability or alienability is one of the conditions for the possibility of reconnection between knowability and ownability, which turns such knowledge into an exchangeable commodity.

We will be using the notion of ownership of knowledge in a slightly broader sense, as we examine struggles over the ownership of a form of artistic practice, Carnatic music, or South Indian classical music, by focusing on four specific performers. We wish to foreground two perspectives. First, we point to the central role of practice as an authoritative way of knowing. The creation of new spaces for performance during the twentieth century, and institutional boundaries for access to such spaces, show how practice becomes a mode of legitimately owning knowledge. Second, for Carnatic music, ownership of knowledge is not just about the use or exchange, but also about a certain social value of the music, which enables entry or membership into a specific community. Thus, ownership is also about being part of a community. In the instances examined in this chapter, the act of owning can change both the owner and the owned. What we examine here is the key point of the constitution of the relation between what is known and what can be owned as knowledge. In short, we must think about ownership of knowledge through performance as a form of relation that is constitutive of these categories.

We will examine knowledge related to a specific set of practices associated with the framework of *raga* in Carnatic music and interrogate how ownership of knowledge is embedded in these practices. While the system of *raga* is a framework that is operationalized in the performance, and in the practices of producing and consuming music, we find that when the relevant knowledge system becomes codified as a *raga* system, it becomes theory that is potentially detachable from the practice of Carnatic music. Though codification has played an increasingly important role since the turn of the twentieth century,[3] detachability does not mean that the theoretical framework is independent of practice; rather, as we shall see, the two remain in a dialectical relationship. Both change over time, since performance continues to be the site for experimentation through improvisational practices and the evaluation of musical production. However, once this *raga* system is codified, it requires actors to produce and consume it in this mode to ensure its reproduction and transmission. The *raga* system, now abstracted, has a technical vocabulary of its own and thus becomes a site for explicating and owning musical knowledge. It begins to function as a conceptual framework that has explanatory power over practice, with a knowledge role disembedded from the performance of the music itself. Given the nature of this codification, it opens up the sphere of Carnatic music to a newly defined knowledge community made up of two groups of people—performers and listeners (or *rasikas*, connoisseurs) of Carnatic music. In short, knowledge is defined in terms that legitimize the communal ownership of a newly emergent unique group who endeavor to distinguish themselves as being a knowledgeable elite by mediating the cultural and/or symbolic value associated with Carnatic music as explicable knowledge, rather than practice alone.

We use the example of Carnatic music to make a wider point about knowledge systems that are owned as practice. Practice, theorized as arrays of human activities, highlights embodied, materially mediated, nonpropositional knowledge;[4] within social theory and ethnology it speaks against deterministic grasps of social structures and systems,[5] and among cultural theorists it is used to depict language as discursive activity as against abstract discourse.[6] Despite the diversity, practice accounts in theory agree that as a field, practices constitute knowledge. Thus, the mind itself is constituted within practices; further, such knowledge is no longer even the property of individuals but instead a feature of groups.[7] We use the performance lens to follow technical practices and practical skills, allowing us to use performance as practice of craft skills and improvisation unfolding in time and space.[8] From this perspective, we follow Nicholas Cook in focusing on how music is performed. Cook writes:

The experience of live or recorded performance is a primary form of music's existence, not just the reflection of a notated text. And performers make an indispensable contribution to the culture of creative practice that is music. My claim is that in order to build this deeply into our thinking about music—in order to think of music *as performance*—we need to think differently about what sort of an object music is, and indeed how far it is appropriate to think of it as an object at all.[9]

This point is perhaps even more true of Carnatic music, where lovers of the art form are much more interested in going to a concert to hear a specific singer, without even knowing what they will perform. Through an analysis of various figures in Carnatic music, we explore how the practice of singing has the potential to signify beyond the social categories that have been imposed on it. Here, while performance is connected to performativity, they are not the same; even as performativity is not an extension of discourse theory, bodies speak without necessarily uttering.[10] Thus, in analyzing practice (both as doing and knowing) as performance we bring together two approaches: the first, performing as *showing doing*,[11] and the second, performing as *doing knowing*.[12] Practice-based research is the default approach in artistic performance; here, we propose that performance itself can produce and validate knowledge, connecting knowing to owning knowledge. Shifting focus from propositional content of knowledge, performance as performing knowledge stresses the co-presence of actors and audiences, temporality, and spatiality. Such a performance situation, where knowledge is presented in person, becomes a place to experiment with new modes of public knowledge, where production of knowledge is inseparably bound up with its reception, as well as the intermediality (sound, visual, material) implied in performance practices in knowledge production.[13] Thus, audiences constituted as communities of practice as well as publics remain important in the discourse about ownership, not only as social determinants, but because they arbitrate the performance of knowledge.

In general, Carnatic music can be placed into a longer historical trajectory of examples where musical forms are reconstituted by capitalism and the nation-state, which we can see in other musical traditions as well. For example, Hermano Viana has masterfully discussed how, in Brazil, samba became a symbol of the nation.[14] Perhaps even more closely related to our project, Fred Lau shows how the concept of modern Chinese music emerged in the early twentieth century as part of a nationalist project.[15] The story of the movement of Carnatic music from palaces to concert halls is similarly very much tied to new forms of identity associated with the emerging Indian nation-state and with market forces. However, there are some additional issues we need to keep in mind when discussing the transformations of Carnatic music. First, it self-identifies as classical music as opposed to, say, samba, or the Chinese folk music of which Lau writes. Second, although there are Carnatic singers, such as M. S. Subbalakshmi, who eventually represented the nation, to do so they needed to overcome regional divides. In other words, Carnatic music is usually understood to mean South Indian music in particular, not Indian music in general. Finally, Carnatic music became intimately enmeshed in a hierarchical social structure—namely, the caste system, which comprised another obstacle to labeling Carnatic music "national music." This is issue not unique to India, since the problem of elite discourses on music becoming dominant is also evident in the Brazilian and Chinese examples. However, in the Indian case, because even the languages of the south differ significantly from those of the north, we find, for example, Tamil nationalism influencing Carnatic music at least as much as Indian nationalism. T. M. Krishna has argued that Carnatic music is fundamentally exclusive because it is monopolized by the Brahmin caste in India.[16] This framing of Carnatic music enacts a barrier between the music and wider society that is almost impossible to negotiate for anyone who is not hereditarily Brahmin. From this viewpoint, membership of a social elite acquired by birth is the only way to know Carnatic music.

In examining the issue from a knowledge ownership perspective, we show how the evolving relation between the *theoretical concept* and *performance* of *raga*, as organizing practices of singing, playing a musical instrument, and listening, potentially allows or excludes participation beyond institutional boundaries of caste and class. A key factor here is the importance of improvisation in the performance of Carnatic music, which goes beyond mastery. Practice or *sadhana* is often done in private and comprises training for performance. During this practice, musicians will often work on various aspects of Carnatic music and, once they gain mastery, will begin to improvise. This private practice, guided by a guru, will eventually become a public performance, which will enable a large audience to experience the music. During the performance, listeners in the audience can demonstrate their knowledge by, for example, identifying *ragas*, and this grants them membership in a community of listeners. Connoisseurs belonging to this community, referred to as *rasikas*, can then become gatekeepers, both of the boundaries of what constitutes valid improvisation in the music itself and of who can be granted entry to become knowledgeable audiences and patrons.

It is through attaining and expressing knowledge of *ragas* that an individual affirms their belonging in a community, either as a listener or as a musician. We begin with an outline of the basic features of *ragas*, followed by a brief discussion of the change in social contexts of performance—of how the centers of music have moved from temples and palaces to concert halls or *sabhas*. This implies a shift in ownership practices, since potentially it has given more people access to Carnatic music, but at the same time, earlier hierarchies related to caste have been reconstituted and become more pervasive.

Specifically, the twentieth century saw Brahminization and modernization emerge as twin phenomena. Once we have set this scene, we will examine various themes and four singers who have influenced *ragas* and Carnatic music more generally in their own ways from the early twentieth century to the present. The four artists who form the subject of our case studies are: G. N. Balasubramaniam (GNB) (1910–1965), who might be called a Carnatic pop star, the most famous female singers of the twentieth century; M. S. Subbalakshmi (MS) (1916–2004); T. M. Krishna (TMK) (1976–), a well-known contemporary critical Carnatic musician; and Vidushi R. Vedavalli (RV) (1935–), a contemporary female musician in her eighties.

The shift from palaces to concert halls brings us to a discussion on the opportunity and the impulse to universalize classical music. In a palace, the king was at the center of musical performance, but music was not played for the public. In contrast, GNB shaped the knowledge of Carnatic music so as to emphasize its universality. After this, we move on to discuss the religious value of devotion in Carnatic music, which is evidenced by Subbalakshmi. MS belonged to the marginalized community of *devadasis*, women who performed in temples and courts to great acclaim, before their practice was deemed promiscuous by the law and lost respectability. She went on to become one of the major figures in Carnatic music and consequently provides a prime example of a person from outside the community of Brahmins who makes Carnatic music their own through negotiating both the social boundaries and the knowledge framework of Carnatic music, including the nuances of *ragas*. In her practice, musical knowledge was connected to a virtuous life, which was accepted by people beyond the Brahmin community. Next, we move on to what could be conceived as a reaction to the above emphasis on devotion, Brahminization, and virtue—an attempt to make Carnatic music "modern" in a different way. Krishna is emblematic of this trend and could be called the rebel star of Carnatic music. He has recently berated the Carnatic tradition as being insular and has attempted to rejuvenate it by radically altering its form, while at the same time rethinking and appropriating the framework of *ragas*. The final section of this chapter deals with the trajectory of traditional Carnatic music and the manner in which, by highlighting the gap between that knowledge and practice, Carnatic music can result in traditionally informed innovation that is distinct from the global mass culture that threatened to marginalize classical music in the 1960s and 1970s. Vedavalli, along with others in the community, has noted how *ragas* have changed over time, by comparing contemporary practice to notation texts written in the late nineteenth century, which were early attempts to codify *ragas*. RV sees herself as very much rooted in the tradition of Carnatic music, but by sticking to this tradition firmly, she changes it. Together, the various sections in this chapter show how Carnatic music is owned and transformed within a particular kind of experimental epistemic culture that self-identifies as traditional and classical music.

RAGA AS A KNOWLEDGE FORM

Raga is a concept that is unique to Indian classical music and notoriously difficult to define, but in this section, we will briefly examine the conceptual framework that people now refer to as a *raga*. A *raga* is a melodic framework or a generative mechanism consisting of rules, which appear in the form of musical scales and phrases. The concept of *raga* emerged before colonial influence and was first elaborated in a classical treatise on music in Matanga Muni's *Brhaddeshi*, usually dated between the sixth and eighth centuries. However, this and later texts do not use *raga* in the same way that we use the term today. Perhaps most importantly, music was not a distinct subject of knowledge at that time—the pre-thirteenth-century literature on music also covered art, aesthetics, beauty, dance, and theater.[17]

Initially, there was no difference between Carnatic music, which originated in South India and Hindustani, and North Indian classical music, and they continue to share a similar structure of *raga*.[18] Carnatic (and Hindustani) musicians have used something analogous to the solfège to map the musical scale of *ragas*. In short, the notes do, re, mi, fa, so, la, ti in Western classical music correspond to the notes or *swaras* sa, ri, ga, ma, pa, da, ni in Carnatic music.[19] The intervals between the notes can change, just like in Western classical music. Although a certain type of notation existed before Western colonialism and influence, as we shall see in the next section, in the twentieth century there was a conscious effort to make an analogy to the Western system, to claim that Carnatic music was also a type of classical music, through a focus on producing texts, some of which were notated. We argue that this codification implied a certain emphasis on theory, but as such, theory associated with the *raga* was continually performed or displaced through practice. We argue that this shift to describing what was previously known primarily through performance sets the stage for the current discussion on Carnatic music as a regime of ownership of knowledge that must accede to particular classificatory rites of passage if it is to become a modern knowledge system.[20] As music became more of an independent artistic practice, the concept of the *raga* also evolved.

In this chapter, we will focus primarily on the concept of *raga* in the twentieth century. In general, a specific combination of *swaras* is the foundation of any given *raga*. For example, the major scale in Western classical music corresponds to the Carnatic *raga Sankarabharanam*.[21] *Ragas* are more than merely scales. For instance, although much baroque music uses the same notes as the *raga Sankarabharanam*, baroque music might not invoke this *raga* for practitioners and listeners of Carnatic music. Rather, there are often also set phrases that evoke the moods and emotions of a particular *raga*. Certain notes in a *raga* are played or sung with embellishments or *gamakas*, which are essential to the *raga*'s mood. A mark of a knowledgeable listener of Carnatic music is that they are able to identify which *raga* is being sung.

To enable the listener to identify a *raga*, the performer has to sing certain key notes in phrases. To evoke the mood of a *raga* and make it identifiable, a performer often plays or sings a melodic phrase with what is called an oscillation of notes or a *gamaka*, a concept that we find in the early texts on Indian music.[22] Similarly, any given *raga* will have certain key phrases that enable the audience to recognize it in the initial stages of a song, or *kriti*. This description shows that although a complex melody can elegantly evoke a *raga*, a *raga* is at a higher level of abstraction than a song or specifically composed melody. One can insert notes and string phrases of a specific *raga* in many possible patterns and combinations. To some extent, this creativity or innovation is the work of the composers, who write songs in any given *raga*. The framework of *ragas* gives the singer a large degree of creativity as well, and to grasp this, we will need to introduce a bit more detail about the manner in which a song is presented.

Typically, in a concert, a singer will not jump straight into a complex *kriti* or composition—that will be preceded by what is called an *alapana*. In this section, the singer sings phrases of a specific *raga*, but the *alapana* is improvised without any lyrics or composed rhythmic structure. The singer sings using syllables such as tha, da, ri, and na, among others. During this phase, the singer brings out the essence of the *raga* and explores various possibilities within it. This is then followed by a violin solo, where the violinist presents their version of the *alapana*, partly mimicking the vocalist.

When these two improvised portions are finished, the singer launches into the *kriti* or song, which largely follows the composed version and includes a specific rhythmic pattern. We say "largely" because—toward the end of this composition—there are two improvised sections, which can be distinguished from the *alapana* because they are performed while conforming to the rhythmic structure in which the *kriti* is composed. These two sections are called the *neraval* and *kalpana swaras*. In the *neraval*, the singer takes a phrase from the lyrics in the *kriti* and improvises around it, this time within the strictures of the rhythm. The *kalpana swara* then continues this type of improvisation, but instead of using lyrics, the singer directly sings the notes sa, ri, ga, ma, pa, da, ni in different patterns at increasing speeds, again exploring the possibilities of the *raga*. In these various periods of improvisation, there are certain aesthetic criteria that govern the manner in which the singer can develop the *raga*. It is here that the community of listeners and connoisseurs plays a role, sanctioning such experimentation as being legitimate or not.

We have gone into some detail about the various modes of improvisation because this illustrates the dynamic nature of the *raga* in performance. In other words, as a knowledge system or framework for practice, *ragas* may be described as comprising a combination of scales, phrases, and *gamakas*; however, because of the high level of improvisation the musical form demands, the initial framework is being both constantly reproduced and gradually transformed based on aesthetic criteria that are not easily codified during practice and performance. We argue that this internal dynamic is connected to what it means to own or master a *raga*. In short, one cannot possess a *raga* like an object, and so when we discuss ownership in this context, we are talking about belonging to a community of listeners and musicians at the same time; owning knowledge of *raga* equates to being recognized as a part of the community. There is a constant tension between three elements—the theory, the musician's performance, and the social mediation that makes up this epistemic culture. There is a dialectic between the structure of the *raga* and the various reproductions and transformations that occur with the performance of improvisations based on various aesthetic criteria.

The training required to learn a *raga* then takes us beyond usual forms of ownership or belonging, because owning or appropriating a *raga* involves a transformation of the subject doing the owning. The repetitive, imitative, technical, and meditative practices required to learn Carnatic music, along with their religious overtones, often call for overcoming the self and following the logic of the *raga* itself. While this is not the place to go into the complex world of Carnatic pedagogy, typically a neophyte will begin by repeating various exercises while imitating their teacher. At that point, students are introduced to forms of codification, especially the notes and the various *ragas*, which have been increasingly formalized in the past two centuries and form the basis of the various exercises. The exercises eventually grow in complexity and the *swara* patterns become increasingly difficult. Such imitation and repetition appear to be opposite to improvisation, but in fact they are creating the conditions for the possibility of creativity.[23] At the final stages of a student's training, teachers will expect moments of spontaneity to emerge without going out of the framework of the *raga*. At this level, the learner goes beyond both codification and mere imitation, but in a manner that retains what they have learned at the previous levels.

Musicians will debate whether *ragas* can be owned at all, as in their framework, ownership is about having the responsibility to perpetuate the art without compromising its essential creative character. At the same time, when a musician is an accredited master of a particular *raga*, it becomes attached to his name, as, for example, in the case of Todi Sitaramayya, who was known for his mastery of the *raga* called *Todi*. One could also say that rather than possessing the *raga*, at that point it should be almost as if the singer is *possessed by* the *raga*, which seems to undermine one of the key conditions of ownership—namely, self-possession. This element of Carnatic music could represent a moment of self-transcendence through self-discipline. Popular fiction about Carnatic music often alludes to this moment of transcendence in Carnatic music. For example, in the famous Kannada novel the *Swan's Song* (*Hamsa Geethe*, 1952), the protagonist's singing teacher tells him that in order to sing, he needs to transcend himself. "Did I not tell you before: Advaita. Everything is me. Everyone's pain and happiness is in my mind. The mind is the ontological whole [*aham braham asmi*], implying that the mind must come to full maturity."[24] Interestingly, in such stories, the protagonist has to eventually disown everything else to learn Carnatic singing, including their own self, and devote their entire being to practice. Then finally they are able to learn improvisation, which begins to come "naturally." There is an analogous narrative for the listener, where true *rasikas* could be found in unlikely places, like the driver of the bullock cart who could converse knowledgeably about the previous night's performance while ferrying the musician of the day to her concert.

These idealized representations of Carnatic music do say something about the practice, and we see contemporary musicians drawing on such ideas as well. By combining musical practice with an understanding of ultimate reality, such texts indicate the universal significance of Carnatic music. Understanding and being able to practice Carnatic music implies understanding the secrets of the world—or even the universe. This idealized vision is inextricably connected with social mediation that puts people in certain subject positions that denote class, caste, gender, and numerous other social relations. Like all art forms, Carnatic music is socially mediated, so we must consider the social hierarchy that conditions its practice. To shed some light on this aspect, we will now briefly explain the transformation of the social context of Carnatic music and the Brahmin hegemony that surrounds its institutional performance spaces.

UNIVERSALIZING MUSIC: FROM IMPERIAL COURTS TO MADRAS CONCERT HALLS

In the late nineteenth and early twentieth centuries, an important social transformation affected the social spaces in which Carnatic music was performed and presented as knowledge. These transformations help us to understand the changing conditions for the production of knowledge. Before the twentieth century, Carnatic music was not unified.[25] Just as music itself was considered in relation to other performing arts, there were various types of performers of Carnatic music. In particular, the two non-Brahmin practitioners, the *devadasis* and those playing the instrument, the *nagasvara*, were able to coexist with the Brahmin musicians and scholars, and all contributed to music. The *devadasi* is unique to the Indian context, but could be understood as the Indian counterpart to the Japanese geisha. A way of distinguishing the gender roles that emerged in Carnatic music, and which needed negotiation in the twentieth century, was that most musician composers, *vaggeyakaras*, were Brahmin and male, and they simultaneously acquired knowledge of language, *shastras* (sacred texts), *agamas* (Hindu devotional scriptures), and musical theory. In contrast, women of the *devadasi* community, though well versed in the sixty-four arts, were seen predominantly as performers. Although they would later be associated with prostitution, their actual artistic identity was much more complex. Strictly speaking, *devadasis* were women who were dedicated to worship and performed various artistic ritual offerings in the temple through song and dance. Precisely because they became identified with prostitution after India's independence in 1947, the government later outlawed them.[26]

Throughout the nineteenth and early twentieth centuries, singers would perform in temples and imperial palaces. Tanjore, a city in Tamil Nadu state in southeast India, was a center for art, and as the center moved to Madras, a Brahmin stream of music came to dominate the *devadasi* practitioners. *Devadasis* had a rich repertoire. including songs of *Shringara* (sensuousness) such as *padams* and *javelis*, but also, at times, larger *raga*-centered improvisation. As the Brahmins came to dominate patronage organizations and the framework of the *raga* became codified as abstract knowledge, it also became increasingly associated with Brahmin music.

This perception of the Brahmin way of singing embodied a contradiction. On the one hand, the emphasis on abstraction and improvisation suggested an openness, tending toward universality. Anyone who learned the rules and the aesthetic principles could participate in this music and potentially be accepted into the community of listeners and practitioners. On the other hand, the training and culture associated with such music was eventually limited to upper-caste and specifically Brahmin individuals, as this knowledge was transmitted within their homes and organizations. Even without caste-based restrictions, the amount of time and practice required to understand and appreciate Carnatic music meant that it was difficult to access.

We see here a type of contradictory development of Carnatic music that is at once open and closed. The same contradictions can be seen in relation to the emergent public. In particular, the concert space moved from the Tanjore palace and temples to public concert halls or *sabhas* in Madras, which were theoretically open to a larger public. The concert halls implied a structural transformation of the spaces where music was performed and appreciated. In place of a palace or temple where music was performed for a specific person, such as a king, performance halls were built for everyone, and included a stage that was separated from the audience's seating area. Other changes accompanied this transformation, such as the introduction of microphones and amplifiers that would enable large numbers of people to listen. But, as we have already mentioned above, this opening up occurred simultaneously with a Brahminization of the music, which in reality tended to make it the property of one particular caste. This exclusivity was sometimes aided by the religious nature of the music, which largely represented Hindu imagery, reinforcing religious boundaries.

At India's independence, people had gained legal freedom—they were increasingly officially free from bonds of caste, but the hierarchies of caste continued to operate in the private domain, and at times grew stronger, while at the same time creating new forms of identity and exclusion. This again fits with recent discussions of caste. For instance, Gill Navyug has used Karl Marx's "On the Jewish Question" (1844) to understand caste in India.[27] In that article, Marx argued that the transition from feudalism to capitalism officially abolished existing hierarchies, but in reality it allowed inequalities to continue in the private sphere. Marx is, of course, most concerned with the power of private property and the market, which is definitely an important part of this story as well. After all, the move to Madras must be understood in relation to the history of the emergence of capitalism in India. This had contradictory effects on music. On the one hand, it had the potential to lead to universalization, since theoretically, anyone who bought a ticket could listen to the concerts, which were open to the public. But on the other hand, this could result in the watering down of Carnatic music, as musicians began to cater to an uninformed public in order to broaden their appeal and therefore increase ticket sales.

As this process developed in the 1950s and 1960s, as we will see, the famous Carnatic singer GNB criticized the commodification of music. Lakshmi Subramanian has helpfully outlined the consequences of these transformations on listening culture, in ways that suggest a connection to the problem of ownership and also to the reproduction of hierarchy. She writes about the importance of a growing middle class in Madras, successful professionals for whom the

> consumption of music was not only a matter of aesthetic pleasure or a marker of status and culture but an articulation of what Emile Durkheim would call a collaborative need to formalize togetherness by adhering to common symbols and practices. . . . However, what distinguished the engagement of the Brahmin elite in Madras was the enhanced symbolic significance they attached to the practice of listening and appreciating music, thereby participating in the construction of a sense of community with exclusivist overtones.[28]

We see here the construction of a new form of identity related to the emergence of the modern nation-state. Indeed, it was largely because of the narrative of the nation-state that it was deemed that if Carnatic music was to remain respectable, the *devadasis*, who were now associated with prostitution, could not be its main performers. In this context, earlier caste distinctions did not disappear but instead were reconstituted in a different form and, in some ways, exacerbated. To understand the reconstitution of the culture of South Indian music, we must also briefly examine the confluence of two aspects of capitalist modernity in the Indian context—namely, colonialism and education. Subramanian explains:

> The emergence of a new Brahmin elite was a direct outcome of the spread of western education and professional service that was evident from around the 1860s, when Tamil and Telugu Brahmin lawyers began to attain prominence, wealth and status as a professional group.[29]

These professionals, who had been educated in a Western style, then began to reframe the relationship between Carnatic music and Indian tradition. Elites would reconstitute Carnatic music in a contradictory dynamic between their colonial education and the imagination of a national culture. Many were Brahmins who stressed that Carnatic music, like the Indian nation, was pure and distinct from the tawdry practices of *devadasis* and other nonelites.

Before moving on to our case studies of four musicians, we should mention one more aspect of this transformation that is intimately connected to tradition—namely, religion. As we have mentioned above, the imagery of Carnatic music is primarily Hindu. However, the transformations associated with shifting the center of music to the Madras concert halls "fostered a genuine listening habit among an urban audience that responded to music not as a ritual experience, but as a kind of hybrid personal experience, that helped to negotiate the new professional life detached from the older moral economy."[30]

THE MAKING OF A CARNATIC POP STAR: G. N. BALASUBRAMANIAM

GNB became an icon in Carnatic music and is considered one of its most innovative musicians. He was born in 1910, at the very height of the transition from the Tanjore court to the Madras concert halls (*sabhas*). His life and career exemplify many of the changes that we have discussed but also show that innovation is possible and sustainable in the framework of modern Carnatic music.

He was born into a lower-middle-class Brahmin family. His father was a schoolteacher who was also the secretary of a concert hall in Triplicane called the Parathasarthy Swami Sabha. Because of his father's post, the young GNB was exposed to a great number of well-known musicians, and an early anecdote about this tells us something about the nature of *ragas* in Carnatic culture. When GNB was about three years old, a number of musicians and his father were discussing the music of a well-known singer, Maha

Vaidyanatha Iyer. These types of discussions are an essential part of Carnatic music, since it is one of the places where listeners of Carnatic music participate in and perform the knowledge of *ragas*. GNB's father praised Vaidyanatha Iyer's ability to sing *kalpana swara* in the *raga Hamsadwani*. The young GNB then interjected that he could do this as well. His father first dismissed GNB's words as the idle boast of a toddler, but the musicians convinced him to give his son a chance. GNB went on to improvise *swara* combinations in the *raga Hamsadwani* in four-speed cycles. Extremely impressed, the musicians urged GNB's father to give him proper music lessons and forget about his school education.

The point of this story for our purposes is that the anecdote shows the manner in which *raga* functions as knowledge and in connecting a community. Learning the *raga* form opens Carnatic music newcomers up to the community. In this case, there is nothing especially unusual about a Brahmin boy who excels at improvising in Carnatic music. But at this level of abstraction, the community is seen as being able to potentially include newcomers who can grasp materially the abstract *raga*, as this group did by recognizing GNB's mastery. GNB would go on to become immensely popular among the young upper-class people of Madras, as an innovator famous for the speed of his singing and his use of a specific *gamaka*, the *brigga*. GNB himself described his gift for visualizing *swaras* thus: "Without so much as any basic training, I acquired swara gnana. . . . Whenever I listened to good music, I had an inner feeling that I could visualize it in the imagery of swaras."[31]

GNB's fast-paced *gamakas* or *briggas* are important not only because they took the music world from the 1930s to the 1960s by storm, but also because their genealogy went against Brahmin dominance, and he would eventually connect such things to a type of universality. Specifically, GNB contended that he had learned to sing *briggas* by imitating the music of the *nagaswaram*, a wind instrument that was primarily played by non-Brahmins. This instrument could be used to play notes at an extremely fast pace, and GNB incorporated this style into his own vocal music. But his incorporation was something like a Hegelian *Aufhebung*, or sublation, since in GNB's musical practice, the techniques of the *nagaswaram* maestros were appropriated and transported to a new space where complexity and innovation in developing *ragas* were key.

GNB graduated with a BA honors degree in English, which might be one reason why people called him an intellectual. Some suggest that because of his intellectual and creative proclivities, he would often push the limits of what was possible in terms of the accepted norms of a *raga*. Consequently, some of the older generation of Carnatic singers and listeners were initially skeptical about how he treated Carnatic music.

In addition to exploring the patterns within *ragas* that people already knew, an important facet of GNB's innovation lay in his singing and exploring *ragas* that people rarely sang. To do this, he drew on rare compositions by famous composers, especially Tyagaraja and Dikshitar, but during the periods of improvisation, he would have to create new melodic phrases that conformed to the aesthetics of Carnatic music. The audience and experts were satisfied with this, and GNB soon went on to become an icon of Carnatic music, partly because of his good looks, but also because he popularized a new style of singing that had an emphasis on creativity and intellectual virtuosity. He not only added to their listening pleasure but also expanded their knowledge base by adding new or rare *ragas* to their repertoire.

In some ways, he epitomized the new mode of Carnatic music in the Madras concert halls, and toward the end of his life, during the late 1950s and early 1960s, he began to reflect on the current state of music and promoted a universal vision of music. His reflections expressed the abstraction related to this genre but attempted to make more general statements about Carnatic music as art at a universal level. In an essay titled, "Art: Its Dawn and Future Role," GNB made the following comment:

> Music is the most universal and least sensual in its appeal. If painting is two-dimensional, sculpture three, one can say that music is four-dimensional, having its basis in tone, colour, rhythm and melody. The artist and the listener are free of the bonds of time and the logic of circumstances. It is enjoying and creating beauty without responsibility. Really, there is no language for sounds or music. Yet it is the most universal language.[32]

GNB's appropriation of Carnatic music makes it universal and accessible to anyone and, to some extent, echoes the ontological claims made about music in new interpretations of Hinduism. According to this, music becomes completely transcendent—a language beyond language. You cannot help hearing in these remarks something like Adorno's claim that "music is language of a completely different type. Therein lies its theological aspect. . . . It is demythologized prayer."[33] With such cryptic remarks, Adorno and GNB turn music into a new type of religion without religion, which perhaps implies a new way of knowing that is based not only on grasping an object but also on a transformation of the subject, which intimates a different relation to time. In the same essay, GNB strikes a messianic tone, which gives meaning to music and its knowledge:

> When all nations of the earth are worn out in their fight for world supremacy, when there is a desperate cry for universal peace, Music will be the Messiah for the golden age, uniting all in one common language and religion of sound at once sensuous and intellectual, exciting to calm, stimulating to appeasement and marshalling all the powers for Goodness, Truth and Beauty to work in unison in a spontaneous, disciplined and organized manner, towards the achievement of the common weal of all mankind.[34]

Music here gestures toward an ideal future where ownership within a community will seem to become irrelevant. This passage suggests that while music might currently be a framework that excludes some, it is seeking a universality that would go beyond social distinctions. By stressing the disciplined, organized, and yet spontaneous manner of Truth and Beauty, GNB emphasized the nature of Carnatic music itself, which involves repetitive training structure and exercises, only to eventually break these and create new patterns through improvisation. For him, the practice of Carnatic music and music in general could also have a meaning beyond India and beyond even the various religions and nations, offering a means for humankind to ultimately achieve harmony. But such a harmony would require Carnatic music to continue to be itself, while gesturing toward the universal. During the same period, GNB was concerned about the "annihilation of art," which was largely connected to the problems of modernity and the social mediation of artistic production. He feared precisely that art and Carnatic music would lose their identity in the throng of mass culture, writing:

> The greatest threat to the life and growth of all classic art is regimentation, which is to a certain extent unavoidable when the dissemination of art is mechanised, commercialised and discriminate. It clogs the springs of imagination and all creativity, forcing the artist into a dull, cold and inane level of achievement, at once tiring him and satiating the listener.[35]

GNB highlights mechanization and commercialization, both of which are connected to the processes associated with capitalist modernity. He had witnessed how commercialization was affecting Carnatic music in the concert halls of Madras. The problem of ownership there was complex, because as the public began to appropriate Carnatic music, it might cease to be itself by losing its creative moment. If the audience no longer had this creative subjective element, there was a risk they might end up treating *ragas* just as something that is purely formal or mechanical. As Hegel warned with respect to the process of thinking,

> The Idea [*Idee*], which is true enough for itself, in fact remains ensnared in its origin as long as its development consists in nothing but the repetition of the same old formula. Having the knowing subject apply the one unmoved form to whatever just happens to be present [*dem Vorhandenen*] and then externally dipping [*eingetaucht*] the material into this motionless element contributes as much to fulfilling what is demanded as does a collection of purely arbitrary impressions about the content.[36]

The analogy with Hegel emerges precisely because with the global reinterpretation of religion, there was an emphasis on interiority or subjectivity, which grounds creativity.[37] But in both philosophy and music, this interiority cannot just do whatever it likes; it must follow the logic of the inherent content—in the case of Carnatic music, the movement of the *raga* itself. The danger was that without creativity and improvisation, *raga* might dwindle into formalism without any aesthetically interesting content. In this way, GNB warned the Carnatic listener to focus on more than just the familiar framework of the *raga*, to include the moment of creativity that cannot be completely codified. This is why the role of practice is so important. With music, repeated practice gives rise to a moment that cannot be completely codified, which is why a *raga* can never be reduced to just its notation or rules. You cannot merely externally apply the framework of a *raga* to an existing voice or instrument. Rather, an artist must follow the aesthetic logic of the *raga* to create music. Because we are dealing with an aesthetic logic, the parallel with Hegel's phenomenology breaks down. There are moments of uncertainty in the improvised portions of Carnatic music, and creativity sometimes lies therein. A further exploration of the parallels might be interesting. If we interpret Hegel from a Carnatic perspective, we could say that he requires a *sadhana* or a practice of thinking, which is needed to follow the logic of concepts.[38]

We will now consider another tendency associated with the development of Carnatic music as knowledge—namely, the idea of a virtuous knower, which implies an innate attribute of the musician that cannot necessarily be codified.

THE MAKING OF A VIRTUOUS KNOWER: M. S. SUBBALAKSHMI

In dealing with the problem of knowledge and virtue, we will draw on the life and work of MS to illustrate how someone outside the community can be seen as a Brahmin by showing that they have the requisite type of musical knowledge. Similar to the GNB case, this section tells the story of how someone became accepted into the Carnatic scene. GNB was a Brahmin who learned the *raga* framework and attempted to go beyond it. The case of MS is more complex because she was not from a Brahmin family; instead, she was born into a family of *devadasis* and received the requisite training in singing and dancing. So, in addition to learning the framework, she had to transform her identity in order to become accepted. MS's virtuosity was not just abstract; she also expressed devotion and virtue, which made her appeal universal. Moreover, part of the transformation of her identity involved constructing her persona as someone virtuous and at the same time knowledgeable about Carnatic music.

One of the ways in which *devadasis* could move up in terms of class was through marriage, so her mother found wealthy suitors for her. However, MS was determined to follow a singing career in Carnatic music and twice fled to Madras at a young age. The second time was in 1936, when she was interviewed by a prominent Brahmin called Sadhasivam for a Tamil magazine. Sadhasivam sponsored MS and would eventually become her husband, which afforded her acceptance by the Brahmin community as well as respectability in her performance.

This story fits with the narrative that Lakshmi Subramanian provides with respect to the move of *raga* performances to the concert halls in Madras, since Sadhasivam was also a prominent nationalist, and part of the nationalist project was to emphasize progress and the move away from *devadasi*-like practices. MS began her musical career by acting in films, and in 1940, the year she married Sadhasivam, she also acted in the film *Shakuntalai* opposite GNB and fell in love with him. During this period, GNB's influence on her music was evident, as she also experimented with fast-paced *gamakas*. It is said that Sadhasivam did not want to let MS act in any more movies, but he made an exception for the film *Meera* (1945), where MS played the heroine. In the film, she plays a woman who is so devoted to Krishna that her love for the deity goes beyond everything else, including her devotion to her husband, a king. The film was made in Tamil and dubbed into Hindi as well, and Jawaharlal Nehru and Lord Mountbatten attended special screenings. The freedom fighter Sarojini Naidu introduced MS to Hindi speakers from North India, saying, "You will cherish her. You will be proud, that India in this generation, has produced so supreme an artist."[39]

MS was then characterized by this type of devotional knowledge, and after India gained independence, Sadhasivam's influence was crucial in turning MS into both an image of the ideal Brahmin housewife and a symbol of the nation, which again required devotion. These two sides of MS went together, since being a good Brahmin housewife implied a certain sublime spirituality that negotiated religiosity and Indian nationalism. Gandhi adored MS's music and Nehru called her the Queen of Music. As her career progressed, MS became a public figure who was known for singing religious songs in languages from both North and South India; she also famously represented India when she sang at the United Nations (UN) in 1966.

Although some commentators have argued that maintaining her social persona affected her music, and that singing a large number of religious songs sometimes restricted her opportunity for improvisation,[40] she continued to sing elaborate *kritis*, *alapanas*, and other forms of improvisation. She was well known for her adherence to *shruti* (pitch) and her ability to sing a pure note with almost a sublime therapeutic effect on the listener. Becoming one with the *shruti* indicates a knowledge that goes beyond written words and echoes a theme in the film *Meera*, where Meera eventually becomes one with her god, Krishna. As part of the narrative around perfection, it became extremely important for MS to sing flawlessly, which some observers say restricted her attempts at creativity. Yet, being a woman subsumed in a marriage to a Brahmin man, the identity she asserted was that of Brahmin piety, attaching spirituality to the community whose membership had been the means for her to gain ownership of music.

MS's case illustrates how the added dimension of gender may have shaped the ability of individuals to know and own Carnatic music by social status. Although GNB had faced some criticism in the initial stages of his career, he eventually became known as an intellectual musician and received the Sangita Kalanidhi, the Madras Music Academy's highest award, in 1958. MS also received this award, in 1968, and became even more of an icon than GNB, but her case is more complex. Her acceptance into this Brahmin community was aided by her ability to pronounce Sanskrit correctly; although she was not known for her intellectual authority, her devotional spirit was lauded.

Partly due to her performance at the UN and her connections to the Indian nationalist movement, she furthermore became the first Indian musician to receive the Ramon Magsaysay Award[41] in 1973, and to become known globally. Such fame came at a price—namely, that she had to conform to the specific role of a religious Brahmin wife, which placed constraints on the individuation that, in modernist terms, is seen as the basis for musical development.

MS's story shows that in addition to its formal rules and various aesthetic criteria, the practice and reception of Carnatic music are mediated by various social structures including gender, capitalism, and the nation-state. In her life we can see the contradiction between something like the abstract elements of Carnatic music, such as intellectual discussion and improvisational experimentation of the *raga*, and the other side of her musical appeal—namely, the portrayal of virtue and devotion through flawless performance of the selfsame *raga*.

THE MAKING OF MODERN CARNATIC MUSIC: T. M. KRISHNA

There are numerous books stressing that Carnatic music as we know it is modern.[42] And by this, scholars usually mean that Carnatic music was fashioned as a genre of classical music that was in competition with Western classical music. Earlier in this chapter, we also showed how Carnatic music was conditioned by the social transformation that India shared with most parts of the world—the transition to capitalism and a larger consumer base—as well as India's particular circumstance of becoming a nation-state. However, between devotion and abstraction, Carnatic music seems to have eluded becoming "modern art," which has implications for the type of knowledge it becomes. In short, modern art, as in forms of painting such as abstract expressionism, is freed from the constraints of tradition and form. It becomes an art that is intimately connected to knowledge. Western music has certainly done this with famous compositions, such as John Cage's *4'33"*. In that piece, the music is not just about sounds, which vary with each performance; rather, the goal of the piece is to inspire a reflexive meditation on what music is. While the context of Carnatic music is obviously different, we suggest that TMK, the youngest musician we will consider, similarly attempts to make the audience reflect on what Carnatic music actually is.[43] He has also attempted to highlight the abstraction of Carnatic music by freeing it from the fetters of caste. TMK's mother was a Carnatic music graduate, and his grand-uncle, T. T. Krishnamachari, was one of the founding members of the Madras Music Academy, which could be called the center of Carnatic music orthodoxy. In the 1990s, TMK was one of many young musicians who attempted to delink Carnatic music from religion to make it more palatable to the younger generation. He rose to stardom in the late 1990s and early 2000s, eventually studying with Semmangudi Srinivas Iyer, a major figure in the Carnatic music scene who was considered the traditionalist rival of GNB in the 1950s and 1960s.

Around 2010, he took the Carnatic music world by storm by openly denouncing its hierarchies with respect to caste and gender. By doing so, TMK to some extent stepped out of his own context, and his explanation for how he could do that echoes the protagonist's guru in the novel mentioned above, *Hamsa Geethe*. TMK contends, "My art has given me a gift. A gift of experience, a gift of empathy. A gift to sense life beyond my limitations."[44] Like the guru in *Hamsa Geethe*, TMK suggests that he is able to experience the suffering of various castes and classes and thereby relativize his own existence. Similar to GNB, TMK asserts that there is something in Carnatic music that tends toward the universal and enables criticism of exclusionary practices. It implies a logic of sharing beyond exclusive ownership. For his practice and his writings on such subjects, including a groundbreaking book on Carnatic music,[45] TMK was awarded the Ramon Magsaysay Award in 2016.

TMK's political commentary about Carnatic music is intimately connected to social issues and could be linked to our discussion of MS. In order for MS to become accepted in the Carnatic music world, she had to conform to a certain ideal, which TMK believes affected her music. TMK asserts that instead, he would like to recapture that pure art form that goes beyond the prevailing social structures. He aims to liberate the framework of Carnatic music from Brahmin hegemony, but to do so, he believes that he must change the practice of Carnatic music in certain ways. He states that he wants to rescue the pure aesthetic and abstract kernel of Carnatic music from the dross of caste, class, and gender oppression. Not only that, he furthermore believes that these forms of oppression have also seeped into the manner in which Carnatic music is presented.

We will describe two examples of how TMK is seeking to rescue Carnatic music from the constraints of its form by commenting on the type of knowledge it entails. First, as we mentioned in our discussion of *raga*, the standard rendering of a *kriti*, a song, begins roughly with an *alapana* or improvisation without lyrics, followed by a song with lyrics, and then ends with a *kalpana swara* or improvisation, explicitly singing the notes from the solfège. At a recent performance, however, TMK decided to sing a concert consisting only of *alapanas*, which bewildered the audience.[46] By doing so, he created something like a John Cage effect, which made the listeners ponder what Carnatic music is and how far you can change the format of Carnatic music while still preserving the art form. Second, TMK has experimented with changing the lyrics of Carnatic music to combat their religious basis, which could be associated with caste.

TMK also experimented with topical choices. While Carnatic music lyrics are generally largely about gods and Hindu mythology, with the help of contemporary Tamil poets, TMK has composed and sung lyrics about the environment and other social problems, such as "Poramboke" by the poet Kaber Vasuki, which is about the environmental crisis.[47] This represents a radical break from earlier practices of Carnatic music. This particular poem, and TMK's interpretation of it, are extremely germane to this chapter, since they fundamentally deal with the problem of ownership. The term *poramboke* in Tamil has three meanings: originally it meant common land, and later it came to mean illegal land and also became a derogatory term applied to people or places. TMK's video begins by asking how the meaning changed from positive to negative. The poem itself grapples with ownership and begins with the lines, "Poramboke is not for me and not for you, / but it is for the earth and for the nation." The poem pushes the work from a situation of exclusion based on laws to a larger inclusive perspective, which was implicit in the earlier interpretations of the term. TMK's rendition of this poem might be the first attempt to make a Carnatic music video, or even Carnatic performance art, since the musicians begin by wearing masks at Ennore Creek. TMK made the video and sang the song as part of an effort to save the backwater,[48] which demonstrates how he mobilizes Carnatic music for political aims and, consequently, potentially makes it available to everyone. TMK is saying that Carnatic music is not just for you or me, but potentially universal.

In "Poramboke," and more generally, TMK continues to use the *raga* to organize his music, which shows that he is still invested in the project of Carnatic music, but has propelled the genre into new contexts. For TMK, knowing about Carnatic music means not just knowing its framework; it also requires an understanding of its relation to caste oppression and other social and political issues. He attempts to bring the knowledge of modern intellectuals into the discourse of Carnatic music. Through this, he hopes to turn Carnatic music away from the logic of exclusive ownership toward a logic of sharing.[49]

THE TRADITION OF INNOVATING IN CARNATIC MUSIC: R. VEDAVALLI

We will now consider how innovation is possible within the matrix of Carnatic music. Sometimes, it is precisely by emphasizing tradition that someone achieves innovation. RV is a contemporary musician and scholar who provides an interesting contrast to TMK, since she sticks to classicism but at the same time explores new vistas within this framework. She is also engaged in expanding the body of knowledge about Carnatic music and thereby renewing it. She herself noted in an interview that when people suggest that she is doing something new, she always retorts that she is merely carrying on the tradition of her gurus.[50] Her practice exemplifies the Confucian dictum, "to know the new through practicing the old."[51]

Her biography is typical of several well-known female Carnatic musicians.[52] She was discovered at the age of five by Madurai Srirangam Iyengar, who heard her voice as he was passing her house. He then insisted to her parents that she take on formal music training. After a few years, her family moved to Madras, where she studied with Mudicondan Venkatrama Iyer, a well-known scholar and musician who won the annually awarded title Sangeetha Kalanidhi in 1948. In the following years, RV embarked on numerous projects to enlarge the tradition from within. For example, she studied *padams* and *javelis* with the teacher Muktha, who was from the *devadasi* tradition. She was an early participant in the Madras Music Academy's musical conferences, at the time, one of few women and performers—as opposed to male patrons and teachers—who used the mode of a lecture-demonstration to describe and transmit musical knowledge to audiences. Performing during a time in the 1960s when musical connoisseurship was not exclusively the remit of the elite, she related that someone as ordinary as the rickshaw puller collecting her from the train station of a small rural town to take her to her concert would share his opinion of the quality of the previous night's performing artiste. Rather than being a space of aspiration, her experience is of the downward trend that followed, as the interest of ordinary people in Carnatic music declined in favor of film songs. She stressed the importance of *ragas* to resist easy appropriation by film music. Film music could be based on *ragas*, but it would not explore a particular *raga* in detail. One needed a certain training to appreciate and understand Carnatic music, and RV attempted to promote the conditions of such an understanding.

From the perspective of the construction of knowledge, perhaps her most interesting contribution lies in her rethinking of *ragas*. From the late 1990s up to today, she has been returning to classical texts to reconstruct *ragas* through the compositions as they were sung before the twentieth century. Consequently, when RV and her students sing the so-called traditional versions of *ragas* in a contemporary concert, they often strike listeners as sounding strange and new. In short, singing *ragas* in an older form, and thus going against the grain of contemporary practice, causes something of a feeling of strangeness, which might also cause people to question what they understand as knowing Carnatic music.

This provides us with an example of two key points in this chapter. First, although there is a shared understanding of the basic framework of a *raga*, it changes by being constantly reproduced through practice. There is an important dialectic between the a priori framework—that is, the notes and phrases associated with a *raga*—and the a posteriori practice of improvising in a concert. It is the a priori framework that appears detachable and that is owned as knowledge by people in the Carnatic music community. But the a priori framework is a result of a dialectic that has gone on in the history of the theory and practice of Carnatic music. The historical framework of the *raga* confronts the singer, learner, and listener as something already given—it is a congealed practice that returns to live in every concert.

Second, Vedavalli's use of notation texts from the late nineteenth and early twentieth centuries demonstrates precisely the manner in which she uses tradition to produce something new. Her practice shows us that tradition is not just one thing, that there will always be a gap between how people imagine the tradition and the various ways in which the tradition is recorded and transmitted. By highlighting the gap between the imagined or presupposed tradition and the manner in which it is actually being documented and transmitted, RV potentially shakes up the practices of Carnatic music from within. Stressing the importance of oral history, and transmission through oral practices of learning and teaching, she uses textual notations of the past to authenticate her own practice, where it has diverged from popular parlance. This is an interesting illustration of epistemic culture in innovating tradition; she renders compositions that are new to the audience and yet are authenticated as traditional versions by her use of older annotated texts.[53]

A comparison with TMK's practices is instructive here. TMK spirits Carnatic music beyond its limits by, for example, singing a concert consisting only of *alapanas*. RV, on the other hand, often uses her scholarship to defend the importance of the nonimprovised parts of a concert, such as the *kriti*, and shows how this portion teaches us the contours of a *raga*, which forms a basis for improvisation. In other words, improvisation obtains its meaning in relation to the nonimprovised compositions that have been handed down by famous composers. But then, she simultaneously reinterprets *kritis* by showing how they would sound if the singer stayed true to the oral tradition of transmission, defending such a refusal to converge with popular taste by referring to classical annotated texts. In this manner, she destabilizes the *kriti* in the hope of expanding the aesthetic horizon of the Carnatic world. She shows the community of listeners that they need to be open to the uncanniness of the tradition that they think they possess, or to which they think they belong. By disrupting what appears to be given as a tradition of the community of listeners, she potentially unsettles their self-understanding, thereby opening them up to new possibilities. Yet, this presupposes the audience's knowledge of Carnatic music and their ability to make experimental aesthetic choices, albeit ones framed as tradition. In pushing this goal, she uses codified texts from the past to authenticate such experimentation in the present, in order to frame it as conserving tradition.

CONCLUSION

We have briefly examined the contours of the *raga* and examined different ways of negotiating through a body of knowledge and practices associated with Carnatic music, using performance as a lens. Carnatic music points beyond exclusive ownership to narratives of inclusion; however, it is embedded in a system that systematically excludes people based on their caste and class. This context allows us to better understand GNB articulating a messianic moment in Carnatic music and TMK's political practice and transformation. MS illustrated the manner in which a woman from outside the Brahmin community could negotiate the social system and become accepted far beyond the Carnatic world, to eventually become a symbol of the nation. Finally, RV shows the resources that the Carnatic tradition has available to go beyond itself and raises the possibility that for change, it should look within its own resources.

The discussions of GNB, MS, TMK, and RV above each demonstrate different determinants of the epistemic culture of Carnatic music, and changes in them over time. GNB popularized classical music after the institutional shift from temples to *sabhas*. By using speed and innovating the *raga* framework, he enabled new young audiences who were seeking entertainment rather than spirituality in their musical experience to identify and appreciate Carnatic music, which made them want to own it and become part of that musical community.

MS popularized classical music by embodying a nationalistic identity of Eastern spirituality, which by stressing her flawlessness and purity in performance became an assertion of the new Brahmin identity that was also seen as a depiction of national identity. While the shift to Brahminical processes of socialization gave her access to knowledge and respectability in the emerging phase of Carnatic music, it was through her performance, both of the music and as a quintessential Brahmin wife, that she came to own it. In doing so, she undeniably shaped the music itself, bringing the focus back to the aesthetics of the sublime, in the affective songs she performed, drawing on her *devadasi* ardor; virtue and virtuosity both become exemplary performances.

TMK stresses liberating himself from the social determinants of musical knowledge—as operationalized through caste, commodification of music, and constraints of institutional authority—yet there is an underlying tension between his attempt to maintain a "pure and abstract" creative art form that embodies individual creative freedom and his aim of liberating Carnatic musical practice on its own terms. This is reflected in the ongoing tension between his musical performance that aims to speak for itself and his reflexive theorizing of his practice through text and lectures.

Finally, RV operationalizes knowing and owning knowledge through an internalist discussion. Rather than separating the cultural and conceptual frames, she focuses on the interaction between the world of practice and the world of words. *Raga*, in this emerging epistemic culture, now has two modes of becoming knowable: one where the concept could be detached from the practice and debated in words, and a second, which is more important for the musicians—the aesthetic creativity that is transmitted orally and evaluated in performance. In order to keep practice authoritative in owning knowledge, rather than text, she holds that the working place for producing and evaluating knowledge and culture—the grammar of the music—is in the musical practice, while contrarily using the abstract discussion space of *ragas* to point to the gap between theoretical texts and popular practice. She experiments with singing *ragas* differently from contemporary practice, while framing her practice as tradition, to see how audiences will respond. Yet, this means that rather than attempting to change the narrative, she resolves the ongoing tension between tradition and innovation only in performance.

In order to understand the embedded nature of Carnatic music knowledge in its community of practice, we could think of Carnatic music as an epistemic culture[54] that is experimental.[55] What is at stake is the fixation of specific epistemic conditions under which it is still possible for processes occurring within the musical community to become manifest outside it, and thus to become accessible to analytical investigation. Experiment in Carnatic music has to do with the practice itself, where virtuosity is evaluated as the ability to improvise within the constraints of the *raga*, and the question of whether an experiment works or not depends on its evaluation by the musical community and the audience. Thus, the *condition for such experimentation* is precisely its history. By historicizing epistemological cultures, we analyze the cognitive history of experimentation in Carnatic music; in this chapter we study how rather than being independent, social determinants are produced in and through the musical practice of the *raga* and the discourse around it. We explore how patterns of change in evaluation of current practice depend on local histories of past practice.[56] Practices of *raga* change, and the treatment and understanding of the conceptual frames of *raga* undergo transformation over time. Furthermore, the correlation between the two suggests that cultures change in relation to conceptual work as much as concepts change in response to how singers practice their art.[57]

Thus, the resistance to opening *raga* up to influences from outside the community is not just social but is also linked to the epistemic culture, to the nature and values embedded in the music itself, and to fears of diluting a conceptual framework that is stabilized by the boundaries of the community of practice—in this case, enforced as caste. Yet, through historicizing different musicians' efforts to keep knowability as well as ownability open, this chapter has focused on Carnatic music as experimental culture rather than its forms of institutionalization, on processes of knowledge production that allow for new knowledge to be created, in which unprecedented things can happen. Foregrounding practice as an authoritative way of knowing and owning—in producing, listening, and evaluating Carnatic music performance—reveals that there is a generative aspect of Carnatic music held by a community of practitioners that goes beyond the gaze of caste and culture. Indeed, the pursuit of universality in Carnatic music is part of a larger search for a new universality that is not based on exclusion. Such a community of practice brings knowledge into manifestation and transforms it into ownable knowledge. They are concretions—not abstractions—in which epistemic, technical, and social moments are inextricably intertwined.[58]

Discussions about epistemic cultures are necessary, as they can destabilize such boundaries and help liberate Carnatic music from itself, thus creating the conditions for including those outside the community who may desire to know and own it legitimately. A study of epistemic cultures of music could lead the way for music to have a different social effect, one that is inextricably connected to creating a world beyond unjust hierarchies. We believe that we have taken the first step toward such a project by showing how a particular genre of music produces logics of ownership and exclusion.

ACKNOWLEDGMENTS

The European Research Council project PENELOPE (ERC funding HORIZON 2020 number 682711) at the Deutsches Museum in Munich and the Deutsche Forschungsgemeinschaft (DFG) project 435681850 at the Technische Universität Berlin have supported Annapurna Mamidipudi's research into ownership of knowledge in craft and hand weaving. She is grateful to the PIs of both projects, Ellen Harlizius-Klück and Friedrich Steinle for taking a leap of faith into the world of traditional Indian crafts and immersing themselves in understanding its naming, performance, and use in modern times.

Notes

1. Although this picture of ownership is prevalent in the modern capitalist world, there were clearly different forms of ownership in precapitalist societies. In such societies, ownership may have been collective and the boundaries between owners and those who did not own an object may have been blurred. Moreover, as many observers have argued recently, capitalism does not do away with all noncapitalist forms of ownership and production. See Massimiano Tomba, *Marx's Temporalities* (New York: Haymarket, 2014); Harry Harootunian, *Marx after Marx* (New York: Columbia University Press, 2017). This might be especially true for certain artistic forms which, although they have been thoroughly commodified, also retain earlier forms of practice.

2. See, for example, Bruno Latour, "Visualization and Cognition: Thinking with Eyes and Hands," *Knowledge and Society: Studies in the Sociology of Culture Past and Present* 6 (1986): 1–40.

3. The paper follows a chronology from the turn of the twentieth century to the turn of the twenty-first century.

4. For philosophical practice thinkers, see Ludwig Wittgenstein, *Philosophical Investigations*, trans. G. E. M. Anscombe, 3rd ed. (New York: Macmillan, 1958); Hubert L. Dreyfus, *Being-in-the-World: A Commentary on Heidegger's Being and Time, Division I* (Cambridge, MA: MIT Press, 1991).

5. For social theoretical thinkers see Pierre Bourdieu, *Outline of a Theory of Practice*, trans. Richard Nice (Cambridge: Cambridge University Press, 1977); Anthony Giddens, *Central Problems in Social Theory: Action, Structure, and Contradiction in Social Analysis* (Berkeley: University of California Press, 1979).

6. For cultural theory, see Michel Foucault, *The Archaeology of Knowledge and the Discourse on Language*, trans. A. M. Sheridan Smith (New York: Pantheon Books, 1972). For a detailed discussion on practice theorists, see Theodore R. Schatzki, Karin Knorr-Cetina, and Eike von Savigny, *The Practice Turn in Contemporary Theory* (London: Routledge, 2005).

7. This thesis unites a broad collection of thinkers of science studies. Karin Knorr-Cetina, *The Manufacture of Knowledge: An Essay on the Constructivist and Contextual Nature of Science* (Oxford: Pergamon Press, 1981); David Bloor, *Wittgenstein: A Social Theory of Knowledge* (New York: Columbia University Press, 1983); Ian Hacking, *Representing and Intervening: Introductory Topics in the Philosophy of Natural Science* (New York: Cambridge University Press, 1983); Bruno Latour, *Science in Action: How to Follow Scientists and Engineers through Society* (Cambridge, MA: Harvard University Press, 1987); Andrew Pickering, ed., *Science as Practice and Culture* (Chicago: University of Chicago Press, 1992); Hans-Jörg Rheinberger, *Toward a History of Epistemic Things: Synthesizing Proteins in the Test Tube* (Stanford, CA: Stanford University Press, 1997).

8. For a detailed discussion on performance lens as a conceptual and methodological contribution to studying ecologies of practices—in this case, farming—see Dominic Glover, "Farming as a Performance: A Conceptual and Methodological Contribution to the Ecology of Practices," *Journal of Political Ecology* 25, no. 1 (2018): 686–702.

9. Nicholas Cook, *Beyond the Score: Music as Performance* (Oxford: Oxford University Press, 2014), 1 (emphasis in original).

10. For a detailed discussion, see Mieke Bal's work on performance as both rehearsal and repetition, as well as the theatrical performance aspect of performativity. Mieke Bal, *Travelling Concepts in the Humanities: A Rough Guide* (Toronto: University of Toronto Press, 2002).

11. Richard Schechner, *Performance Studies: An Introduction* (London: Routledge, 2012), 28.

12. For discussion on performing knowledge as doing, see Tim Ingold, *Making: Anthropology, Archaeology, Art and Architecture* (London: Routledge, 2013); Pamela H. Smith, "In the Workshop of History: Making, Writing, and Meaning," *A Journal of Decorative Arts, Design History, and Material Culture* 19, no. 1 (2012): 4–31; Karamjit S. Gill, "Hermeneutic of Performing Knowledge," *AI & Society* 32, no. 2 (2017): 149–156.

13. Mary Helen Dupree and Sean B. Franzel, *Performing Knowledge, 1750–1850* (Berlin: De Gruyter, 2015).

14. Hermano Viana, *The Mystery of Samba: Popular Music and National Identity* (Raleigh: University of North Carolina Press, 1999).

15. Fred Lau, "Nationalizing Sound on the Verge of Chinese Modernity," in *Beyond the May Fourth Paradigm: In Search of Chinese Modernity*, ed. Kai-Wing Chow et al. (Lanham, MD: Lexington Books, 2008), 209–229.

16. He notes that the connection between caste and music is a recent formation. Perhaps the most famous work to underscore the modernity of the concept of caste is Nicholas Dirks, *Castes of Mind: Colonialism and the Making of Modern India* (Princeton, NJ: Princeton University Press, 2001). Carnatic music in the twentieth century was coeval with numerous new forms of religious subjectivity.

17. T. M. Krishna, *A Southern Music: The Karnatik Story* (Delhi: HarperCollins India, 2013), 439.

18. A full discussion of the relationship between Hindustani and Carnatic music is beyond the scope of this chapter.

19. We find these notes in early texts such as the *Brhaddeshi* by Matanga Muni, which is usually dated between the sixth and the eighth century. See Lewis Rowell, *Music and Musical Thought in Early India* (Chicago: University of Chicago Press, 2015), 144–179.

20. See Geoffrey C. Bowker, *Memory Practices in the Sciences* (Cambridge, MA: MIT Press, 2005).

21. See S. Bhagyalekshmi, *Ragas in Carnatic Music* (Chennai: CBH Publications, 2010).

22. For the mood of a *raga* to emerge, the manner in which the individual notes are treated is just as important as the actual notes that are sung. We could say a great deal about this here, but we will limit ourselves to one central aspect—namely, the distinction between plain notes and notes with oscillations, since specific *ragas* require the practitioner to oscillate certain notes. Notes in Carnatic *ragas* often entail oscillations, or *gamakas*.

23. Annapurna Mamidipudi, *Towards a Theory of Innovation for Handloom Weaving in India* (Maastricht: Maastricht University, 2016).

24. T. R. Subbha Rao, *Hamsa Geethe* (1952; Bengaluru: Hemantha Sahithya, 1984), 100, our translation.

25. T. M. Krishna describes this period in almost utopian terms. Krishna, *Southern Music*, 315.

26. For an overview of the history, see Hari Krishnan, "Bharatanatyam," section 5, "Social Reform and the Disenfranchisement of Devadasis," Accelerated Motion: Towards a New Dance Literacy, website produced by the Wesleyan University Press and the Academic Media Studio, 2009, accessed November 20, 2019, http://www.oberlinlibstaff.com/acceleratedmotion/dancehistory/bharatanatyam/section5.php.

27. Gill Navyug, "Limits of Conversion: Caste, Labor and the Question of Emancipation in Colonial Punjab," *Journal of Asian Studies* 78, no. 1 (2019): 3–22.

28. Lakshmi Subramanian, *From the Tanjore Court to the Madras Music Academy: A Social History of Music in South India* (Oxford: Oxford University Press, 2006), 42–43.

29. Subramanian, 45.

30. Subramanian, 45, 47.

31. Lalitharam, *The Prince of Music: A Biography of G. N. Balasubramaniam*, trans. V. Ramnarayan (Chennai: Wordcraft, 2018), 14.

32. G. N. Balasubramaniam, "Art: Its Dawn and Future Role," in *Gandharva Ganam: G. N. Balasubramaniam Centenary—Commemorative Volume*, ed. Lalitha Ram and V. Ramnarayan (Chennai: G. B. Bhuvaneswaran and Mahesh G. Bhuvaneswaran, on behalf of the GNB family, 2009), 113–119, 116.

33. Theodor W. Adorno, "Music, Language and Composition," *Musical Quarterly* 77, no. 33 (1993): 401–414.

34. Balasubramaniam, "Art," 118.

35. Balasubramaniam. "The Annihilation of Art," in Ram and Ramnarayan, *Gandharva Ganam,* 130.

36. Georg Wilhelm Friedrich Hegel, *The Phenomenology of Spirit* (Cambridge, UK: Cambridge University Press, 2018), 11.

37. The paradigm case is Protestantism, but other religions followed suit, stressing an individual's inner feeling or subjective experience. For a discussion of this issue in relation to Judaism, see Leora Batnitzky, *How Judaism Became a Religion: An Introduction to Modern Jewish Thought* (Princeton, NJ: Princeton University Press, 2013). For a discussion of the Indian case, see A. R. Mohapatra, *Social Philosophy of Swami Vivekananda* (Delhi: Readworthy Publications, 2009).

38. Because we are dealing with an aesthetic logic, this is where the parallel with Hegel's phenomenology breaks down. There are moments of uncertainty in the improvised portions of Carnatic music, and creativity sometimes lies therein. A further exploration of the parallels might be interesting. If we interpret Hegel from a Carnatic perspective, we could say that he requires a *sadhana*, or a practice of thinking, which is needed to follow the logic of concepts.

39. Karan Bali, "The Making of MS Subbulakshimis *Meera*: Her Final and Finest Film," Scroll.In, September 16, 2016, https://scroll.in/reel/816654/the-making-of-ms-subbulakshmis-meera-her-final-and-finest-film.

40. See T. M. Krishna, "MS Misunderstood: The Myths and Misconceptions around MS Subbulakshmi—India's Most Acclaimed Musician," *The Caravan*, October 1, 2015, https://caravanmagazine.in/reportage/ms-understood-ms-subbulakshmi.

41. The Ramon Magsaysay Award is an annual award established to perpetuate former Philippine president Ramon Magsaysay's example of integrity in governance, courageous service to the people, and pragmatic idealism within a democratic society. It is considered the Asian equivalent of the Nobel Peace Prize.

42. Amanda Weidman, *Singing the Classical, Voicing the Modern: The Postcolonial Politics of Music in South India* (Durham, NC: Duke University Press, 2006).

43. T. M. Krishna, interview by Viren Murthy, December 28, 2019, Chennai.

44. T. M. Krishna's acceptance speech can be viewed here: S. Harihanan, "Indira Gandhi National Integration Award for T. M. Krishna," YouTube video, uploaded November 1, 2017, 14:34, https://www.youtube.com/watch?v=f2r0GeJZoR4&t=189s.

45. Krishna, *Southern Music*.

46. T. M. Krishna, "The Argumentative Musician," interview by Sumana Ramanan, *Open Magazine*, January 23, 2014, http://www.openthemagazine.com/article/arts/the-argumentative-musician.

47. TMK's performance of "Poromboke" can be viewed here: Vettiver Collective, "Chennai Poromboke Paadal ft. T. M. Krishna," YouTube video, uploaded January 14, 2017, 9:33, https://www.youtube.com/watch?v=82jFyeV5AHM.

48. The poem touches on the building of power plants near Ennore Creek and the problem of pollution. The lyrics say that the floods have come and gone, but this has not changed people's attitudes to the creek or the environment. The problem was especially exacerbated during the Chennai floods of 2015.

49. Concert given by T. M. Krishna at the University of Chicago's Logan Center, October 14, 2018, attended by one of the authors, Viren Murthy.

50. Vidushi R. Vedavalli, interview by Annapurna Mamidipudi and Viren Murthy, December 27, 2017.

51. Confucius, *Lunyu yizhu* [Translation and annotations of the Analects], ed. Yang Bojun (Beijing: Zhonghua Shuju, 2002).

52. The famous Carnatic singer M. L. Vasantakumari (1928–1990) was similarly discovered when GNB overheard her voice when he was walking past her house.

53. In a two-CD recording, *Pramanam*, she sang compositions that illustrate the changes that time has wrought on well-known pieces.

54. Karin Knorr-Cetina, *Epistemic Cultures: How the Sciences Make Knowledge* (Cambridge, MA: Harvard University Press, 1999).

55. Hans-Jörg Rheinberger, "Cultures of Experimentation," in *Cultures without Culturalism: The Making of Scientific Knowledge*, ed. Karine Chemla and Evelyn Fox Keller (Durham, NC: Duke University Press, 2017).

56. Donald MacKenzie, "On Invoking Culture in the Behaviour of Financial Markets," in Chemla and Keller, *Cultures without Culturalism*.

57. Karine Chemla, "Changing Mathematical Cultures, Conceptual History, and the Circulation of Knowledge," in Chemla and Keller, *Cultures without Culturalism*.

58. Rheinberger, "Cultures of Experimentation."

Bibliography

Adorno, Theodor W. "Music, Language and Composition." *Musical Quarterly* 77, no. 33 (1993): 401–414.

Balasubramaniam, G. N. "The Annihilation of Art." In Balasubramaniam, *Gandharva Ganam*, 128–132.

Balasubramaniam, G. N. "Art: Its Dawn and Future Role." In Balasubramaniam, *Gandharva Ganam*, 113–118.

Balasubramaniam, G. N. *Gandharva Ganam: G. N. Balasubramaniam—Commemorative Volume*. Edited by Lalitha Ram and V. Ramnarayan. Chennai: G. B. Bhuvaneswaran and G. Bhuvaneswaran, on behalf of the GNB family, 2000.

Bal, Mieke. *Travelling Concepts in the Humanities: A Rough Guide*. Toronto: University of Toronto Press, 2002.

Bali, Karan. "The Making of M. S. Subbulakshmi's *Meera*: Her Final and Finest Film." Scroll.in, September 16, 2016, https://scroll.in/reel/816654/the-making-of-ms-subbulakshmis-meera-her-final-and-finest-film.

Bhagyalekshmi, S. *Ragas in Carnatic Music*. Chennai: CBH Publications, 2010.

Bloor, David. *Wittgenstein: A Social Theory of Knowledge*. New York: Columbia University Press, 1983.

Bourdieu, Pierre. *Outline of a Theory of Practice*. Cambridge: Cambridge University Press, 1977.

Bowker, Geoffrey C. *Memory Practices in the Sciences*. Cambridge, MA: MIT Press, 2005.

Butler, Judith. *Gender Trouble*. London: Routledge, 2002.

Chemla, Karine. "Changing Mathematical Cultures, Conceptual History, and the Circulation of Knowledge." In Chemla and Keller, *Cultures without Culturalism*, 352–398.

Chemla, Karine, and Evelyn Fox Keller, eds. *Cultures without Culturalism: The Making of Scientific Knowledge*. Durham, NC: Duke University Press, 2017.

Confucius. *Lunyu yizhu* [Translation and annotations of the Analects]. Edited by Yang Bojun. Beijing: Zhonghua Shuju, 2002.

Cook, Nicholas. *Beyond the Score: Music as Performance*. Oxford: Oxford University Press, 2014.

Dirks, Nicholas. *Castes of Mind: Colonialism and the Making of Modern India*. Princeton, NJ: Princeton University Press, 2001.

Dreyfus, Hubert, L. *Being-in-the-World: A Commentary on Heidegger's* Being and Time, *Division I*. Cambridge, MA: MIT Press, 1991.

Dupree, Mary Helen, and Sean B. Franzel. *Performing Knowledge, 1750–1850*. Interdisciplinary German Cultural Studies 18. Berlin: De Gruyter, 2015.

Foucault, Michel. *The Archaeology of Knowledge and the Discourse on Language*. Translated by A. M. Sheridan Smith. New York: Pantheon Books, 1972.

Giddens, Anthony. *Central Problems in Social Theory: Action, Structure, and Contradiction in Social Analysis*. Berkeley: University of California Press, 1979.

Gill, Karamjit S. "Hermeneutic of Performing Knowledge." *AI & Society* 32, no. 2 (2017): 149–156.

Glover, Dominic. "Farming as a Performance: A Conceptual and Methodological Contribution to the Ecology of Practices." *Journal of Political Ecology* 25, no. 1 (2018): 686–702.

Hacking, Ian. *Representing and Intervening: Introductory Topics in the Philosophy of Natural Science*. Cambridge: Cambridge University Press, 1983.

Harootunian, Harry. *Marx after Marx*. New York: Columbia University Press, 2017.

Hegel, Georg Wilhelm Friedrich. *The Phenomenology of Spirit*. 1807. Translated and edited by Terry Pinkard. Cambridge: Cambridge University Press, 2018.

Ingold, Tim. *Making: Anthropology, Archaeology, Art and Architecture*. London: Routledge, 2013.

Knorr-Cetina, Karin. *Epistemic Cultures: How the Sciences Make Knowledge*. Cambridge, MA: Harvard University Press, 1999.

Knorr-Cetina, Karin. *The Manufacture of Knowledge: An Essay on the Constructivist and Contextual Nature of Science*. Oxford: Pergamon Press, 1981.

Krishna, T. M. "The Argumentative Musician." Interview by Sumana Ramanan. *Open Magazine*, January 23, 2014. http://www.openthemagazine.com/article/arts/the-argumentative-musician.

Krishna, T. M. "MS Misunderstood: The Myths and Misconceptions around MS Subbulakshmi—India's Most Acclaimed Musician." *The Caravan*, October 1, 2015. https://caravanmagazine.in/reportage/ms-understood-ms-subbulakshmi.

Krishna, T. M. *A Southern Music: The Karnatik Story*. Delhi: HarperCollins India, 2013.

Lalitharam. *The Prince of Music: A Biography of G. N. Balasubramaniam*. Translated by V. Ramnarayan. Chennai: Wordcraft, 2018.

Latour, Bruno. *Science in Action: How to Follow Scientists and Engineers through Society*. Cambridge, MA: Harvard University Press, 1987.

Latour, Bruno. "Visualization and Cognition: Thinking with Eyes and Hands." *Knowledge and Society: Studies in the Sociology of Culture Past and Present* 6 (1986): 1–40.

Lau, Fred. "Nationalizing Sound on the Verge of Chinese Modernity." In *Beyond the May Fourth Paradigm: In Search of Chinese Modernity*, edited by Kai-wing Chow, Tze-ki Hon, Hung-yok Ip, and Don Price, 209–229. Lanham, MD: Lexington Books, 2008.

MacKenzie, Donald. "On Invoking Culture in the Behaviour of Financial Markets." In Chemla and Keller, *Cultures without Culturalism*, 29–48.

Mamidipudi, Annapurna. *Towards a Theory of Innovation for Handloom Weaving in India*. Maastricht: Maastricht University, 2016.

Navyug, Gill. "Limits of Conversion: Caste, Labor and the Question of Emancipation in Colonial Punjab." *Journal of Asian Studies* 78, no. 1 (2018): 3–22.

Pickering, Andrew, ed. *Science as Practice and Culture*. Chicago: University of Chicago Press, 1992.

Rheinberger, Hans-Jörg. "Cultures of Experimentation." In Chemla and Keller, *Cultures without Culturalism*, 278–295.

Rheinberger, Hans-Jörg. *Toward a History of Epistemic Things: Synthesizing Proteins in the Test Tube*. Stanford, CA: Stanford University Press, 1997.

Rowell, Lewis. *Music and Musical Thought in Early India*. Chicago: University of Chicago Press, 1992.

Salih, Sara. *Judith Butler*. London: Routledge, 2002.

Schatzki, Theodore R., Karin Knorr-Cetina, and Eike von Savigny, eds. *The Practice Turn in Contemporary Theory*. London: Routledge, 2005.

Schechner, Richard. *Performance Studies: An Introduction*. London: Routledge, 2012.

Smith, Pamela H. "In the Workshop of History: Making, Writing, and Meaning." *Journal of Decorative Arts, Design History, and Material Culture* 19, no. 1 (2012): 4–31.

Subbha Rao, T. R. *Hamsa Geethe*. 1952. Bengaluru: Hemantha Sahithya, 1984.

Subramanian, Lakshmi. *From the Tanjore Court to the Madras Music Academy: A Social History of Music in South India*. Oxford: Oxford University Press, 2006.

Tomba, Massimiano. *Marx's Temporalities*. New York: Haymarket, 2014.

Viana, Hermano. *The Mystery of Samba: Popular Music and National Identity*. Raleigh: University of North Carolina Press, 1999.

Weidman, Amanda. *Singing the Classical, Voicing the Modern: The Postcolonial Politics of Music in South India*. Durham, NC: Duke University Press, 2006.

Wittgenstein, Ludwig. *Philosophical Investigations*. Translated by Gertrude Elizabeth Margaret Anscombe. 3rd ed. New York: Macmillan, 1958.

5
IMITATING CRACKLES: MATERIAL MIMESIS IN STONES AND TEXTILES

Marjolijn Bol

This chapter proposes to study the ways in which processes of making can perform in claims to knowledge and ownership from the perspective of imitation. It focuses on a special kind of imitation, "material mimesis," here defined as the phenomenon in which artisans used one or more materials to imitate the characteristics of another material.[1] Today, still, our daily lives are pervaded by practices of material mimesis. We walk on laminated floors made to look like wood and navigate the digital world through computer screens imitating the appearance of office spaces, and modern medicine can replace the materials of our bodies on almost every scale. Examples of material mimesis can be found across societies, in almost any period, and it occurs throughout a great variety of art forms. Ancient ceramic vessels, for instance, some nearly four millennia old, were often glazed to look like metals, or imprinted to look like straw; bronze and copper objects, in turn, were worked to assume the appearance of leather (figure 5.1). Medieval artisans gave wood the appearance of solid gold, and the sophisticated application of colored plaster was used to metamorphose seventeenth-century furniture into quasi-marble structures (figure 5.2).

In each of these examples, the mimetic materials or objects are appropriated or rejected through a wide variety of uses ranging from ritualistic, ethical, and aesthetic considerations to practices of deceit, social distinction, health, learned inquiry, adaptation to change, and play. Imitation fur that looks too much like real fur, for instance, may be rejected by those who have ethical reasons for not wanting to wear fur obtained from animals (or because they fear they may be judged by those who do have ethical concerns). To reject imitation fur on the basis of looking "too real" implies sophisticated material knowledge on the side of the users (i.e., what does real fur look like?), and this, in turn, plays a prominent role in shaping the materials and processes used to produce imitation fur (i.e., how to keep it fur-like while downplaying the resemblance to natural animal fur).

Many, if not most, material mimetic objects raise similar fundamental questions about how knowledge and ownership (1) were performed through the different social

Figure 5.1
Flask (*bianhu*), Eastern Zhou dynasty, Warring States period (475–221 BCE). Bronze inlaid with copper, probably modeled after leather flasks. Lucy Maud Buckingham Collection, Art Institute Chicago (CC0 Public Domain Designation).

functions of practices of material mimesis and (2) shaped the materials and processes used to produce the imitations. This chapter explores these issues by focusing on two case studies in which the practice of material mimesis was defined by making, knowing, and owning "the crackle." I will study the ancient practice of imitating precious stones by crackling crystals and Western attempts at reproducing the crackle of batik textiles from the East between the eighteenth and twentieth centuries. In both cases, the dual conditioning of kn/own/ables is at work. What was known about the material characteristics of precious stones and traditional batik, as well as the processes by which these objects were imitated, determined their value and prestige and hence

Figure 5.2
The Warwick Castle table, attributed to Baldassare Artima and Diacinto Cawcy, England, ca. 1671. The top of slate is decorated with scagliola, the frame of pine and beech faced with scagliola, imitating marble, and inlaid pietra dura. © Victoria and Albert Museum, London.

the ways they were desired and owned. The case of imitation batik also shows that when the social context in which it was consumed changed, the lack or unimportance of practical knowledge (or that actors assigned to it) eventually gave the textiles an entirely new meaning. This transformed the imitation material into a different kind of kn/own/able distinct from the "original."

CRACKLED CRYSTALS

Today, diamonds, sapphires, rubies, and emeralds can be created in the laboratory in such a manner that they have the same physical, chemical, and optical characteristics as gems produced by geological processes.[2] Sharing the same chemical and physical properties with their natural analogs, modern lab-created gems are just as hard, transparent, and brilliant as gems produced by nature. As a result, these synthetic gems can be distinguished from their natural counterparts only by trained experts using sophisticated equipment. Notwithstanding their physical correspondence, there are three crucial differences between natural and lab-created minerals. The first has to do with time: in nature it takes billions of years to create a precious stone, but in the lab the same mineral can be produced in the space of a few weeks. This means that lab-created diamonds, sapphires, rubies, and emeralds are less rare than their natural counterparts. Another important difference is that unlike gems produced by nature, lab-created gems are not procured by mining. The purchase of lab-created minerals therefore is not complicated by the same ethical issues that often surround mined minerals.[3] And, finally, stones produced by nature almost inevitably include other elements and defects that make each gem unique. In the controlled conditions of the modern laboratory, such inclusions and defects can almost be eliminated.[4] This means that synthetic diamonds, emeralds, sapphires, and rubies are purer than almost any of their natural counterparts. To find this kind of purity in a gem today therefore indicates that one is dealing with either an extremely rare mineral, or, quite the opposite, a lab-created stone material.

From this we can extrapolate that lab-created imitations of precious minerals become more desirable when consumers know and value the fact that their material makeup is identical to, or, in terms of purity, better than natural precious minerals, and/or when consumers know and care about the societal impact of the mining industry. Precious minerals produced by nature, on the other hand, become more desirable when consumers know and value rarity (also in terms of a natural stone's "uniqueness") over the reasons for wanting to own a lab-created imitation gem.

But what is it about "rarity" that makes something the object of our desire to "know" and "own" it? The answer to this question is not simply relative value, or economic worth. Our desire for the rare is deeply intertwined with the social-cultural prestige tied to owning something that, because of its scarcity, cannot easily be owned by others. To describe how rare materials were used as an expression of status, the renowned archaeologist Grahame Clark introduced the phrase "symbols of excellence"—"a quality which stems from aesthetic awareness but the striving for which lies at the very root of the civilizations created by man."[5] To explain this concept, Clark refers to Thorstein Veblen's economic theory of symbolic substances. In his seminal book *The Theory of the Leisure Class* (1899), Veblen coined the term *conspicuous consumption* to describe the act of buying symbolic objects to display one's social status. Since symbolic objects used as expressions of status are typically useless for purposes of daily life, Veblen terms them "conspicuous waste." Veblen points out that, often without realizing it, consumers appreciate the superior article not, as they like to think, because it has more intrinsic beauty, but because it is more rare and therefore more honorific. Veblen explains this idea with the example of why a handmade silver spoon typically has a higher value than a spoon made in a factory or of a base metal.[6] All three spoons are equally "serviceable" and, Veblen points out, do not differ much in terms of beauty. Still, the handmade silver spoon is more valuable because it is not so readily available as the base metal spoon or the one mass-produced in a factory; to own the handmade silver spoon, therefore, expresses status. To appreciate how an object expresses status in certain social contexts, however, one also needs to know something about the object's material properties and how the object is made. Indeed, only if the handmade silver spoon can be distinguished from its base metal or factory-produced counterparts can it be recognized as a symbolic object expressing the status of its owner. This leads us to consider two important questions that I will examine in what is about to follow: If the rare is typically the most honorific, in what instances can imitations of the rare become the subject of desires for ownership as well? And when this happens, how does it affect and/or how is it affected by consumers' knowledge about the materials and making of these imitations as opposed to their rarer counterparts?

We have seen that in the case of natural precious minerals, knowledge of the ethical aspects of the mining industry may increase consumers' desire to own lab-created minerals. This is quite similar, in fact, to why some prefer to own items made from imitation fur or leather over skins obtained from a certain species of animal. In the past, material, social, ethical, and cultural know-how of material mimetic objects likewise played an important role in defining their meaning and value as "ownable things" *and* deeply influenced how they were made and with what materials. Let us turn to a more ancient practice of making imitations of precious stones to study this in more detail.

In the Latin West, the practice of making imitation gems was first described by Seneca the Younger (ca. 1 3CE–65 CE) in his *Epistles*, or *Moral Letters to Lucilius* (*Ad Lucilium epistulae morales*). Seneca's letters to his friend Lucilius describe the history of civilization as it developed from a period called the Golden Age, during which mankind lived without distinctions of ownership or social status, to an age in which mankind exchanges this common ownership for the pursuit of luxury goods, something which Seneca deeply condemns.[7] In his nineteenth letter, Seneca deals with the argument of Posidonius of Apamea, who argues that philosophy was the inventor of the arts. To defend the thesis that wisdom comes from artisanal practice, Posidonius mentions several philosophers to whom important artisanal inventions were attributed. In response, Seneca introduces one of the alleged inventions of the pre-Socratic philosopher Democritus (460–370 BCE), who was said to have discovered how to turn "a pebble into an emerald by baking it, a procedure by which even today we color stones found to respond to it."[8]

Seneca argues, however, that such inventions by philosophers are not proof that the arts are the province of philosophical knowledge:

> The sage may well have discovered these things, but not by virtue of his wisdom. In fact, he does many things that we observe quite unwise people doing just as well or even with greater skill and ease.⁹

For Seneca, philosophical and artisanal knowledge are two very different things. Yet, he does appear to consider the art of gemstone imitation to be one of such great skill that it could lead someone like Posidonius to confuse artisanal knowledge with philosophical knowledge.

Written at around the same time, Pliny the Elder's *Historia naturalis* contains similar comments about the remarkable skill with which gems were imitated. In Book 37, which deals with the natural history of gemstones, Pliny points out that "to distinguish genuine and false gemstones is extremely difficult, particularly as men have discovered how to make genuine stones of one variety into false stones of another."¹⁰ Pliny adds that he knows of treatises that describe how to make such imitations:

> There are treatises by authorities, whom I at least shall not deign to mention by name, describing how by means of dyestuffs emeralds and other transparent coloured gems are made from rock-crystal. . . . And there is no other trickery that is practised against society with greater profit.¹¹

Like Seneca, Pliny condemns the pursuit of luxury goods for the sole purpose of expressing status or wealth, and this is a recurring theme throughout the different books that make up the *Natural History*.¹² But Pliny condemns even more sharply the imitation gems produced for the sole purpose of cheating the buyer. Thus, he views the knowledge that helps detect such frauds as something that may empower the consumer and that should therefore be recorded and shared: "I, on the other hand, am prepared to explain the methods of detecting false gems, since it is only fitting that even luxury should be protected against deception."¹³

No treatises detailing the practice of gemstone imitation have survived from Pliny's time or earlier, but a later echo can be found in a collection of recipes dating to the fourth century CE, the so-called Stockholm Papyrus. The Stockholm Papyrus contains more than seventy recipes for the imitation of precious stones, including ruby, beryl, amethyst, and sunstone.¹⁴ This document, written in Greek, was possibly copied as a funerary gift around 200 to 300 CE, but its recipes are believed to have belonged to a much older tradition.¹⁵ Significantly, the recipes included in the Stockholm Papyrus do not focus on trying to approximate the supposed perfection of natural precious stones, but rather on recreating their typical flaws to make a convincing imitation. As mentioned above, natural precious minerals are often marked by elemental inclusions and internal cracks and fissures. It was these cracks and fissures that provided an opportunity for creating imitation gems, and translucent precious minerals in particular.

The papyrus recipes explain that to imitate a precious stone, a less rare, colorless, transparent mineral such as rock crystal (a clear quartz variety) had to be "opened up" so that it could become receptive to the colorants that were supposed to transform its appearance into that of a rare translucent ruby, emerald, or sapphire. The recipes are various, but the main procedure typically involves crackling transparent minerals by heating them and then quickly cooling them down in a liquid:

> Put the stones in a dish, lay another dish on it as a cover, lute the joint with clay, and let the stones be roasted for a time under supervision. Then remove the cover gradually and pour alum and vinegar upon the stones. Then afterward color the stones with the dye as you wish.[16]

As a result of the quick decline in temperature, the stones crackle inside, with some of the fissures reaching the surface of the stone (figure 5.3). The main challenge would have been to keep the stone from breaking, because either the direct heat from the fire or the quick change in temperature could cause the crystal to shatter rather than crackle. Some of the recipes in the papyrus address this concern. A recipe for the "preservation of crystal" calls for heating the stone in a fig to prevent the stone from breaking under the intense heat of the fire:

Figure 5.3
Quench-cracked quartz crystals. Pieces of quartz were heated and then quenched in cold water. The photo shows the numerous small fissures that are the result of this treatment. This network of fissures (reaching up to the surface of the crystal) allow the dye to penetrate the stone when it is heated again and quenched in the dye liquid (see figure 5.4). Photo by Marjolijn Bol.

Preservation of Crystal.
In order that small stones which are prepared from crystal do not break into pieces, take and open a fig, put the stone therein, and lay the fig upon the coals to roast.[17]

After the stone was successfully crackled, it had to be heated again—thus slightly opening the newly created cracks—and then it was dipped into a dye bath. The recipes suggest that these dye baths typically consisted of a natural resin (a viscous substance exuded by trees and other plants)—made liquid through the application of heat—in which a colorant (a dye or a pigment) was dissolved:

Boiling of Stones.
If you wish to make ruby from crystal, which is worked to any desired end, take and put it in the pan and stir up turpentine balsam and a little pulverized alkanet there until the dye liquid rises; and then take care of the stone.[18]

When certain colorants, such as the dye alkanet (extracted from the roots of *Alkanna tinctoria* (L.) Tausch) in the recipe above, are mixed with liquid resin, they become dissolved in it.[19] When the previously crackled crystal is reheated and then dipped in this substance, it soaks up the warm, colored resin mixture through the phenomenon of capillary action before the cracks "close" again due to the sudden change in temperature.

With the colored resin solution now inside the crystal, the stone assumes the appearance of a colored, translucent mineral (figure 5.4). Even though its physical and chemical properties are markedly different, such dyed quartz may appear quite similar to precious minerals produced by nature. This is especially true when we consider that most of the gems available in the postclassical and early medieval period would not have been of the highest purity and clarity. The greater number of available precious stones would have been marked by inclusions, internal cracks, and fissures. Thus it

Figure 5.4
Imitation gems colored with copper green and alkanet (red dye) dissolved in hot resin. Photo by Marjolijn Bol.

appears that the knowledge of these natural imperfections not only inspired the processes used to imitate precious minerals with quartz, but also influenced the success of the imitation. If the consumer was aware of the natural flaws in gems, such knowledge of the natural material could make the imitation gem more convincing. But what, then, did "success" in terms of these imitation gems entail?

Historical sources suggest that an important use of dyed quartz was for committing fraud. As we have seen with Pliny, sources stress the difficulty of discovering fake gems among the natural stones, and they provide the reader with a variety of methods for their identification.[20] Similarly, in his seventh-century *Etymologiae*, Isidore of Seville writes about the skill with which imitations of gems—in this case, *smaragdi* (emeralds)—were produced:

> As a substitute for that most precious stone, the *smaragdus*, some people dye glass with skill, and its false greenness deceives the eyes with a certain subtlety, to the point that there is no one who may test it and demonstrate that it is false.[21]

However, precious stones were imitated not only to deceive innocent consumers into thinking they were buying a natural gem; gemstone imitations were also produced for aesthetic, ritualistic, religious, and pecuniary purposes. In these instances, the material mimetic gems were known to be imitations and were appreciated as such. The more convincing they were, the more valuable they became. Again, this implies a rather specific kind of knowledge on the side of the consumer, not just about the natural material but also about what constitutes a skillfully made imitation.

The most obvious reason for wanting to own an object that imitated rare and desirable materials is that they were typically cheaper than objects made from solid gold or silver and decorated with the rarest of natural gems. Owning an object that looked like a solid golden piece could, for instance, help a less wealthy monastery or church acquire a set of liturgical objects that fulfilled the same purpose as the objects owned by the wealthiest of religious institutions. Yet, objects characterized by material mimesis not only served as cheaper substitutes for the rare; they were also ordered and owned by the richest of patrons. Examples of the latter can be found in instances where imitations of precious stones were used as grave gifts, a practice found in some of the earliest known societies. In these practices, material mimetic objects were used to "replace" certain materials in the deceased's journey to the afterlife.[22] Imitations of gems also played an important role in the decoration of a variety of medieval religious objects. Reliquaries, for instance, were often set with gemstone imitations made from colored, translucent glass. In addition to being decorated with imitations of precious stones, religious objects were often entirely material mimetic. They were typically made of wood decorated with gold and silver leaf set with imitations of enamels and precious stones so that they appeared to be made of solid gold or silver and studded with gems. On a much larger scale, the embellishment of medieval churches with colored glass windows needs to be understood in a similar light. The saturated colors of blue, red, green, and yellow glass were meant to give worshippers the impression that the House of God was illuminated by translucent sapphires, rubies, emeralds, and gold. In these glass imitations, however, the crackle did not play a role. Since natural precious stones occur only in small sizes, it is impossible to glaze an entire building with them, even for the richest of patrons. The purpose of stained glass windows was thus to imitate only the best qualities of natural precious stones—their saturated color and their ability to transmit the visible light—on a scale that would have been impossible were natural minerals used. To this end, the goal was to procure only the clearest of colored glass. The crackle, a fundamental technique for giving color to gemstone imitations with crystal, was not required to give color to stained glass windows. But what is more, the crackle was undesirable in this case because it would have been considered unfitting to enlarge a defect of a natural material to embellish a religious building. Indeed, in the case of the stained glass window, it was the imitation of the flawless nature of the rare that became the subject of desire for ownership.

To be able to identify a material mimetic object and distinguish it from the materials it was meant to imitate *and* to know the creative skill necessary to produce it—that is, to have an explicit, practical knowledge of how this was done, or to be aware that this skill was being applied—played an important role in the desire to own material mimetic objects. This becomes especially clear in the second example discussed in this chapter: European attempts at imitating handpainted textiles from the East.

CRACKLED CLOTHS

The production of cloths by means of a dye-resist method goes back to ancient Egypt (at least), and over the course of the centuries, it developed independently in various geographical regions. The art form itself may be an ancient kind of material mimesis, as it likely developed from the desire to create alternatives for the more laborious practices of weaving and embroidering. To decorate a textile by means of a resist method, patterns are applied by painting or printing designs on the cloth in a dye-resistant substance such as wax or a paste (e.g., rice, gums, clay, etc.), or by using the more modern discharge method, in which a chemical agent resists penetration of certain dye types. Another well-known method is to tie (e.g., tie-dyeing) or stitch a pattern, or to create a patterned stencil that can be attached to the cloth during its immersion in the dye bath. When a cloth is dyed by one of these methods, it remains uncolored in the areas where the resist was applied. To obtain multiple colors and patterns, the process can be repeated several times by changing or increasing the resist areas and dipping into different colors.[23]

The history of the West's engagement with resist-dye textiles goes back to the sixteenth and seventeenth centuries, when Europe started importing textiles from the East.[24] The sophistication of the colorful painted and printed patterns of these cloths was completely new to Western consumers (figure 5.5). Up til then, the European textile industry had been concerned mostly with embroidering and weaving silk and wool. Compared to the East, it was much less advanced in its knowledge of creating designs on cloth, particularly on cotton, by means of handpainting or printing. We know that carved woodblocks were used to imitate patterns of velvet and damask on linen since at least the beginning of the fifteenth century, when the technique was first described by the Italian painter Cennino Cennini. In his *Il libro dell'arte*, Cennini provides us with recipes for the production of all kinds of painting, including painting on walls, panels, glass, and, indeed, printing colors on cloth. Figure 5.7 shows an example of such an early printed textile produced by a block next to the type of velvet cloth it was meant to imitate (figure 5.6).

According to Cennini, block-printed textiles made from hemp cloth are good for the clothing of young boys and "for particular lecterns in churches." He explains that a walnut or pearwood block, which ought to have the dimensions of a "terracotta block or brick," has to be carved with a continuous pattern on four of its sides: "Any kind of silk drapery that you like should be drawn on it, whether with leaves or animals." The block has two uncarved sides; one is fitted with a handle to be able to apply the pattern without disturbing the ink and the other is used for resting the block. Cennini explains that the preferred binding medium for the pigments used to create the patterns is "liquid varnish," a viscous mixture of linseed oil and natural resin (also used in the practice of gemstone imitation described above) cooked together on a fire. A cloth is stretched on a frame so that the print can be applied systematically by "rolling" the block over the textile. The pattern may be decorated further by handpainting it with other colors mixed with an oil-resin binder, the color palette being limited to a variety of yellow, red, green, blue, black, and white pigments and their mixtures.[25]

European recipe treatises since the fifteenth century suggest that in addition to this block printing, fabrics such as silk and linen and even paper were handpainted with pigments ground with linseed oil or varnish.[26] After having been painted in this manner, such fabric or papers were drenched with oil or varnish to make them see-through and, stretched on a frame, were used to imitate colored, translucent glass. In both Cennini's printed cloth and the handpainted screens, the applied colors are pigments mixed with a binder. This means that unlike in the mordant-dyed textiles produced in

Figure 5.5
Quilt, Coromandel Coast (made for the European market), painted and dyed cotton chintz, ca. 1700. Nr. IS.121–1950, © Victoria and Albert Museum, London.

Figure 5.6
Silk fragment, lampas weave, Italy, ca. 1375–1399. Nr. 1941.391, Cleveland Museum of Art.

Figure 5.7
Block-printed linen, woven (originally black), Germany (probably), 1350–1400. Nr. 1745–1888, © Victoria and Albert Museum, London.

the East, there is no chemical bond between the color and the fabric's fibers. Cloths printed or painted in this manner will not withstand washing as a mordanted fabric would, and this greatly reduced their practical appeal.

In addition to their lack of knowledge concerning the direct application of dyes, Europeans also did not have the range of permanent and radiant dyes known to cloth dyers in the East.[27] In 1689, the Englishman John Ovington remarks on this fact when he writes in his travel notes,

> In some things the Artists of *India* out-do all the Ingenuity of Europe, *viz.* in the painting of Chites or Callicoes, which in *Europe* cannot be parrell'd, either in the brightness and life of the colours, or in their continuance upon the Cloath.[28]

Before their first attempts at exporting printed cloths in imitation of Eastern textiles, Europeans had to acquire expertise in permanent dyes and their application by means of mordanting. It is significant that these efforts were largely focused on improving methods for the printing of cloths; the manual painting techniques practiced in the East were not explored further. Two additional innovations were also crucial for the development of the printing of cloth: the use of engraved copper and faster methods for printing the patterns.

In the eighteenth century, it was discovered that with the use of engraved copper plates as opposed to the high-relief woodblocks that Cennini had also described, high-quality prints full of detail could be achieved. But because everything was printed by hand, the process was still quite laborious. In terms of cost, too, these European printed textiles could not compete with textiles in local markets in the East.

It was not until the implementation of more efficient rotary and block-printing methods in the textile industry that fabrics could be produced quickly and cheaply enough to become a viable export product.[29] By the middle of the nineteenth century, factories in Great Britain, Belgium, and the Netherlands were competing to produce imitations of textiles imported from the East, with a special focus on batik textiles from Indonesia. Maxine Berg, in her work on the import and imitation of luxury goods from India, China, and Japan, points out that a century earlier, the West's efforts to imitate Eastern commodities for their own markets were largely fueled by taste and style. The knowledge of production of these traditional textiles remained in the East, and as a result, the ensuing technological advances that produced the Western imitations were far removed from their Eastern counterparts in terms of materials and technique. According to Berg, European imitations derived their prestige not from the materials with which they were made, "but from the craftsmanship that so effectively replicated the natural world—in the case of printed calicoes, the vividly imitated European and exotic flowers and gardens."[30] It is remarkable, therefore, that by the nineteenth century, when textile printing techniques had improved to the extent that European fabrics could now be exported to Eastern markets (and the Indonesian Archipelago in particular), it was not just the patterns of the cloths but also their material qualities for which local consumers judged them and, in some cases, rejected them.

The resist-dye process of Indonesia, referred to as *batik*, involves the application of hot melted wax, typically mixed with resin, to create a patterned resist.[31] Via freehand drawing, a bamboo or copper-spouted stylus (canting), or copper stamps (cap), the resist is applied on both sides of the fabric so that after dyeing, the textile is decorated with the same pattern on each side. The techniques of handpainting and canting are laborious, and depending on the complexity of the pattern and the number of colors used, it may take several days to produce a batik cloth. For the most exquisite designs, a relatively small piece of cloth may take several months to produce (e.g., a *sarung* is about 2.5 meters). To own these more sumptuous batik garments was a privilege reserved for the upper strata of society.

European textile manufacturers assumed a potential market for industrially produced imitations that would undercut the price of traditional batik textiles.[32] Thomas Stamford Raffles (1781–1826), lieutenant governor of Java and neighboring islands between

1811 and 1816, was one of the first to point out the potential of selling batik imitations produced by the British textile industry on the island of Java. Raffles sent traditional batiks to England to be studied by British manufacturers for the purpose of imitating them. Yet, the British imitations that were produced on the basis of his samples were not immediately successful.[33] In his *History of Java*, Raffles argues that this was because after one washing, "the natives had discovered that the colours would not stand," and this was "a disadvantage which all the British printed cottons labour under."[34] He wonders:

> Would it not tend greatly to the improvement of the British manufacture, and consequently greatly extend the export, if the enquiries of scientific men in India were directed, in a particular manner, to an observation of the different dye stuffs used in Asia, and to the manner followed by the natives in different parts, for fixing the colours and rendering them permanent?[35]

In the next decades, with the invention of an increasing number of synthetic dyes, these issues with dye permanency would largely be dealt with. Manufacturers would soon find out, however, that the greater durability of imitation batik did not guarantee its success in local markets in the East. Traditional Indonesian batik is characterized by small crackles over the surface of its pattern, as can be seen in the eighteenth-century shawl or belt today kept in the Art Institute of Chicago (figure 5.8). These crackles are the result of the wax slightly breaking during drying. The broken wax allows the color to penetrate the resist in the dyebath, resulting in an accidental pattern of colored crackles over the surface of the cloth. Traditionally, artisans would have tried to prevent the wax from breaking, but even with the greatest care, crackling can never be completely avoided.[36] It is therefore significant that when the Westerners first started exporting their industrially produced batik imitations to Indonesia, they quickly found that their cloths did not meet with approval because the crackle was missing from their factory-produced imitations. Indeed, rather than being considered a mistake or error, the craquelure patterns of traditional batik had become one of the most important signifiers for attributing value and ownership to this textile. To know the crackle was to understand the material characteristics of traditional batik cloth and the methods by which it was produced. And knowing the crackle helped consumers in the East distinguish European imitation batiks from traditionally produced cloths. In fact, the crackle had become so important to the identity of traditional batik that imitating it became a prerequisite for Western imitation batiks to be successful in Eastern markets.

As a result, printers in the West became ever more inventive in trying to imitate the crackled surface of traditional batik. An early solution involved engraving a pattern of crackles in the copper plates or the rollers of the presses used for printing batik imitations.[37] But the engraved crackles looked quite different from the organic

Figure 5.8
Selendang (shawl) or belt, batik dyed, Java, 1775–1825, 168.2×103.1 cm. Nr. 1938.241.1–2, gift of Mrs. Charles H. Worcester. Art Institute Chicago (CC0 Public Domain Designation).

crackles produced by the accidental breaking of the wax. To consumers in the East, the engraved crackle patterns appeared rough and unnatural; consumers immediately recognized these cloths as imitation batiks and did not appreciate their appearance.[38] European printers, most notably the Belgian printer Jean Baptiste Theodore Prévinaire (1783–1854), active in the Dutch city of Haarlem, therefore searched for new methods to improve the material mimesis of batik. In 1844, Prévinaire patented a method that he insisted "was similar in all respects" to the batik produced in the East. He proposed a method in which resin mixed with wax had to be applied while warm. During the drying process the resin cracked, and this generated the sought-after pattern of fine veins that traditional batik was known for.[39] Although this method was not new to the textile industry, Prévinaire was the first to realize its potential for enhancing his batik imitations with unique crack patterns. With this, Prévinaire appropriated an important aspect of the traditional production process of batik, and by way of his patent, he turned this knowledge into an economic asset.

In addition to the crackle, European cotton printers also tried to imitate another visual characteristic typical of traditional batik: the flow effect that appeared when using brown dyes. In the traditional batik process, the brown dyes, unlike the other colors, had to be applied in a warm bath. This meant that the wax remained somewhat liquid from the heat and, as a result, the dye paint penetrated the painted resist wax pattern, causing some flowing out of the color. European printers discovered ways to mechanically reproduce this effect and eventually started using it with colors other than brown as well.[40] Such reengineering of a visual effect seen in the original textile can be considered a form of new knowledge (i.e., a new method of production), but depending on how the new material mimetic fabric was put into the market, it could also sideline the original makers and owners of batik textiles (i.e., the new technology could take over local batik production). Interestingly, and as we will see in the next section, this did not happen in the case of imitation batik. The material nature of locally produced batik was so important to Eastern consumers that Western attempts at visual imitation of the textile were not enough to guarantee its success.

Indeed, European printers discovered that in addition to the visual characteristics of traditional batik cloths, the scent of traditional batik was crucial to the reception of their imitations. At first, Western textile printers did not know where the particular scent of batiked textiles came from. They speculated that it could be attributed to a special kind of wax, to certain local dyes, or even to a particular step in the batik process unknown to Western printers. Perhaps close observation of the traditional batik process would provide an answer? An 1855 report written by a Dutch agent of the Nederlandsche Handel-Maatschappij (NHM) based on the island of Java describes the outcome of his investigations into "the peculiar aromatic scent of Java-batiks."[41] The Dutch NHM agent meticulously observed the materials and processes used to produce batik. Based on this research and some further experimentation of his own, he concludes in his report that it must be the wax-resin mixture that gives batiks their particular aromatic scent. The cloths, as he describes, were infused with this scent during the process of washing off the resist after dyeing. To remove the resist from the fabric, batiks were washed in boiling water baths. The cloth was plunged into the boiling water several times, and this caused it to come into contact with the wax-resin mixture that was floating at the surface of the water bath. It was this treatment of immersing the entire cloth into the wax-resin mixture, the NHM agent argues, that was the "fundamental and first cause" of the scent of batik.[42] When Prévinaire learned that his imitations were lacking in this respect, he decided to add this step to obtain the special scent for his batik imitations.

Paradoxically, the reception of the European imitation batiks depended on how closely they resembled traditional batik textiles with respect to a material characteristic that the Eastern dyers traditionally were mostly trying to avoid: the crackle. As a result of their being confronted with European textiles that were able to imitate traditional batiks ever more closely, the crackle became a crucial signifier for batik's local identity. Eventually, locally produced batik imitations, produced with copper stamps, successfully challenged European imitation batiks, which quickly lost their place on the Eastern markets.[43]

It is well known that Western manufacturers of imitation batiks survived only because they discovered new markets for their products in Central and West Africa. The shortcomings of Dutch resist-dye textiles that were problems in the Eastern markets now became the textile's greatest strengths—on the African continent there was nothing better on the market than Prévinaire's extremely refined batik imitations, and because they displayed their colorful patterns on both sides, Western imitation batik was perfect for African wrap-style fashion.[44] Indeed, the "Veritable Java Print," "Guaranteed Dutch Java Hollandis," "Genuine Amsterdam," "Vlisco True Original"—just a few of the names by which these imitation batiks are known today—became highly prized commodities through which the African elite could express their social status. In Africa, the numbered designs, most of which are nowadays produced by the Vlisco company in a factory in the Dutch city of Helmond, were given local meaning by salesmen on the marketplace and by the consumers who purchased the textiles to transform them into clothing.[45] Imitation batik was so successfully adopted by the African elite that it is difficult to imagine African dress without the colorful textiles that have such an intricate history as material mimetic objects. As a result, the Dutch imitation batiks became the subject of a rather confusing tale of appropriation in which the last word has not yet been said.

There is an additional twist to this story that should not be left out in the context of this volume. Ironically, Vlisco was almost put out of business in the 1990s because cheap Asian copies of their textiles flooded the market. The company survived because it reinvented itself as a luxury fashion brand, working together with both established and aspiring designers, including Jean Paul Gaultier and Yinka Shonibare. To thwart counterfeiting, Vlisco now brands all its textiles and has a disclaimer on their website that explains the uniqueness of their products and educates the public about the nature of "counterfeited" fabrics. In this case, consumers are informed about the production process of Dutch Wax Cloth to learn what constitutes an original Vlisco.

Interestingly, the "crackle," once so important in the East because of its connection to the traditional batik process, did not lose any of its prominence in Vlisco's reappropriation of the textiles. The "crackle," in fact, became one of the most important attributes of Vlisco's finest quality fabrics.[46] The process of purposely breaking wax ensured that no meter of Vlisco fabric is the same, and this became one of its main selling points. Again, as in the original batik process, the crackle helped impart a sense of uniqueness to each of Vlisco's textiles, even though it is produced mechanically. Vlisco's fabrics are therefore a case of material mimesis in which their history as imitation batik is embraced, but the changed social function of these fabrics on the African continent gave them new value as something unique as well. Here, the crackles became indicative of a quality and unique piece of imitation batik.

During the first decades of the twentieth century, artists of the arts and crafts movement again assigned new meaning to the crackle when they begin exploring handpainted batik. They argued that the subtle presence of crackles in batik textiles is a sign of artisanal skill that should not be recreated by mechanical means or other shortcuts. Pieter Mijer, a Dutch artist who worked in New York, was a great advocate of the art of batik in America. In 1919 he published a treatise, *Batiks, and How to Make Them*, in which he presents the technique of producing a batik cloth by handpainting. Mijer laments the fact that due to its popularity, making batiks had become a fashionable pastime comparable to "peasant wood-carving, burnt-wood work or sweater knitting," and that people also "tried to produce the effect [of batik] without work" (i.e., by producing imitations) in a so-called secret process that "enjoyed considerable vogue."[47] Mijer explains that the general public believed that these substitutes were real batik because the material had been dipped and some wax had been used, but

> anyone who knew anything about the genuine process was not fooled and recognized that stencils and various other fake methods had been utilized. The unlimited patience of the native worker was unknown, and unsung was the thoroughness of the painstaking craftsman. At this period the watch-word was "speed" and the results showed it.[48]

Mijer continues that the importance placed on the "crackle effect" in batik imitations produced for the American and Dutch markets should be seen as part of the same development:

> Crackle certainly has its place in the beauty of batik, but the indiscriminate use of it as a complete motive of decoration in itself, is to be regretted. It would be used less, probably, if examples of the best native and European work were studied, in which real design and colour are the arresting features.[49]

Once again, the crackle had been appropriated. For Mijer, having thorough knowledge of the crackle in the traditional batik technique, and its proper application in contemporary handpainted batik art, helps distinguish a good artist from one with less skill, or even an amateur.

CONCLUSION

With a focus on the crackle, the two cases discussed in this chapter showed how the phenomenon of material mimesis prompts a negotiation of ownership between makers and users. In the case of imitating precious stones, colored resins were used to penetrate the crackles of colorless crystals in an attempt to come closer to the appearance of colored translucent gems—using a precious mineral's natural flaws to make the imitation and to make it more convincing. In the case of batik, resin likewise played a crucial role. Here, resin was the material responsible for the crackles in the handpainted batik cloth. These crackles in batik cloth were initially considered a defect, something the artisan tried to avoid in the final product as much as possible. But in conversation with Western attempts at mechanically imitating handpainted batik textiles, the sophisticated local material knowledge of a traditional product helped the crackle gain in importance. It became the means to *know* and *own* traditionally made handpainted batik textiles; to *know* and *own* its imitations and to assess their quality; and finally, instigated by their new role in African fashion, to *know* and *own* these imitations as something with a value and uniqueness independent of what they originally were meant to imitate.

The practice of material mimesis thus involved a complicated interplay between appropriation and ownership on the side of the consumer, the artisan, and the scholar. Artisans tried to transform one material to appear to be another as convincingly as possible either to show off their skill or, in the case of fraud, to disguise their craftsmanship to the best of their abilities. The context in which material mimetic objects were offered to the consumer was therefore crucial for their reception as either a marvel of artistic skill or a deceitful piece of work. But in both instances, consumers had to first know—that is, learn to recognize and understand—the creative mastery that went into the production of the material imitations in order to avoid accidentally buying a consciously fraudulent material mimetic product, and second, to be able to evaluate the skill and materials with which the mimetic object had been made. In the practice of "crackling," these practices of kn/own/ables thus come together in a unique way. The value of "crackled" objects in the sphere of economy (their use) is intertwined with practices of use and naming that hide or embrace the way in which the objects were produced as mimesis (the performance of this knowledge), or that designate a new object in a commercial process of rebranding independent of this knowledge. Hence, the example of the crackle in the imitation of both precious stones and batik shows that the consumer's knowledge about materials and production deeply impacts the ways in which artisan, industry, and later "materials science" shape and value the practice of material mimesis.

ACKNOWLEDGMENTS

This chapter is partly the result of a project that has received funding from the European Research Council (ERC) under the European Union's Horizon 2020 research and innovation program (Grant agreement No. 852732—DURARE). I would like to extend my sincere thanks to the editors and the anonymous reviewers for their valuable feedback on this chapter.

Notes

1. Marjolijn Bol and Emma C. Spary, eds., *The Matter of Mimesis: Studies on Mimesis and Materials in Nature, Art and Science* (Leiden: Brill, forthcoming).

2. This fact has been the subject of a fair amount of discussion in recent years; see, e.g., Harriet Constable, "The Sparkling Rise of the Lab Grown Diamond," BBC, February 10, 2020, https://www.bbc.com/future/article/20200207-the-sparkling-rise-of-the-lab-grown-diamond.

3. Because of the huge amount of energy required to produce a lab-created mineral, the environmental benefit is less obvious. See, e.g., Rob Bates, "Just How Eco-Friendly Are Lab-Created Diamonds?," *JCK*, March 29, 2019, https://www.jckonline.com/editorial-article/lab-created-diamonds-eco-friendly/.

4. Features such as inclusions and defects can, of course, also be purposefully added to control the properties of the mineral.

5. Grahame Clark, *Symbols of Excellence: Precious Materials as Expressions of Status* (Cambridge: Cambridge University Press, 1986), 3.

6. Thorstein Veblen, *The Theory of the Leisure Class* (1899; New York: Macmillan, 1912), 126–128.

7. Christos P. Baloglou, "The Tradition of Economic Thought in the Mediterranean World from the Ancient Classical Times through the Hellenistic Times until the Byzantine Times and

Arab-Islamic World," in *Handbook of the History of Economic Thought: Insights on the Founders of Modern Economics*, ed. Jürgen Backhaus (New York: Springer, 2012), 51–52.

8. Lucius Annaeus Seneca, *Letters on Ethics: To Lucilius*, ed. and trans. Margaret Graver and A. A. Long (Chicago: University of Chicago Press, 2015), 332 [Letter XC.33].

9. Seneca, *Letters*, 332 [Letter XC.33].

10. Pliny the Elder, *Natural History*, ed. and trans. D. E. Eichholtz, Loeb Classical Library 419 (Cambridge, MA: Harvard University Press, 1962), 10:227 [XXXVII.lxxv.197].

11. Pliny the Elder, 327 [XXXVII.lxxv.198].

12. Eugenia Lao, "Luxury and the Creation of a Good Consumer," in *Pliny the Elder: Themes and Contexts*, eds. Roy K. Gibson and Ruth Morello (Leiden: Brill, 2011), 35–56.

13. Pliny the Elder, *Natural History*, 327 [XXXVII.lxxvi.199–200].

14. The other half of the papyrus is kept in Leiden, and for this reason it is referred to as the *Leiden Papyrus*. The most extensive critical transcription and translation to date is Robert Halleux, ed. and trans., *Les alchimistes grecs*, vol. 1, *Papyrus de Leyde—Papyrus de Stockholm—Recettes*, Collection des Universités de France (Paris: Belles Lettres, 1981). For the English translation of the two papyri cited here (based on Otto Lagercrantz's German translation), see Earle Radcliffe Caley, "The Leyden Papyrus X: An English Translation with Brief Notes," *Journal of Chemical Education* 3, no. 10 (1926): 1149–1166, and Earle Radcliffe Caley, "The Stockholm Papyrus: An English Translation with Brief Notes," *Journal of Chemical Education* 4, no. 8 (1927): 979–1002.

15. Marcellin Berthelot suggests that the papyri may have been preserved in the mummy case of an Egyptian chemist, and Otto Lagercrantz argues that the papyri were a luxury copy (internal evidence shows the manuscripts are copied from another source) made for the purpose of entombment. This would also explain why they have been preserved after Diocletan's 296 CE decree banning all treatises dealing with alchemy. See Marcellin Berthelot, *Introduction à l'étude de la chimie des anciens et du moyen age* (Paris: G. Steinheil, 1889), 5; and Otto Lagercrantz, ed., *Papyrus graecus holmiensis (P. Holm.): Recepte für Silber, Steine und Purpur* (Uppsala: Akademiska Bokhandeln, 1913), 13:55.

16. Caley, "Stockholm Papyrus," 987 [nr. 54].

17. Caley, 987 [no. 24].

18. Caley, 985 [no. 31].

19. See also Marjolijn Bol, "Coloring Topaz, Crystal and Moonstone: Gems and the Imitation of Art and Nature, 300–1500," in *Fakes!?: Hoaxes, Counterfeits and Deception in Early Modern Science*, ed. Marco Beretta and Maria Conforti (Sagamore Beach, MA: Science History Publications, 2014), 108–129; and Marjolijn Bol, "The Emerald and the Eye: On Sight and Light in the Artisan's Workshop and the Scholar's Study," in *Perspective as Practice: Renaissance Cultures of Optics*, ed. Sven Dupré (Turnhout, Belgium: Brepols Publishers, 2019), 71–101.

20. For a discussion on these identification methods, see Bol, "Coloring Topaz," 121–124.

21. Isidore of Seville, *The Etymologies of Isidore of Seville*, ed. and trans. Stephen A. Barney et al. (Cambridge: Cambridge University Press, 2006), 328 [XVI.xv.27].

22. Such practices have been extensively discussed in the field of archeology, where material mimesis is typically referred to as *skeuomorphism*. See, e.g., Michael J. Vickers and David Gill, *Artful Crafts: Ancient Greek Silverware and Pottery* (Oxford: Oxford University Press, 1994); and Michael Vickers, *Skeuomorphismus oder die Kunst, aus wenig viel zu machen* (Mainz: Philipp von Zabern, 1999). On the imitation of leather and wood in ceramic urns in which Irish Early Bronze Age cremations were deposited, see, e.g., T. G. Manby, "Skeuomorphism: Some Reflections of Leather, Wood and Basketry in Early Bronze Age Pottery," in *Unbaked Urns of Rudely Shape: Essays on British and Irish Pottery for Ian Longworth*, ed. Ian Kinnes and Gillian Varndell (Oxford: Oxbow Books, 1995), 81–88. And see also, for funerary ceramics impressed with textiles and other organic material culture, Linda Hurcombe, "Organics from Inorganics: Using Experimental Archaeology as a Research Tool for Studying Perishable Material Culture," *World Archaeology* 40, no. 1 (2008): 83–115, esp. 106–107. For the role of the skeuomorph in ritualistic practice, see also Jeroen Stumpel, "The Vatican Tazza and Other Petrifications: An Iconological Essay on Replacement and Ritual," *Simiolus-Netherlands Quarterly for the History of Art* 24, no. 2–3 (1996): 111–127.

23. For an overview of these techniques with references, see Gerald W. R. Ward, ed., *The Grove Encyclopedia of Materials and Techniques in Art* (New York: Oxford University Press, 2008), 696–705.

24. Anthony Reid, "Southeast Asian Consumption of Indian and British Cotton Cloth," in *How India Clothed the World: The World of South Asian Textiles, 1500–1850*, ed. Giorgio Riello and Tirthankar Roy (Leiden: Brill, 2009), 40.

25. Cennino Cennini, *Cennino Cennini's "Il libro dell'arte": A New English Translation and Commentary with Italian Transcription*, ed. and trans. Lara Broecke (London: Archetype, 2015), 232–238 [nos. 208–216].

26. This was still common in the nineteenth century. See, e.g., the English (expanded) edition of Pierre François Tingry, *The Painter's and Colourman's Complete Guide* (London: Sherwood, Gilbert, and Piper, 1830), 295–297.

27. Beverly Lemire and Giorgio Riello, "East & West: Textiles and Fashion in Early Modern Europe," *Journal of Social History* 41, no. 4 (2008): 898.

28. John Ovington, *A Voyage to Suratt in the Year 1689 Giving a Large Account of That City and Its Inhabitants and of the English Factory There* (London: Printed for Jacob Tonson), 282.

29. G. P. J. Verbong, "Katoendrukken," in *Geschiedenis van de techniek in Nederland: De wording van een moderne samenleving 1800–1890*, ed. H. W. Lintsen (The Hague: Stichting Historie der Techniek; Zutphen, Netherlands: Walburg Pers, 1992–1995), 3:59–61.

30. Maxine Berg, "In Pursuit of Luxury: Global History and Consumer Goods in the Eighteenth Century," *Past & Present* 182 (2004): 126–127.

31. Rens Heringa and Harmen C. Veldhuisen, eds., *Fabric of Enchantment: Batik from the North Coast of Java—From the Inger McCabe Elliott Collection at the Los Angeles County Museum of Art* (Los Angeles: Los Angeles County Museum of Art, 1996), 16, 224–230.

32. See, e.g., Rens Heringa, "Javaanse katoentjes," in *Katoendruk in Nederland*, ed. Bea Brommer (Tilburg, Netherlands: Textielmuseum; Helmond, Netherlands: Gemeentemuseum, 1989), 131–156.

33. Maria Wronska Friend, "The Early Production of Javanese Batik Imitations in Europe (1813–1840)," in *Glarner Tuch Gespräche: Tagungsband Internationale Fachtagung vom 2./3. Juni 2016 im Hänggiturm Blumer & Cie., Schwanden zum Thema "Kunst und Geschichte des Glarner und europäischen Zeugdrucks,"* ed. Reto D. Jenny (Schwanden (Kanton Glarus), Switserland: Comptoir of Daniel Jenny & Cie, 2017), 50–51.

34. Thomas Stamford Raffles, *The History of Java: In Two Volumes* (London: Printed for Black, Parbury, and Allen, 1817), 216–217.

35. Raffles, 216–217.

36. For some colors there were exceptions; cloths dyed brown, for instance, were often purposefully crackled by crumpling the fabric before dyeing. See Verbong, "Katoendrukken," 73–74.

37. Verbong, 59–61.

38. Verbong, 73–74.

39. G. P. J. Verbong, "Technische innovaties in de katoendrukkerij en ververij in Nederland 1835–1920" (PhD diss., Technische Universiteit Eindhoven, 1988), 347 (appendix 13).

40. Verbong, "Katoendrukken," 73–74.

41. G. P. Rouffaer and H. H. Juynboll, *De Batik-Kunst in Nederlandsch-Indië en haar geschiedenis* (Haarlem, Netherlands: H. Kleinmann, 1899), I (appendix 1).

42. Rouffaer and Juynbolll, *De Batik-Kunst*, IV (appendix 1).

43. See, e.g., C. H. Krantz, "De Export van in Nederland bedrukte katoen naar het Verre Oosten en Afrika," in Brommer, *Katoendruk in Nederland*, 111–130.

44. Krantz, 115–130.

45. See, e.g., Nina Sylvanus, "The Fabric of Africanity: Tracing the Global Threads of Authenticity," *Anthropological Theory* 7, no. 2 (2007): 201–216; and Danielle Bruggeman, "Vlisco: Made in Holland, Adorned in West Africa, (Re)appropriated as Dutch Design," *Fashion, Style & Popular Culture* 4, no. 2 (2017): 197–214.

46. See, e.g., the description of Super-wax on the company website: "Super-wax is of the finest quality in wax print fabrics. For this product we use an extra densely woven, fine cotton fabric. Super-wax has a recognizable design signature as it always features two blocking colours showing a natural and unique crackling effect." "Product Information," Vlisco (website), accessed December 10, 2020, https://www.vlisco.com/support/product-information/.

47. Pieter Mijer, *Batiks, and How to Make Them* (New York: Dodd, Mead, 1919), 25–26.

48. Mijer, 26.

49. Mijer, 26–27.

Bibliography

Baloglou, Christos P. "The Tradition of Economic Thought in the Mediterranean World from the Ancient Classical Times through the Hellenistic Times until the Byzantine Times and Arab-Islamic World." In *Handbook of the History of Economic Thought: Insights on the Founders of Modern Economics*, edited by Jürgen Backhaus, 7–92. New York: Springer, 2012.

Bates, Rob. "Just How Eco-Friendly Are Lab-Created Diamonds?" *JCK*, March 29, 2019. https://www.jckonline.com/editorial-article/lab-created-diamonds-eco-friendly/.

Berg, Maxine. "In Pursuit of Luxury: Global History and Consumer Goods in the Eighteenth Century." *Past & Present* 182 (2004): 126–127.

Berthelot, Marcellin. *Introduction à l'étude de la chimie des anciens et du moyen age*. Paris: G. Steinheil, 1889.

Bol, Marjolijn. "Coloring Topaz, Crystal and Moonstone: Gems and the Imitation of Art and Nature, 300–1500." In *Fakes!?: Hoaxes, Counterfeits and Deception in Early Modern Science*, edited by Marco Beretta and Maria Conforti, 108–129. Sagamore Beach, MA: Science History Publications, 2014.

Bol, Marjolijn. "The Emerald and the Eye: On Sight and Light in the Artisan's Workshop and the Scholar's Study." In *Perspective as Practice: Renaissance Cultures of Optics*, edited by Sven Dupré, 71–101. Turnhout, Belgium: Brepols Publishers, 2019.

Bol, Marjolijn, and Emma C. Spary, eds. *The Matter of Mimesis: Studies on Mimesis and Materials in Nature, Art and Science*. Leiden: Brill, forthcoming.

Bruggeman, Danielle. "Vlisco: Made in Holland, Adorned in West Africa, (Re)Appropriated as Dutch Design." *Fashion, Style & Popular Culture* 4, no. 2 (2017): 197–214.

Caley, Earle Radcliffe. "The Leyden Papyrus X: An English Translation with Brief Notes." *Journal of Chemical Education* 3, no. 10 (1926): 1149–1166.

Caley, Earle Radcliffe. "The Stockholm Papyrus: An English Translation with Brief Notes." *Journal of Chemical Education* 4, no. 8 (1927): 979–1002.

Ceninni, Cennino. *Cennino Cennini's "Il libro dell'arte": A New English Translation and Commentary with Italian Transcription*. Edited and translated by Lara Broecke. London: Archetype, 2015.

Clark, Grahame. *Symbols of Excellence: Precious Materials as Expressions of Status*. Cambridge: Cambridge University Press, 1986.

Constable, Harriet. "The Sparkling Rise of the Lab Grown Diamond." *BBC*, February 10, 2020. https://www.bbc.com/future/article/20200207-the-sparkling-rise-of-the-lab-grown-diamond.

Halleux, Robert, ed. and trans. *Les alchimistes grecs*. Vol. 1: *Papyrus de Leyde—Papyrus de Stockholm—Recettes*. Collection des Universités de France. Paris: Belles Lettres, 1981.

Heringa, Rens. "Javaanse katoentjes." In *Katoendruk in Nederland*, edited by Bea Brommer, 131–156. Tilburg, Netherlands: Textielmuseum; Helmond, Netherlands: Gemeentemuseum, 1989.

Heringa, Rens, and Harmen C. Veldhuisen, eds. *Fabric of Enchantment: Batik from the North Coast of Java—From the Inger McCabe Elliott Collection at the Los Angeles County Museum of Art.* Los Angeles: Los Angeles County Museum of Art, 1996.

Hurcombe, Linda. "Organics from Inorganics: Using Experimental Archaeology as a Research Tool for Studying Perishable Material Culture." *World Archaeology* 40, no. 1 (2008): 83–115.

Isidore de Seville. *The Etymologies of Isidore of Seville.* Edited and translated by Stephen A. Barney, W. J. Lewis, J. A. Beach, and O. Berghof. Cambridge: Cambridge University Press, 2006.

Krantz, C. H. "De export van in Nederland bedrukte katoen naar het Verre Oosten en Afrika." In *Katoendruk in Nederland*, edited by Bea Brommer, 111–130. Tilburg, Netherlands: Textielmuseum; Helmond, Netherlands: Gemeentemuseum, 1989.

Lagercrantz, Otto, ed. *Papyrus graecus holmiensis (P. Holm.): Recepte für Silber, Steine und Purpur.* Uppsala: Akademiska Bokhandeln, 1913.

Lao, Eugenia. "Luxury and the Creation of a Good Consumer." In *Pliny the Elder: Themes and Contexts*, edited by Roy K. Gibson and Ruth Morello, 35–56. Leiden: Brill, 2011.

Lemire, Beverly, and Giorgio Riello. "East & West: Textiles and Fashion in Early Modern Europe." *Journal of Social History* 41, no. 4 (2008): 887–916.

Manby, T. G. "Skeuomorphism: Some Reflections of Leather, Wood and Basketry in Early Bronze Age Pottery." In *Unbaked Urns of Rudely Shape: Essays on British and Irish Pottery for Ian Longworth*, edited by Ian Kinnes and Gillian Varndell, 81–88. Oxford: Oxbow Books, 1995.

Mijer, Peter. *Batiks, and How to Make Them.* New York: Dodd, Mead, 1919.

Ovington, John. *A Voyage to Suratt in the Year 1689 Giving a Large Account of That City and Its Inhabitants and of the English Factory There.* London: Printed for Jacob Tonson, 1696.

Pliny the Elder. *Natural History.* Vol. 10, *Books 36–37*, edited and translated by D. E. Eichholtz. Loeb Classical Library 419. Cambridge, MA: Harvard University Press, 1962.

Reid, Anthony. "Southeast Asian Consumption of Indian and British Cotton Cloth." In *How India Clothed the World: The World of South Asian Textiles, 1500–1850*, edited by Giorgio Riello and Tirthankar Roy, 29–51. Leiden: Brill, 2009.

Rouffaer, G. P., and H. H. Juynboll. *De Batik-Kunst in Nederlandsch-Indië en haar geschiedenis.* Haarlem, Netherlands: Kleinmann, 1899.

Seneca, Lucius Annaeus. *Letters on Ethics: To Lucilius.* Edited and translated by Margaret Graver and A. A. Long. Chicago: University of Chicago Press, 2015.

Seneca, Lucius Annaeus. *Epistles 1–65.* Vol. 1. Edited and translated by Richard M. Gummere. Loeb Classical Library. Cambridge, MA: Harvard University Press, 1917.

Stamford Raffles, Thomas. *The History of Java: In Two Volumes.* London: Printed for Black, Parbury, and Allen, 1817.

Stumpel, Jeroen. "The Vatican Tazza and Other Petrifications: An Iconological Essay on Replacement and Ritual." *Simiolus-Netherlands Quarterly for the History of Art* 24, no. 2–3 (1996): 111–127.

Sylvanus, Nina, "The Fabric of Africanity: Tracing the Global Threads of Authenticity." *Anthropological Theory* 7, no. 2 (2007): 201–216.

Tingry, Pierre François. *The Painter's and Colourman's Complete Guide.* London: Sherwood, Gilbert, and Piper, 1830.

Veblen, Thorstein. *The Theory of the Leisure Class.* New York: Macmillan, 1912.

Verbong, G. P. J. "Katoendrukken." In *Geschiedenis van de techniek in Nederland: De wording van een moderne samenleving 1800–1890*, edited by H. W. Lintsen. 6 vols. The Hague: Stichting Historie der Techniek; Zutphen, Netherlands: Walburg Pers, 1992–1995.

Verbong, G. P. J. "Technische innovaties in de katoendrukkerij en ververij in Nederland 1835–1920." PhD diss., Technische Universiteit Eindhoven, 1988.

Vickers, Michael J. *Skeuomorphismus oder die Kunst, aus wenig viel zu machen.* Mainz, Germany: Philipp von Zabern, 1999.

Vickers, Michael J., and David Gill. *Artful Crafts: Ancient Greek Silverware and Pottery.* Oxford: Oxford University Press, 1994.

Ward, Gerald W. R., ed. *The Grove Encyclopedia of Materials and Techniques in Art.* New York: Oxford University Press, 2008.

Wronska-Friend, Maria. "The Early Production of Javanese Batik Imitations in Europe (1813–1840)." In *Glarner Tuch Gespräche: Tagungsband Internationale Fachtagung vom 2./3. Juni 2016 im Hänggiturm Blumer & Cie., Schwanden zum Thema "Kunst und Geschichte des Glarner und europäischen Zeugdrucks,"* edited by Reto D. Jenny, 49–58. Schwanden (Kanton Glarus), Switzerland: Comptoir of Daniel Jenny & Cie, 2017.

6

EDUCATIONAL INEQUITIES AND THE DISTRIBUTION OF TECHNICAL KNOWLEDGE: THREE INSTRUMENTS

Amy E. Slaton

This is the story of a banana in a shoebox. The cardboard box has its cover on, but light peeps in through a series of small holes that have been poked in the lid at regular intervals. The numbered holes form a grid into which a thin wooden barbeque skewer, itself calibrated at regular intervals like a ruler, can be dipped. Gathered around the shoebox in a sixth-grade US classroom, children of eleven or twelve are being taught to reveal its contents with repeated thrusts of the skewer. They proceed methodically along each line of holes, recording the depth of each dip onto a waiting piece of graph paper, and as they complete their suspenseful prodding the readings gradually reveal that it is a banana that lurks inside the box—not a potato, not a shoe, not empty space. The banana is, in short, a scientific specimen; the work going on is that of remote-sensing; and the shoebox is a very cheap, very easily maintained scanning probe device: a $4 atomic force microscope (AFM).[1]

How this particular scientific apparatus has come to be known as such, and not as trash or plaything, and the banana known as a research object, not as a misplaced part of someone's lunch, is a worthwhile analytic project for those interested in the ownership of knowledge. Science education in the United States represents a complex project of staged mastery, where learners are meant to encounter increasingly complex, precise—and not least important, costly—versions of "real" instruments as they move in their schooling toward the accumulation of ever more remunerative and prestigious skills.[2] After the shoebox AFM, once in high school, students may encounter a $40 scanning probe instrument made up of LEGO blocks and the sort of pocket laser pointer normally used by carpenters or public speakers; the sample under study might itself be a configuration of LEGO blocks. This more elaborate AFM is accompanied by basic imaging software instead of pencil and graph paper.[3]

Beyond both experiences may come an encounter with a $25,000 AFM in the well-appointed college classroom or industrial training facility, along with the instrument maker's proprietary software, or with a still more expensive system installed in a high-level university or commercial research and development laboratory. Among all the AFMs, only those costing in the tens of thousands of dollars involve the literal mapping of electrons to reveal surface forces of materials at the atomic scale, but in the world of US science education, each of these successive learning experiences moves the aspirant closer to what is understood to be atomic-scale expertise, closer to the comprehensive "sense-making" that is identified as science.[4] But this experience of forward or upward movement, of accreting skill, is not assured to all those present. This chapter departs from other historical studies of science education in suggesting that no singular process of scientific training is occurring in any given classroom. The idea that science education primarily derives from programs of occupational or disciplinary reproduction is challenged as well. Instead, I want to consider, historically, the social instrumentality of occupations and disciplines among *other* possible ways of organizing the epistemic commonalities we know as "science."[5]

THE BANANA IN THE SHOEBOX: MERIT AND THE PRODUCTION OF KNOWERS AS OWNERS OF KNOWLEDGE

The promissory character of contemporary science education for young people—the directionality and continuity that each stage of mastery implies about individuals' potential movements from kindergarten through high school (i.e., progress through the "K–12" system)—reflects deep commitments to the vision of the United States as a meritocratic society.[6] That vision maintains that through the actualization of proficiencies, the student steadily gains access to the next stage of education and, ultimately, to remunerative employment. Meritocratic commitments are, above all, cast in education policy discussions as an automating mechanism of US social life: an individual's innate capacities will differentially yet inevitably lead to achievement, and achievement will yield distributed prosperity.[7] Some students will naturally reach greater life attainments than others, but all have some enhanced future awaiting via education—that is, via actualized competence.[8] The particular actualization of achievement and knowing undertaken by US technical education includes both the enlistment and disciplining of future scientific workers and also, foundationally, the determination of who is eligible for such work at what level—for example, as a manufacturing plant worker, technician, or research scientist, to invoke one customary ladder-like trope. In prevailing discourse around US science and engineering education (the two occupational destinations are conjoined in most recent literature on K–12 education), curiosity about the natural world is cast as something that is to be detected and satisfied in learners through the provision of staged, age-appropriate curricula, with individuals expected to drop out of the science-learning process as they reach the limit of their natural abilities.[9] As Secules et al. incisively capture, each classroom must have a "worst" student.[10]

The twinned ideas that the objects of scientific knowledge await knowing and that such knowledge exists prior to its acquisition together produce a dehistoricizing understanding strongly challenged by this volume. Predicated on a model of education as a process of transmission, or of the "banking" of knowledge in students, as Freire frames it,[11] the proposition of meritocratic pedagogy is especially consonant with the conception of Western science as an enterprise of value-neutral individuals seeking to learn about the world—that is, as a cognitive disposition of "timeless dualist detachment."[12] If there is a singular cosmos to be known, then each person can be understood to know less or more of it, and each individual's particular capacity—how much they know—can thus be gauged and compared with others'. We could say that science makes of the world an instrument for calibrating scientists.

More particularly, empiricism in the context of science education, just as in mature scientific practice, is a technique for sorting *inquiring subjects*, not only or primarily their objects of inquiry, to perhaps extend Barad's formulation regarding scientific discovery to the process of learning to be scientific.[13] This ascription of differing value to individuals' differing cognitive labors does a great deal to hide the ontological operations of capitalism in the United States. In this chapter I will account for the ways in which the twenty-first-century science classroom cospecifies the activity of effectual remote sensing (in contrast to improper conduct) and the effectual remote sensor (versus the improper actor).[14] Such sorting of learners is required in a setting such as the US, where work and economic security, for which education prepares young people, are comprehensively stratified; the outcomes of teaching and learning must map individuals onto a range of capabilities, must establish divergent and legibly hierarchical personal endowments, if the labor needs of capital are to be warranted as democratic.[15]

In the United States, the history of scientific merit is thus a history of the belief in the multiplicity of human intellectual endowments, a timeline of change and continuity in distributive understandings of racial, gender, and other forms of identification. Scholars have shown that this multiplicity of endowments historically has justified differential life circumstances as people of different ascribed identities proceed through schooling into adulthood and working lives, part of a broader Euro-American take-up of heritability in depictions of human cognitive attainment now understood as a racialized, if not fully eugenic, approach.[16] In recent years, as conceptions of "inclusion" and "diversity" have conformed educational and hiring policies, the existence of differences in ascribed intellectual endowment, or "potential," has also been cast as a matter for pluralistic celebration.[17] It is this recent form taken by merit—as driver of, explanation for, and *commendation* of social stratification—that a close look at the sociomaterialities of contemporary K–12 science education can articulate for us.

As Ferguson writes, after World War II, *yes* finally became "a word attached to minority difference" in America. The activism and legal provisions of the civil rights era redirected educational resources toward historically disadvantaged communities and eventually eroded the acceptability of explicit race-based discrimination in many settings; reformist projects along lines of identities based on gender, ethnicity, disability, and sexuality followed. In education, precepts of inclusion, and by the 1990s, so-named diversity initiatives, elevated a notion of identity-blind pedagogy in which the new stance of welcome would be accompanied by selectivity based on "excellence." Affirmative action and other compensatory approaches declined in the face of this broad neoliberal project to "maintain excellence" in places of US learning and employment. Science, technology, engineering, and mathematics education and work were now rearticulated as the "STEM" sector, a vital source of national prosperity and international competitiveness in the "globalizing" world, and one in which no rigor need be sacrificed to the aim of diversity and inclusion.[18] Yet, these efforts installed what Ferguson has called "a political economy that deploys minority affirmation to rebuttress institutional power."[19] The turn at the end of the century toward pluralism as an ostensible support for a more just and democratic polity embodied majority impulses to preserve structural inequities.

The three AFMs, as fabricated by educators and embedded in US science curricula in the early twenty-first century, embody this history. Science education—tracked and stratified—enacts excellence in opposition to whatever is not excellent. This necessarily involves a determination of *whomever* is not excellent; the three instruments themselves detect and specify human differences of these kinds, facilitating the passage of some students from rudimentary to advanced educational standing. Following the historical ownership of knowledge among individuals and groups of actors (such as youngsters encountering staged scientific experiences) can support our efforts to historicize participation in American science not as individuals' experience of knowing the world, but as the individual coming to be seen (by others, and by the self) as a knower of the world.[20]

In this way, we follow the demarcation of "excellent" scientific learners from others as deriving from the priorities of majority society (including those priorities supporting capitalist labor and production), while obscuring that derivation. The possession of knowledge is not a matter of acquisition alone, but of ascribed potential for acquisition; a central point of this chapter is that these are inseparable attributions. In particular, we can emphasize how understanding knowledge as constituting ownership resonates with the framing of whiteness (or other majority identifications such as masculinity, heterosexuality, cis-identified, or abled in body or mind) as property and resource; any instance of "belonging [to]" predicates both the thing that belongs and the owner to whom it belongs. Accounts of the historical investment in whiteness as a value proposition have helped us grasp the interested and material (rather than just the attitudinal) features of white advantage.[21]

Foundationally, for the three AFMs to differentiate among individuals as learners of various capacities, or detect them as nonknowers, academic capability and achievement must be seen to reside in the individual student. Merit makes of intellect something expressed as talent or capability, but also a quality imaginable as separate from all other factors in one's life, whether a factor is thought to be personal or societal in nature. It is significant that in the post–civil rights–era US, the meritocratic disposition has been, for some of those concerned with racial, gender, and other forms of education and employment inequity, notoriously subject to violation. Advocates of inclusive education and hiring policies meant to correct "minority underrepresentation" or encourage "STEM diversity" point to, for example, the role of stereotype bias in depriving minoritized people of resources or recognition.[22]

But while those exclusionary practices do occur, we misunderstand them as first causes of minoritization, or of racial and gender essentialization in, say, places of schooling and work. In fact, when analyzed strictly as acts of exclusion, such discriminatory operations—however pernicious they may be—disguise the social function of merit: to render individual intellects as such and provide possible classifications for intellects. In conferring or denying ownership of knowledge, and as the term *ownership* itself implies, attributions of merit distribute material and social benefits. Defined and assessed as a trait of individuals, merit produces individuated actors, each with their own calibrated potential and differential deserved life circumstances, and thereby renders unreasonable the economic redistribution or reparative initiatives associated with social-structural change. That is, belief in merit reinstates the stratified character of learning and working under US capitalism, naturalizing that system's political and economic inequities.

Hacking's conceptualizations of "making up people" suggest that such taxonomic projects comprise the production of spaces of possible action. He tells us that social change brings about new categories into which people may be sorted, but also, as in the case of a census, that "counting is no mere report of developments." That is, Hacking discourages our customary sense of empirical inquiry as an operation distinct from social action; to seek and/or find human attributes in individuals or groups, he might suggest, is to make real, or at least, actionable, differences among those human subjects. In a parallel formulation, we might say that in science education, educators' recognition of meritorious students is no mere revelation of talent. Studying the sociabilities of learning, scholars have begun to conceptualize knowledge "as emerging—simultaneously with identities, policies, practices and environment—in webs of interconnections between heterogeneous things, human and nonhuman."[23] The point, in Fenwick's words, is that "attention to the sociomaterial can help reveal the dynamics that are actually constituting what comprises everyday life, including learning," and that the objects that comprise experiences of learning and work

> might be taken by a casual observer as natural and given—things comprising a "context." But a more careful analysis notes that these objects including objects of knowledge, are very messy, slippery, and indeterminate.[24]

Allowing for this indeterminacy lets us see that one's commendable use of the scientific instrument, the child's approach to the banana as object of detection rather than snack, enacts ownership of knowledge about the banana. This is an example of Barad's "agential cut," whereby one possibility, and not another, is brought into being: what is cut in the elementary, high school, or college science classroom is each knower-and-known-about-thing.[25]

Given the huge body of critical literature on effective and equitable science education as a guarantor of positive futures for minoritized young people in the US, it's perhaps worth clarifying that the aim of this chapter is not to implicate the AFMs in the US production of "good learners"—that is, in the demarcation of eligible achievers—as a biased process. That description implies that standards exist by which an organization, such as US K–12 schooling, may "accurately" compare individuals' abilities, and that we might, with awareness, eliminate bias is such determinations. A more ontological lens is needed, whereby we can see how the ascription of relative technoscientific potential (i.e., merit) is neither anterior nor posterior to the identification of individuals by, for example, race or gender; rather, these characteristics are brought into being at once.[26] In an important sense, this recognition allows us to question the universalist idea of a postracial world in which racism is "a distortion of an otherwise aracial rationality," ready to be dispelled once the truly objective judgment of personal capacities is liberated from distorting forces.[27]

The focus on the ontological character of science teaching and learning—on the making of knowledge about people's talent, and people, as a single operation—also helps us disrupt the idea of educational opportunity as an empty location awaiting arrival or uptake by yet unspecified individuals. The proposition of such emptiness—purporting that any given human might arrive at any institutional or social position, depending only on their innate endowments—falsely claims an essential democratic nature for US economic systems. It is not mistaken notions regarding some racial bases of intelligence, nor gendered ideas of mental discipline, nor ableist conceptions of mind-body relations that falsify those claims, but rather the unified nature of individual intellectual endowment and identity US culture.

Put another way, in keeping with this volume's interrogation of the ownership of knowledge, rather than formulate the democratic ideal of science education as an "even playing field"—an image of figures of diverse identities given the chance to move freely against an unchanging, neutral ground (and the classic analogy for a bias-free society)—we might see that educational opportunity is instead a *location of ownership*. That is to say, opportunity, embodied in the next level of science instruction, in the eligible student's next possible encounter with knowledge of remote-sensing, is specified *with* the eligible actor. It does not vacantly await the meritorious, but instead conditions the meritorious. If the sociabilities of US learning and work—still vastly disadvantageous to minority communities in 2022—concern us, we need to see that a child's future that includes a more expensive remote-sensing apparatus or a more elaborate physics syllabus is not best seen as an experience of more knowledge but as the making of a proper person, a propertied person.

For generations in the United States, science curricula have been elaborated in massive guideline-setting enterprises, most recently 2013's Next Generation Science Standards (NGGS), a collaboration among twenty-six states following a framework supplied by the National Research Council.[28] The NGGS are explicitly based on the premise that the acquisition of scientific knowledge by US students will democratically support personal and collective prosperity.[29] Thus, the science classroom is meant to serve as both driver of student engagements and source of reward and promotion, according to the performance of each student. In this way, science pedagogy guides students in maintaining their motivation to learn about the world "not just in school but throughout life," and to expect rewards for acting on those impulses.[30] This is a braided imaginary of available empirical knowledge, innate endowments, cognitive agency, and postschooling rewards. It requires that the figures of students are seen to move against the ground of a knowable (for some) world. Unpacking this imaginary, the remainder of this chapter follows continuities and ruptures among student experiences of the three "levels" of AFM mastery—from shoebox to LEGOs, to commercially supplied instrument—to see how a world and people who can (and cannot) know it come into being together in US places of science education.

SCHOOLED FUTURES: THE HISTORICAL MAKING OF KNOWERS AND NONKNOWERS

Hands-On Learning: Whose Hands and Whose Learning?

Two broad historical developments in recent decades have intertwined to land the banana inside the punctured shoebox: the renewed centrality of "hands-on" learning in K–12 science education and the rise of atomic-scale science and technology in projections of US economic growth. The integration of "hands-on" and lab-based experiences into US science education has been an identifiable priority for educators since the last quarter of the nineteenth century, sidelined at moments in favor of more theoretical content but never terribly widely or for very long.[31] Within areas delineated for such instruction, modeling and materials characterization have been part of US science education for generations, but students' facility with microscopic imaging and manipulation gained new importance among such experiences with the rise of so-called nanoscience and other atomic-scale applications at the end of the twentieth century. That rise produced strong pedagogical and economic arguments for "nano-related" content in K–12 curricula. It is important to avoid the formulation that classroom experiences of scanning probe microscopy represent rudimentary forms of professional laboratory conducts in any singular or direct sense. We can review the landscape of variable selves and futurities—emplacements through ownership of knowledge—that science schooling in the US has projected since the turn of the twenty-first century. Here, we focus on a set of possible moves toward cognitive and economic self-development for some individuals that constitute nanoscale science learning.

As Rudolph and others have made clear, since the origins of formal schooling in this nation, the education of US adolescents in areas of science and technology has included hands-on experiences. The term has encompassed student experiences loosely defined in opposition to "passive" listening to lectures, watching instructors or aides conduct demonstrations, or reading textbooks. In the late nineteenth century, the take-up of German research ideals in US universities prompted influential high school educators in the country to introduce laboratory work to younger learners. Through the 1910s and 1920s, as Rudolph recounts, widening public high school enrollments and teaching experts' enthusiasm for tenets of Deweyan pragmatism lent still greater value to material engagements with scientific methods.[32] Although laboratory exercises have been understood by many generations of science educators to be a particularly engaging form of pedagogy for learners, we can be very clear that they have also historically shared social instrumentalities with other bodily experiences in science learning, such as fieldwork in botany or zoology, or survey camps for engineering students. All such experiences helped inculcate students' sense of themselves as possessing particular sociabilities: masculinity, whiteness, manifest heterosexuality, ablebodiedness, and other forms of "belonging" in places of science or engineering learning.[33]

This experience of belonging maps onto hierarchical occupational and wage structures that have served US employers' needs for human capital, following lines of race, gender, ethnicity, [dis]ability, and socioeconomic status, among other categories. Science and its applications are customarily seen by educational proponents to categorically promise advances in human welfare, meaning that all science-related or science-derived labor has some purported value, but ladders of opportunity and achievement remain integral to the pedagogy. After the emergence of wide networks of trade and vocational schooling that could follow or replace high school in the 1910s and 1920s, educators formalized programs reflecting stratified technical labor with particular ambition. This pattern was repeated in the 1950s and 1960s with the advent of junior and community colleges offering sub-baccalaureate education to students after they complete high school.[34] Tellingly, "middle-ness" has been delineated in each of these episodes as a permanent condition of some work and some workers. Technicians operate at a "middle" place between lesser and more skilled personnel; in the United States the "middle skilled" represent a frequent target audience for sub-baccalaureate STEM training in the twenty-first century.[35] In this system of discrepant experiences of economic and career mobility, getting one's hands into or onto the materials involved in technoscientific labor has historically never represented one thing, pedagogically. On one level, instructors in different sorts of schools have meant different things by the term *hands-on*. For those taking courses in academic high schools, who are generally considered more likely to reach college, hands-on learning might mean work with laboratory instruments for future scientists, or experiences with testing apparatus for future engineers. For those engaged in vocational instruction, hands-on work meant and still means a sort of manual labor strongly demarcated from cognition—literally manipulating industrial materials or medical, mining, or construction technologies preparatory for paid work of that exact sort.[36] But we also find the making of a multiplicity of life outcomes underway within a single K–12 classroom where all students are engaging in the same exercise, because different hands rest on different matter, even at the same workbench. The student who, dismayingly, eats the banana in the shoebox is handling food and thus fails the exercise; the student who, appropriately, remotely maps the banana is handling a scientific specimen and passes. More subtly distinguished errors and achievements follow to delineate students and futures at subsequent grade levels, perhaps, but the linkages of knowers and the known-about persist (just as, ultimately, the factory quality-control inspector does not examine a microchip as a piece of experimental evidence, and the industrial researcher does not examine it as a piece of inventory fit or unfit for sale).[37] The purported continuity of K–12 science education—its ostensible meritocratic futurity—masks the classificatory function of elementary and secondary school instruction, meant to cull those incapable of college from the overall pool of scientifically literate future workers—that is, to produce nonknowers, or, we might say, nonowners of the particular knowledge associated with particular futures.

Functioning in a promissory mode, scientific literacy appears as an immersion in knowledge that is imagined to be continuous over time and not susceptible to interruption. The planners of the NGGS described their priorities thus:

> The framework is designed to help realize a vision for education in the sciences and engineering in which students, over multiple years of school, actively engage in scientific and engineering practices and apply crosscutting concepts to deepen their understanding of the core ideas in these fields.[38]

The scientific knowledge accumulated by meritorious students during this passage of time is seen to be particularly durable and mobile, cast as a combination of "content," and "practices" that carry "core ideas" from location to location. Both "content" and "practices" are to be "intertwined in designing learning experiences in K–12 science education":[39]

> Student performance expectations have to include a student's ability to apply a practice to content knowledge. Performance expectations thereby focus on understanding and application as opposed to memorization of facts devoid of context.[40]

As was the case a century ago, those students in the US seen to have the most far-reaching potential for performing technical labor are seen today to require something beyond facts. As in past decades, arguments today about the economic utility of schooling and requirements of national security require that a cutting-edge of research practice be delineated and become the basis of teaching in both public and private schooling.[41] Since the mid-1990s, so-called grade-banded techniques have sought with particular focus to instill facility with techniques of detection, imaging, and modeling, and in this way, microscopy has become a central area of student learning for livelihood.

Educating for a Nanoworld

Microscopy and telescopy, and indeed, a great many science techniques including most involved in chemistry and physics, depend on detecting the so-called unseen through instrumentation. Despite this genealogy, aspects of scanning probe microscopy entered K–12 science curricula after 1995 or so as features of what had come to be understood as "next-generation science and engineering practices," seemingly novel in both form and economic import for the nation. This was the point at which practical atomic-scale science and engineering received considerable attention from the National Science Foundation (NSF), which lent its imprimatur to scientific research and enterprise loosely defined by the terms *nanoscience* and *nanotechnology*.[42] The terms referred very broadly to atomic- and subatomic-scale operations emerging in biotech, medical, materials, electronic, energy, and other sectors, and with this governmental recognition of research and development at unprecedentedly small scales came tremendous enthusiasm for projected discoveries and applications, including the creation of mass manufacturing operations claimed to bring wide employment.[43]

The enthusiasm for all things nano- went hand in hand with a view that equated new scientific knowledge with economic expansion. The founding of the National Nanotechnology Initiative (NNI) in 2000, for instance, indicated "focused investment" by multiple federal agencies supporting dozens of individual programs around the country, including both pertinent scientific research and research on science education.[44] The novel character of nanoscale science was established in conjoined technical and political terms; the notion of vast human benefits in health, environmental sustainability, and other arenas to be wrought by the "revolutionary" nature of nanoscience and nanotechnology made sense of a new vision of "nanolabor." This vision of a new prosperity for workers under capitalism was said by proponents to promise millions of new jobs in the "nanosector" within a decade or two.[45] Related allocations of NSF funds for curricular initiatives meant to prepare the nation's future workforce carried outward a general sense of urgency around the current, ostensible nonpreparedness of US students for emergent economic conditions and program development in many schools systems, at all levels. Students in elementary, middle school, and high school soon began to receive instruction on "real-world" applications of nanoscale knowledge, including regarding composites, ceramics, concrete, biosensors, electronics, polymers, pharmaceuticals, and other topics. Educational encounters via nano-focused lab kits, student-grade instrumentation, video and animated materials, board games, and other formats proliferated to prepare students for promised employment opportunities in nanoscale science and engineering fields.[46]

The rapid uptake of nanoscience and nanotechnology as bases for K–12 and postsecondary curricula in the US expressed many sorts of futurity for their proponents, often tied to regional economic concerns but inseparable from beliefs about national security and global economic competitiveness, as in the NSF funding strategies. That relatively few jobs in "nanomanufacturing" currently exist in the United States suggests that the scale of NNI-funded initiatives may have been based on extravagant projections.[47] Moreover, the positive role projected for nanotechnology sectors in the lives of students conflates persuasive pictures of economic growth and of oneself moving upward to accrete skills and knowledge. But as we discuss below, meritocratic concepts of science-based competence are nothing if not adaptable to less fortunate, and even unfortunate futures for some students.

The vision of an economy requiring particular scientific and technological competencies was invoked to support educational reform efforts after 2000 that enlisted empiricism in the service of capital yet again, now through coordinated projects of quantified assessment aimed at individual teachers and classrooms, schools, and the local- or state-level administrative entities that oversee schools. Through national testing initiatives, the gauged effectiveness of K–12 education allowed authorities to compare individual instructors, schools, and districts, just as classroom testing had long allowed comparison among students. Scholars have shown that multiple forms of funding depended on districts' commendable test performance, rather than on evidenced need for greater resources. This pattern strongly suggests that the further marginalization of disadvantaged communities was an acceptable outcome of the national testing initiatives, if not an acknowledged goal.[48] As Antonia Darder describes:

> Just as educational reform efforts of the civil rights era began to reap some promising outcomes in the late 1970s and early 80s, with improvement in educational outcomes for the most impoverished communities and an increase in college and university attendance by historically underrepresented student populations, the conservative antics of the Right revived their bitter campaign to discredit progressive educational efforts, advance the privatization movement, and usher in some of the most Draconian accountability measures in the history of US education. This, in turn, led to the most expansive national high-stakes testing campaign ever, aggressively solidified by the federal passage of *No Child Left Behind* by the Bush administration in 2001 and its transmutation to *Race to the Top* (RTTT) by the Obama administration in 2009.[49]

The perceived reasonableness of a national education system that relegates some communities to much less occupational preparation than others meshes in the US with centuries-old majority understandings that race and other ascribed identifications rightly determine life circumstances.[50] More apposite perhaps is that through these neoliberal formulations, an economy of profoundly stratified wage and mobility structures was firmly associated with responsible, evidenced-based oversight of national educational provisions, and that both were predicated on avowals of accountability. Examining particular moments in which talent and opportunity, as organizing epistemics of twenty-first-century STEM education in the US, have been cospecified in science classrooms can help us see these structural conditions in which knowledge finds its owners.

Claimed Epistemic Continuities, from Skewer to Electron

The three scanning devices considered in this chapter evidently engage with natural phenomena on three different physical scales. A wooden or metal barbeque skewer and a laser pen, moved over the surface of objects under scrutiny, will not reveal characteristics of the same size, and neither of these tools operates at the revelatory scale of the scanning electron probe embedded in the atomic force microscope. But educators claim considerable epistemic continuities for these three instruments, and such continuities are elemental to the promise of personal development made to many students. The near destination of a completed classroom exercise and the far horizon of college, graduate school, and professional employment in the sciences both mean that all cognitive engagements can be cast as learning or its absence—that is, as resulting in the student's acquisition of knowledge, or not. Naturalizing the image of a self moving from having not much demonstrable understanding of the world to having greater such understanding when faced with the possibility of acquiring knowledge, these claims make up the social instrumentality of "mastery" on which staged education in the US is predicated.

This enactment of mastery rests, first, on the assumption that knowledge of the world consists of knowing the world as data. All three scanning devices involve the proposition that scientific knowledge of the world is continuous and thereby subject to accretion, making reasonable the idea of expanding understanding in individuals. The very construction of scientific instruments as yielding "no," "some," or "more" information unites them as objects operating in relation to a singular cosmos, as directly representational in the unmediated sense that historians of science and science studies scholars problematize.[51] Instruments are seen to produce demarcated but not separable bodies of data because the universe they address is, according to the norms of Western science, not a disunified one; what the eleven-year-old finds out about the world is not incompatible with what the PhD molecular scientist finds out—merely, it is maintained in scientific settings, somehow lesser in amount. The stepwise character ascribed to the scientific method, in which observation and testing yield knowledge of that universe, works cognitively for the competent inquiring mind, however calibrated the tool one holds might be. Measurement and other uses of instruments produce one set of "convergent claims of decontextualized truth," as Patel paraphrases Spivak's characterization of Western epistemologies.[52]

In "Exploring the Nanoworld with LEGO® Bricks," a set of K–12 teaching materials on scanning probe microscopy (SPM) technologies produced by the University of Wisconsin in 2012, teachers are told that "by mapping [surface forces of materials on the atomic scale] much can be learned about the surfaces of materials, where many interesting and complex phenomena occur."[53] SPM includes atomic force microscopy, magnetic force microscopy, and lateral force microscopy, all "variations on the same basic principle": "Forces between the surface and a cantilever tip cause the tip to deflect up and down." The play of tip over surface (or substrate) is the meaningful interaction here. Not unlike those moving into high school from middle school, we can likely grasp the significance of such deflection by recalling the calibrated skewer playing over the surfaces of the hidden banana. Here, however, deflections are not transferred from banana to skewer to child operator's hand and eye, but from LEGO substrate to LEGO cantilever to laser-pointer beam, and movements of the beam are then recorded by a photodiode array. The resulting data is then studied by adolescent operators. As with the shoebox apparatus, good students are presumed to grasp, or to be able to learn to grasp, what about the model will carry forward in time as they move onward from their status as learners to workers, and what will remain behind. Instructions clarify that "*as in a real AFM* [my italics], the cantilever is held in place and the surface is moved back and forth underneath the probe." By contrast, neither LEGO bricks nor other elements of the apparatus ("Here, the *refrigerator magnet* at the end of the cantilever interacts with a *refrigerator magnet* taped to the LEGO® surface to alternately attract and repel the cantilever" [my italics]) have any place in the postpedagogical science world. The nature of pedagogy, the aim of personal cognitive growth, and indeed the value of accepting mock-ups and models as stand-ins for more legitimate things are parts of the lesson.[54]

For the authors of "Exploring the Nanoworld," the modularity of LEGO bricks—the possibility that one can build any shape from the pieces—seems to add to the toys' appropriateness for teaching nano-related science. Like atoms, LEGOs may be used to represent any and all imaginable things: "A set of bricks can be used to model structures of matter and the techniques used to study them."[55] Commensurable with atomic models, LEGO bricks imply a unified character to all objects of study; the intended object of representation can present no impediment to the bricks' use as representational medium. But if the standardized, modular features of LEGO bricks reflect a constitutional commonality among all scientific objects of study, what distinctions among these objects are to be revealed, to be attended to, by the modeler? What constitutes the successful use of LEGO bricks as scientific activity?

We will return below to the formulation of bad knowledge about bananas and LEGO bricks and world, but it is important to be clear that once they take the form of a classroom AFM apparatus, LEGO bricks are both autonomous and well protected from any mistaken association with frivolity. The bricks have a complex instructional credibility: they are very mobile actors, at home not only in a toy store or playroom, but also in a STEM classroom, education research program, NSF grant proposal, and many other locations in which US science education initiatives take form and achieve credibility. The LEGO-brand AFM is in fact part of an immense global initiative of corporate engagement with STEM literacy, operating in direct service to industry. Integral to a broad public/private partnership centered on "21st century readiness for every student," LEGO personnel and other corporate educational leaders advocate for federal and state K–12 STEM initiatives closely tied to projected workforce needs.[56] That workforce is of course not one of equally capacitated, or rewarded, workers, and so those who learn with LEGO bricks will somehow be differentiated from one another even as they remain enlisted in forward motion, in learning.

As educators enact this system of teaching and learning for mastery, the conception of knowledge as data correlates with a specific understanding of *what there is to be learned* in the world. The revelatory functions of the shoebox- and LEGO-based AFMs, like that of the $40,000 AFM, orient the capable operator of each toward useful/factual understandings; matter is willing to reveal itself to the right human partner. The three detecting instruments are all premised on some specific shared elements of observation undertaken in the scientific manner—that is, of attention to the world undertaken in support of inferential learning. As Warren et al. point out, experimentation is taught to children as a process of logical inference rather than as one that is open-ended, or geared toward constructing meanings for emerging variables.[57] All three AFMs deploy incremental, systematic description of the unseen in order render the unseen visible. The good student is drawn into an act of temporary faith (an orientation of "stick with it and knowledge will be had") and a willingness to sustain attention. The weak student drifts from the task or otherwise comes to infer nothing and, possessing no recognized knowledge, goes nowhere—or possibly goes somewhere, but not in the direction of science mastery.

The nature of optimized science pedagogy in the United States institutionalizes these binary dispositions of learning and not learning, as in what the NGGS define as "basic understandings about the nature of science," expressed on a matrix of grade-appropriate language for students in grades K–2 (roughly ages 6 to 8), grades 3–5 (ages 9 to 11), middle school (ages 12 to 13), and high school (ages 14 to 18). These "basic understandings" include such directives as "Science Is a Way of Knowing" that demarcate science from other activities and separate knowing from other ways of sense-making (such as pleasure, pain, yearning, art, or love, for example).[58] Another basic understanding, "Scientific Knowledge Assumes an Order and Consistency in Natural Systems," is expressed on the NGGS matrix of grade-appropriate formulations as follows:

[K–2] Science assumes natural events happen today as they happened in the past. . . . Many events are repeated.
[3–5] Basic laws of nature are the same everywhere in the universe.
[Middle school] Objects and events in natural systems occur in consistent patterns that are understandable though measurement and observation.
[High school] Scientific knowledge is based on the assumption that natural laws operate today as they did in the past and they will continue to do so in the future. . . . The universe is a vast single system in which basic laws are consistent.[59]

Two points bear emphasis, as I read this rubric. First, we can note that there is no possibility in the above descriptions to consider that the ways in which such "basic laws" are formulated are themselves historical, changing with our changing understand of nature. Second, the similarity among the statements for six- through eighteen-year-olds here is striking. It suggests that science mastery is both staged and additive (where it indeed occurs); no knowledge is subject to being taken away from the individual who possesses it, even as shoeboxes and LEGO bricks are replaced by laboratory-grade apparatus. The rhetoric of this pedagogy often takes the consistency of knowledge across time and across different levels of learning to be fundamentally grounded in the assumed consistency of the very objects it studies; as one teacher's manual puts it, students shall learn "what a nanometer *is* [my italics]."[60] For the meritorious learner, the world's spatial and temporal continuities assure that a sustained disposition of inquiry cannot result in anything but more knowledge.

Many ontological functions are achieved by this projection of a world waiting to be known. Historians of education and human capital, some following Foucault, have identified the moral and mental disciplining on which modern schooling depends, and historians of science enrich our understanding of research as a further disciplining experience.[61] Nanoscale science pedagogy in no way departs from these regimes of student self-control. Inquisitive individuals not only will take up a focused position regarding inquiry but also remain comfortable with the existence of temporary unknowns; "good science pedagogy" produces not only questions that are not easily answered, but also some that are unanswerable.[62] Observation in places of science should be unimpeded by impatience or other inappropriate impulses that may supersede the intention to learn (such as, say, fatigue, hunger, or rage), and it should be norm-based in all possible respects; there are right and wrong ways to hold the body, the eye, the mind in proximity to the object of study. Only certain "kinds" of people hold the potential to assume these relationships with matter.[63]

With our close study of classroom instruments as expressions of staged mastery we can follow specific techniques for the production of differences supposedly inhering in individuals. In the K–12 science classroom, instructional experiences are organized so that students hear the message over and over: "Here are the necessary actions for a person of your stage in order to become eligible for the subsequent stage. . . . Can you take these up?"[64] Since the system requires incapacity in some individuals, teachers and learners understand that an answer of "no" to that question is always a possibility and exclusion always a possible outcome of effort. At the same time, the filter is meant to let some through; the high schooler is not faced with a research-grade AFM but rather a construction of LEGO bricks that, intentionally, "does not intimidate or frustrate" the competent student unnecessarily.[65] Those who pass through to become college undergraduates are taught via staged presentations of the concepts behind and operational features of research-grade AFMs, with the sensation of challenge again carefully titrated. In other words, science education posits that knowledge cannot circulate apart from its possession by humans, some of whom are incapable, and others of whom are capable, of knowing. No student, no person, evades placement on that totalizing scale of capability. Overall, the purported learnability of scientific techniques and findings establishes individuals deserving of more educational opportunities (those seen to learn in a classroom) and those who manifest disqualifying deficits (those seen not to learn).[66]

The Not-to-Be-Knower and the Not-to-Be-Known

That sorting simply could not happen if only learners were being made in places of science schooling. The to-be-learned must also come into being. Experimentation in school, as in settings of professional science, derives from that possibility of restricted material essences: Students need "the ability to conduct an investigation where they keep everything constant while changing a single variable."[67] The "variable" is of course merely one that is subject to swapping for certain alternatives, not one that varies in its essence. Laboratory enactments included in the process of schooling tell us about individuals directly, signaling their presence in the classroom, their willingness to grow, and their intellectual ability as ingredients of scientific merit. But they thus also tell us about the solidity and predictability of the cosmos to be known by the meritorious.

As critical scholarship on enlightenment ideologies and coloniality have shown, those features of the world are necessary if Western scientific learning is to be posited as independent of culture and maintain its self-confirming political functions.[68] We might now add to such power relations that such behavioral constancy on the part of the world is necessary if students are to be understood as more or less *able* (and with fitting levels of education, *enabled*) to detect the world—that is, if performance metrics for efficacious citizens are to seem reasonable (building on Ashley Taylor's powerful framing of "knowledge citizens"[69]). Biesta, reflecting on Dewey, gives us a basis for seeing this relational character of education as it currently restricts our view of students as empowered subjects of "action and responsibility." For Dewey, student experience and, accordingly, an immediacy or flexibility to all pedagogy supersede in importance student acquisition of particular pieces of knowledge. To this image of dynamic learning, Biesta adds:

> Whereas many would argue that the prime function of schools is to create a common outlook so that future collective action becomes possible, Dewey suggests that schools should instead focus on the creation of opportunities for participation for such a shared outlook to emerge. . . . [Yet the] creation of a shared outlook will not result from simple coexistence or from forms of pseudo-participation in which the activity is set and controlled by others.[70]

Thus, solidifying both knower and known-about in the science classroom involves exclusions of unauthorized subject (i.e., student) activity, of illegitimate knowing.

The science student, swathed in a positivist epistemology, is drawn into the existence of both cognitive possibilities and impossibilities; only that which is deemed potentially seeable, revealable, can possibly exist. As Mukharji makes clear, the rise of modern "oculocentric science" dependent on visualizing and visualization helped displace any knowledge that involved occulted powers and forces—a historical development reflecting the "privileged ontology of the west."[71] This privileged ontology populates the NGGS grid that tabulates the "nature of science" for learners at all levels. In that grid, we find graded expressions of the proposition that "Science Addresses Questions about the Natural and Material World," as follows:

[K–2] Scientists study the natural and material world.
[3–5] Science findings are limited to what can be answered with empirical evidence.
[Middle school] Science limits its explanations to systems that lend themselves to observation and empirical evidence.
[High school] Science knowledge indicates what can happen in natural systems—not what should happen. The latter involves ethics, values, and human decision about use of knowledges.[72]

Uncertainty plays a profoundly political role here. In particular presentations, uncertainty signals the "not yet known," a safe object of attention, and in others, uncertainty indicates that attention is being paid to the problematic, to "not knowable" or impossible worlds. It might be helpful to bring such a notion of "multi-natures," as Viveiros de Castro terms it, to the study of minority marginalization in US technoscientific places.[73] As a very small step in that destabilizing project, it is worth thinking about how nano-focused K–12 education articulates the two distinct versions of uncertainty.

The first version might be titled, "The Bag Remains Closed!" This version of uncertainty concerns the "not-yet-known" and those eligible for reaching knowledge; it projects a valorous disposition towards the accumulation of knowledge. Consider an exercise meant to prepare grade schoolers for their encounter with the punctured shoebox AFM, as described by Jones, Falvo, Taylor, and Broadwell in their "nanoscale science activities for grades 6–12." The instructor is to conceal an object unknown to the students (a model configured of LEGO bricks is recommended) in a sealed, fully opaque plastic garbage bag. Student teams are asked to use as many nondestructive methods as they can to try to determine features of the hidden object; this might include feeling through the bag, or using magnets "to glean material information." The student teams are equipped with a second set of LEGO bricks, outside of the bag, that they can use to make a known model of the (temporarily) unknown. Since "discussion and consensus building" are significant parts of the project, in the next stage of the exercise multiple student teams hold "conferences" to compare and reconcile their findings.

The exercise is meant to imitate the ways in which scientists can come to know the world, and in which they shall contribute to their own or others' well-being through extremely specific sorts of efforts. In a "Helpful Hint" for teachers piecing together the hidden models, the authors note that "it is best to err on the side of making [models] more complicated rather than less. The more challenging the modeling of the unknown, the more effective the exercise." Making the unknowns so complex that students will be "unlikely . . . to reproduce them exactly" is said to produce a still "richer and more interesting exercise, as well as being a better analogy for real scientific work." Here again, patience and fortitude are invoked as valuable characteristics for science learners. But even more interesting, perhaps, is that this educational exercise ends with "The Last Lesson: The Bag Remains Closed!" because "scientists don't get to 'open the bag.'" That is, "We never have complete knowledge of a system under investigation," yet the pursuit of such completion is reasonable and commendable:

> The nanoscale of molecules, viruses, and DNA is a reality of which we have increasingly refined view using the latest technology but there are large gaps in our knowledge. Much remains within the bag. Much remains to discover![74]

The future belongs to explorers: those who not only seek to learn what is in the black bag, and believe its contents to be knowable, but also feel no urge to attack it with a scissors or find some other path of lesser resistance into the future. Keeping the bag closed represents the appropriate, credible condition of investigation in this instance. Surely innovation in science involves "breaking rules" at times, but in each setting of scientific learning or investigation, some standard of rigor (whether centered on objectivity, precision, replicability, or other normative expectation) must operate for activity to be recognized as science (and not, say, fraud, or fantasy), with attendant social effects.

The second version of uncertainty (and thereby what shall constitute scientific certainty), by contrast, points to a less valorous orientation towards the unseen or unexplained. Were our nano-educators to describe this form of nonknowing, it might be headed: "Leaving aside the Black Bag." For example, Jones, Falvo, Taylor, and Broadwell also provide true/false questions on nanoscale science findings to guide students away from any temptations of fantasy or empirical overreach while maintaining their forward-facing stance toward learning: "Gold nano-sized balls can be injected into the body to destroy cancer cells . . . *True*; . . . Scientists have created a nano-sized car that has four doors, tires, and tiny seats and can move around freely . . . *False*."[75] We can put aside for the moment the somewhat fantastic tone that the false statement lends to the true one (awe and wonder are perhaps not as easily controlled as the authors presume). More to the point of the arguments being made here is that the entire dichotomy of real and unreal science marginalizes what Mukharji frames as "potentialities [that] could never be fully or exhaustively witnessed," conditions or experiences that are "unavailable to naturalistic interrogations."[76] The social consequences of this marginalization are multiple. The nature of knowability is narrowed in order to exclude particular people.

Many scholars of indigenous epistemics and cognitive diversity have made this prejudicial function of intellectual credibility abundantly clear; the notion of epistemic rigor itself depends on defining the membership of a given epistemic community, as my own research on the whiteness of US engineering has tried to show.[77] But the exclusion of potentialities has additional worldly effects. The true/false binary, deployed by educators in support of naturalistic interpolation, also rejects the possibility that the surface of a material might be characterized, say, genealogically or politically. The surface of the nanoengineering device, developed for mass production, cannot reasonably be detected in the nano-classroom as involving "concern" and "regret," such that risks to the health of the nano-factory workforce or to those living downstream of the factory become evident. The barbeque skewer cannot find the conditions in which the banana was grown, the pollution caused by pesticides, or the low wages of farmworkers to be "facts" of the banana.[78] Science might provide other devices for such investigative purposes, but for the learner presented with the "necessary tool for the research," the tool determines the research to be done.

This rather simplistic set of examples at least helps us see why, as I said earlier, the idea of expanding STEM participation for minority communities is a problematic goal if what we actually seek is a more just society with fewer differentials in life circumstances. Critical literature on the inequities of US STEM education makes clear that marginality historically has not been incidental or in any helpful sense prior to schooling, but systematically produced by schooling.[79] The study of the three remote-sensing devices helps us see the mechanics by which faith in the possibility of possessing knowledge is in no sense equivalent to eligibility for the possession of knowledge, let alone to possession itself. The idea of "kn/own/ables" grounding this volume captures well how the particular imaginable ways of knowing in a given historical setting preset the possibility of possessing knowledge, which also entails dispossession. The description of learning as incremental, celebrated by Dewey, may in fact hide the role of that incremental nature in demarcating those without merit. That is to say, US schooling and work comprise a system in which facility itself constitutes a technology of oppression, something that is apportioned in ways that naturalize different life attainments. As Patel has written,

Democracy is often conflated to actualizing equity, with assumptions that democratic processes can only result in more equity. This frame thereby situates ruptures of equity and injustice as problems within a society rather than as architecture within the tenets of the society itself.[80]

The production of science knowers and things-known-about that constitutes K–12 education in the US surely represents such architecture.

CONCLUSIONS

A central point of this chapter is that the lack of dependence among learner encounters with the $4, $40, and $40,000 AFMs in US science classrooms in 2022—the foreshortening of this sequence, in many instances—is not well explained by any accounting of missing opportunities for minoritized students, missing people of color in the places of science education, or not-yet-mastered knowledge among certain groups of aspirants. That comforting imaginary of potential learning awaiting actualization is captured in well-worn tropes of the "leaky pipeline" in science and technology education, which depict eligible young learners dropping out of occupational contention due to local issues of underresourcing or discrimination. This picture of STEM fields conveniently elides the ways in which science literacy enacts and necessitates marginalization.[81] Instead, the proposition of existing, possessable knowledge about a singular world, a possession to be obtained in part through science education as we know it, is historically an adjudicating system, an ontological production of scientific minds, but also of "valueless understandings" and thus "nonknowers," or, we might say, nonowners of knowledge.[82]

Perspectives drawn from science and technology studies concerned with the relational nature of technoscientific knowledge, such as we find in Law and Lien, may be helpful here.[83] Specifically, they can help us to recognize insistent indeterminacies regarding human difference with which onto-epistemological projects grapple in the science classroom. We see how US science education is predicated on the making up of people with set and identifiable inclinations, conducts, and prospective bodies of knowledge regarding a fixed and knowable world.[84] This is the premise on which notoriously unjust deficit-based models of student difference are founded, as Freire articulated. It has also lately produced possibly well-meaning, inclusive STEM diversity interventions that for the most part neither interrupt the violence enacted by meritocracies nor address structural conditions generally.[85] But so, too, do notions of knowable individuals as disrupted by queer studies, and ideas of measurable human capacities problematized by disabilities studies, remind us that education and resultant instances of knowledge ownership express much wider cultural commitments in the US to ordering individuals' prospects in service to current distributions of capital, legal security, political influence, and other advantages of the nonminoritized.

With those commitments in mind, we might come to see the relationality that historicizing educational opportunity brings to the surface, yet without falling back on the duality of classified people/apparatuses of classification that I think configures Hacking's outlook on the "making up people."[86] For example, the idea that in a racialist, ableist, sexist, heterosexist, xenophobic society, it is *identity* that constrains *opportunity* badly mistakes the nature of both. Instead, we can grasp that opportunity is not separable from either "potential" or "achievement." Those latter terms exist *in order to* valuate people in light of particular societal aims—again, to cast society as a collection of figures moving against a single, unchanging ground to greater and lesser effect depending on each figure's capacities. In the case of science learning, students are in actuality constructed as figures against the ground of a world that is knowable by some and not others. Identifications such as race, gender, sexuality, and (dis)ability operate throughout, not as a priori determinants of opportunity or post hoc ascriptions following from individuals' perceived relative attainments.

Writing about the interpretive possibilities of actor-network theory (ANT) for a newly critical understanding of learning, Fenwick commends ANT for tracing networks as they produce "force and other effects: knowledge, identities, rules, routines, behaviors, new technologies and instruments, regulatory regimes, reforms, illnesses and so forth." She welcomes ANT's premise that "nothing is given in the order of things, but performs itself into existence" and its constant questioning of any network about "what is holding its system together."[87] I would suggest that in the case of the twenty-first-century United States, the project of emplacement/ownership we have seen in the teaching and learning of remote-sensing is one answer to that question. Generations-old patterns of stratified education and employment would of course be a sturdy enough foundation for any such distributive efforts in 2022, any such difference-making among people. But the location of scientific merit in individuals has recently achieved a new efficacy. Following a period of civil rights reforms that legally prohibited segregation, as the nation turns to the comprehensive criminalization of people of color and the foreign-born, the sorting of individuals—the classificatory projects of racism and related dispositions—continue through these more conciliatory means.

The alternative to the three AFMs as we have met them might be envisioned as a liberatory pedagogy of knowers-who-are-not-knowable, of "caring yet vulnerable and risky relations" between teachers and learners, or employers and employees. This stance might involve crediting all sorts of presently illegible knowledges of the world, such as schoolchildren's everyday knowledge, or presuming competence in order to demolish "archaic distinctions between people with or without learning difficulties."[88] Decentering the humanist achievements ascribed to education in a self-consoling democratic polity also recommends itself; this is required if self-determination is to be honored as an alternative to self-efficacy, to the "grit" of the dedicated STEM learner.[89] Whatever our critical approach, if we are determined not to reproduce the oppressions that the notion of alterity enacts, we can recognize the inescapably ontological character of science learning and knowing—its making of people and what it is they know and can know, all in one operation. The conditions of learning and work, and the world to be learned and worked with, this way stay together in the analytic frame. But the power of the assemblage is no longer denied.

Notes

1. M. Gail Jones et al., *Nanoscale Science: Activities for Grades 6–12* (Arlington, VA: NSTA Press, 2007), 54–57; Yvonne S. Kao, Anthony Cina, and J. Aura Gimm, "Inside the Black Box," *Science Teacher* 73, no. 9 (2006): 46–49.

2. NGSS Lead States, "Executive Summary," in *Next Generation Science Standards: For States, by States*, June 2013, https://www.nextgenscience.org/resources/ngss-introduction-and-overview; Susan Loucks-Horsley et al., *Elementary School Science for the 1990s* (Andover, MA: NSTA Press, 1990), ix; National Research Council, *A Framework for K–12 Science Education: Practices, Crosscutting Concepts, and Core Ideas* (Washington, DC: National Academies Press, 2012); David Kaiser, "Introduction Moving Pedagogy from the Periphery to the Center," in *Pedagogy and the Practice of Science: Historical and Contemporary Perspectives*, ed. David Kaiser (Cambridge, MA: MIT Press, 2005).

3. Kenneth Turner et al., "Seeing the Unseen: The Scanning Probe Microscope and Nanoscale Measurement," *Science Teacher* 73, no. 9 (2006), 58–61.

4. Christina V. Schwartz, Cynthia Passmore, and Brian J. Reiser, *Helping Students Make Sense of the World Using Next Generation Science and Engineering Practices* (Arlington, VA: NSTA Press, 2017); Dean Campbell et al., "Exploring the Nanoworld with LEGO® Bricks," Bradley University, 2012, section 2–1, 17–19, accessed September 4, 2022, https://chem.beloit.edu/edetc/LEGO/PDFfiles/nanobook.PDF. On the history of scanning probe microscopy, see Cyrus C. M. Mody, *Instrumental Community: Probe Microscopy and the Path to Nanotechnology* (Cambridge, MA: MIT Press, 2011). On historiographies of science pedagogy, see Kathryn M. Olsesko, "Science Pedagogy as a Category of Historical Analysis, Past, Present, and Future," in "Textbooks in the Scientific Periphery," ed. Antonio Garcia-Belmar et al., special issue, *Science & Education* 15, no. 7/8 (2006): 863–880, On the "staged mastery" of pedagogy as a modern project, see John Carson, *The Measure of Merit: Talents, Intelligence, and Inequality in the French and American Republics, 1750–1940* (Princeton, NJ: Princeton University Press, 2007), and Kaiser, *Pedagogy and the Practice of Science*.

5. Michael R. Welton, ed., *In Defense of the Lifeworld: Critical Perspectives on Adult Learning* (Albany: State University of New York Press, 1995). I take as exemplary here Simon Schaffer, "Accurate

Measurement Is an English Science," in *Values of Precision*, ed. M. Norton Wise (Princeton, NJ: Princeton University Press, 1995), 135–171.

6. Kristin L. Gunckel and Sara Tolbert, "The Imperative to Move toward a Dimension of Care in Engineering Education," in "A Critical Examination of the Next Generation Science Standards," ed. Troy D. Sadler and David E. Brown, special issue, *Journal of Research in Science Teaching* 55, no. 7 (2018): 938–961; Michael Lachney, "Building the LEGO Classroom," in *Lego Studies: Examining the Building Blocks of a Transmedial Phenomenon*, ed. Mark J. P. Wolf (New York: Routledge, 2014), 174.

7. For example, William C. Symonds, Robert Schwartz, and Ronald F. Ferguson, *Pathways to Prosperity: Meeting the Challenge of Preparing Young Americans for the 21st Century* (Cambridge, MA: Harvard University Graduate School of Education, 2011); National Academy of Sciences, National Academy of Engineering, and National Institute of Medicine of the National Academies, *Rising Above the Gathering Storm: Energizing and Employing America for a Brighter Future* (Washington, DC: National Academies Press, 2007); "Benchmarks for Science Literacy," American Association for the Advancement of Science/Project 2061, accessed August 5, 2018, http://www.project2061.org/publications/bsl/online/index.php.

8. For an overview of the historical function of education as actualizing potential, see Steven Bowles and Herbert Gintis, *Schooling in Capitalist America: Educational Reform and the Contradictions of Economic Life* (Chicago: Haymarket Books, 1976), 4–20; Christopher J. Phillips, *The New Math: A Political History* (Chicago: University of Chicago Press, 2015). Recent work has addressed how instructors and others in authority bring into being scientific interest and attainment in individuals concurrently with racial, gender, or other biological identifications; see Michael S. Dumas, "Against the Dark: Antiblackness in Education Policy and Discourse," *Theory into Practice* 55 (2016): 11–19; Kathryn L. Kirchgasler, "Dangers of 'Making Diversity Visible': Historicizing Metrics of Science Achievement in U.S. Education Policy," in *Handbook of Education Policy Studies*, ed. G. Fan and T. S. Popkewitz (Singapore: Springer, 2020); and Will Letts and Steve Fifield, eds., *STEM of Desire: Queer Theories and Science Education* (Leiden: Brill, 2019).

9. NGSS Lead States, "Executive Summary"; Gunckel and Tolbert, "Engineering Education," 940. Also, Peter Whalley and Stephen R. Barley, "Technical Work in the Division of Labor: Stalking the Wily Anomaly," in *Between Science and Craft: Technical Work in U.S. Settings*, ed. Stephen R. Barley and Julian E. Orr (Ithaca, NY: Cornell University Press, 1997): 23–52.

10. Stephen Secules et al., "Zooming Out from the Struggling Individual Student: An Account of the Cultural Construction of Engineering Ability in an Undergraduate Programming Class," *Journal of Engineering Education* 107, no. 1 (2018): 63.

11. Paolo Freire, "The Banking Concept of Education," in *Thinking about Schools: A Foundations of Education Reader*, ed. Eleanor Blair Hilty (Boulder, CO: Westview Press, 2011).

12. Andrew Pickering, "New Ontologies," in *The Mangle in Practice*, ed. Andrew Pickering and Keith Guzik (Durham, NC: Duke University Press, 2007), 4.

13. Karen Barad, *Meeting the Universe Halfway: Quantum Physics and the Entanglement of Matter and Meaning* (Durham, NC: Duke University Press, 2007), 18–19. Science education is of course

not the only means by which education sorts people; on the recent history of the production of individuals through learning, see, e.g., Gert Biesta, "Why 'What Works' Still Won't Work: From Evidence-Based Education to Value-Based Education," *Studies in Philosophy and Education* 29, no. 5 (2010): 491–503; and Gregory Hollin, "Failing, Hacking, Passing: Autism, Entanglement, and the Ethics of Transformation," *BioSocieties* 12, no. 4 (2017): 611–633.

14. Tara Fenwick, "Re-thinking the 'Thing': Sociomaterial Approaches to Understanding and Researching Learning in Work," in "Selected Papers from the 6th International Conference on Researching Work and Learning, Denmark, June 2009," ed. Henning Salling Olesen, special issue, *Journal of Workplace Learning* 22, no. 1/2 (2010): 104–116.

15. On the historical claims of education as an animating force of capitalist democracy, see Michael Young and Geoff Whitty, eds., *Society, State and Schooling: Reading on the Possibilities of Radical Education* (Ringmer, UK: Falmer Press, 1977); Bowles and Gintis, *Schooling in Capitalist America*; Amy E. Slaton, *Race, Rigor, and Selectivity: The History of an Occupational Color Line* (Cambridge, MA: Harvard University Press, 2010).

16. Slaton, *Race, Rigor and Selectivity*; Carson, *Measure of Merit*; Stephen Jay Gould, *The Mismeasure of Man* (New York: Norton, 1986).

17. See Carnegie Corporation of New York and Institute for Advanced Study, *The Opportunity Equation: Transforming Mathematics and Science Education for Citizenship and the Global Economy*, Commission on Mathematics and Science Education, Carnegie Corporation of New York and Institute for Advanced Study, June 2009, https://production-carnegie.s3.amazonaws.com/filer_public/80/c8/80c8a7bc-c7ab-4f49-847d-1e2966f4dd97/ccny_report_2009_opportunityequation.pdf; Symonds, Schwartz, and Ferguson, *Pathways to Prosperity*; David E. Drew, *STEM the Tide: Reforming Science, Technology, Engineering and Math Education in America* (Baltimore: Johns Hopkins University Press, 2012); Kevin O'Connor, Frederick A. Peck, and Julie Cafarella, "Struggling for Legitimacy: Trajectories of Membership and Naturalization in the Sorting Out of Engineering Students," *Mind, Culture, and Activity* 22, no. 2 (2015): 168–183; Heidi B. Carlone, Julie Haun-Frank, and Sue C. Kimmel, "Tempered Radicals: Elementary Teachers' Narratives of Teaching Science within and against Prevailing Meanings of Schooling," *Cultural Studies of Science Education* 5, no. 4 (2010): 941–964; and Secules et al., "Zooming Out," 56–86.

18. Slaton, *Race, Rigor, and Selectivity*, 171–204.

19. Roderick A. Ferguson, *The Reorder of Things: The University and Its Pedagogies of Minority Difference* (Minneapolis, MN: University of Minnesota Press, 2012), 179; see also Damani J. Partridge and Matthew Chin, "Interrogating the Histories and Futures of 'Diversity': Transnational Perspectives," *Public Culture* 31, no. 2 (2019): 197–214; Sara Ahmed, *On Being Included: Racism and Diversity in Institutional Life* (Durham, NC: Duke University Press, 2012).

20. See Kirchgasler, "Dangers of 'Making Diversity Visible'"; and O'Connor, Peck, and Cafarella, "Struggling for Legitimacy." Critical studies of intellectual disability are immensely helpful here. See, e.g., Ashley Taylor, "Knowledge Citizens? Intellectual Disability and the Production of Social Meanings within Educational Research," *Harvard Educational Review* 88, no. 1 (2018): 1–25.

21. See Daniel Martinez HoSang, Oneka LaBennett, and Laura Pulido, eds., *Racial Formation in the Twenty-First Century* (Berkeley: University of California Press, 2012).

22. See National Academies, "Executive Summary," in *Rising Above the Gathering Storm*. Very helpfully, Kirchgasler historicizes such claims of a one-way causal relationship between stereotype bias and resource distribution; see "Dangers of 'Making Diversity Visible,'" esp. 338–342. Adding further complexity to all such historical accounts are vigorous claims from the right about the fragility of meritocratic systems following the civil rights era—for example, David O. Sacks and Peter A. Thiel, *The Diversity Myth: "Multiculturalism" and the Politics of Intolerance at Stanford* (Oakland, CA: Independent Institute, 1995).

23. Ian Hacking, "Making Up People," in *Reconstructing Individualism: Autonomy, Individuality, and the Self in Western Thought*, ed. Thomas C. Heller, Morton Sosna, and David E. Wellbery (Stanford, CA: Stanford University Press, 1986), 223; Clara O'Shea, "Learning How Kinds Matter: A Posthuman Rethinking Ian Hacking's Concepts of Kinds, Dynamic Nominalism, and the Looping Effect," in *Proceedings of the 11th International Conference on Networked Learning 2018*, ed. Milan Bajic et al. (Zagreb: Zagreb University of Applied Science, 2018), 203; Fenwick, "Re-thinking the 'Thing,'" 104.

24. Fenwick, "Re-thinking the 'Thing,'" 105.

25. Barad, *Meeting the Universe Halfway*, 140.

26. See O'Connor, Peck, and Carafalla, "Struggling for Legitimacy."

27. Gary Peller, *Critical Race Consciousness* (Boulder, CO: Paradigm Publishers, 2011), 6.

28. NGSS Lead States, "Executive Summary," 1. Phillips follows the parallel development of mathematics teaching standards through the Cold War and subsequent political periods in the US. See Christopher J. Phillips, *The New Math: A Political History* (Chicago: University of Chicago Press, 2015).

29. Darren G. Hoeg and John Lawrence Bencze, "Values Underpinning STEM Education in the USA: An Analysis of the Next Generation Science Standards," *Science Education* 101, no. 2 (2017): 278–301, 279.

30. National Research Council, *Inquiry and the National Science Education Standards: A Guide for Teaching and Learning* (Washington, DC: National Academy Press, 2000), xiii; and Loucks-Horsley et al., *Elementary School Science*, ix.

31. Amy E. Slaton, *Reinforced Concrete and the Modernization of American Building, 1900–1930* (Baltimore: Johns Hopkins University Press, 2001), 20–61; see Kaiser, *Pedagogy and the Practice of Science*.

32. John L. Rudolph, "Epistemology for the Masses: The Origins of 'The Scientific Method' in American Schools," *History of Education Quarterly* 45, no. 3 (2005), 352–354. On technical education in particular, see Slaton, *Reinforced Concrete*, 20–61.

33. See Slaton, *Race, Rigor, and Selectivity*; Sharon Traweek, *Beamtimes and Lifetimes: The World of High Energy Physicists* (Cambridge, MA: Harvard University Press, 1988); and Jonson Miller, *Engineering Manhood: Race and the Antebellum Virginia Military Institute* (Ann Arbor, MI: Lever Press, 2020).

34. See Steven Brint and Jerome Karabel, *The Diverted Dream: Community Colleges and the Promise of Educational Opportunity in America, 1900–1985* (Oxford: Oxford University Press, 1989).

35. Mary F. E. Ebeling and Amy E. Slaton, "Promise Her Anything: Education for Work in the U.S. 'Nanoeconomy,'" (unpublished manuscript, 2018); Gunckel and Tolbert, "Imperative to Move," 945; Whalley and Barley, "Technical Work," 32–34.

36. See Mike Rose, *The Mind at Work: Valuing the Intelligence of the American Worker* (New York: Penguin, 2004); Slaton, *Race, Rigor, and Selectivity*; and Justin Carone, "Fixing Value: History, Ethnography, and Material Ontologies of Deservingness in a Philadelphia Repair Shop," *History and Technology* 33, no. 4 (2017): 367–395.

37. See Gunckel and Tolbert, "Imperative to Move"; O'Connor, Peck, and Cafarella, "Struggling for Legitimacy"; and Erin A. Cech et al., "Epistemological Dominance and Social Inequality: Experiences of Native American Science, Engineering, and Health Students," *Science, Technology & Human Values*, 42, no. 5 (2017): 743–774.

38. NGSS Lead States, "Appendix A: Conceptual Shifts," in *Next Generation Science Standards*.

39. National Research Council, *Framework for K–12 Science Education*, 10–11.

40. NGSS Lead States, "Appendix A: Conceptual Shifts."

41. See Christopher Newfield, *Ivy and Industry: Business and the Making of the American University, 1880–1980* (Durham, NC: Duke University Press, 2003); David C. Mowery et al., *Ivory Tower and Industrial Innovation: University-Industry Technology Transfer before and after the Bayh-Dole Act* (Stanford, CA: Stanford University Press, 2004); and Slaton, *Race, Rigor, and Selectivity*.

42. See Mihail C. Roco, Chad A. Mirkin, and Mark C. Hersam, *Nanotechnology Research Directions for Societal Needs in 2020: Retrospective and Outlook* (Berlin: Springer, 2011).

43. See Mody, *Instrumental Communities*; Mary F. E. Ebeling, "Mediating Uncertainty: Communicating the Financial Risks of Nanotechnologies," *Science Communication* 29, no. 3 (2008): 335–361; and Ebeling and Slaton, "Promise Her Anything."

44. See Ebeling, "Mediating Uncertainty"; Judith Light Feather and Miguel F. Aznar, *Nanoscience Education, Workforce Training, and K–12 Resources* (Boca Raton, FL: CRC Press, 2011), 126. Nano-related workforce anxieties took shape as part of the more general worry in the US about global economic competition. For representative arguments, see National Academies, *Rising Above the Gathering Storm*.

45. William Sims Bainbridge, ed., *Societal Implications of Nanoscience and Nanotechnology* (Boston: Kluwer Academic, 2001).

46. See Light Feather and Aznar, *Nanoscience Education*; Campbell et al., "Exploring the Nanoworld"; Turner et al., "Seeing the Unseen," 58.

47. See Ebeling and Slaton, "Promise Her Anything."

48. See Frederick M. Hess and Michael J. Petrelli, *No Child Left Behind: Primer* (New York: Peter Lang, 2006); and Tim Walker, "Despite Progress, the 'Charade' of High-Stakes Testing Persists," *NEA Today*, April 4, 2018, https://www.nea.org/advocating-for-change/new-from-nea/despite-progress-charade-high-stakes-testing-persists.

49. Antonia Darder, foreword to *Neoliberalizing Education Reform*, ed. Keith M. Sturges (Rotterdam: Sense Publishers, 2015), x–xi.

50. See Jacqueline Jones, *American Work: Four Centuries of Black and White Labor* (New York: Norton, 1998); and Nikhil Pal Singh, *Black Is a Country: Race and the Unfinished Struggle for Democracy* (Cambridge, MA: Harvard University Press, 2004).

51. Pickering, "New Ontologies," 5; see Geoffrey C. Bowker, *Science on the Run: Information Management and Industrial Geophysics at Schlumberger, 1920–1940* (Cambridge, MA: MIT Press, 1994). Of special interest for this paper are accounts that analyze identity constructions contingent on such representations, e.g., Traweek, *Beamtimes and Lifetimes*; Tiago Saraiva, *Fascist Pigs: Technoscientific Organisms and the History of Fascism* (Cambridge, MA: MIT Press, 2016); and Carone, "Fixing Value." See also Slaton, *Race, Rigor, and Selectivity*.

52. See Rudolph, "Epistemology for the Masses"; Leigh Patel, "Reaching beyond Democracy in Educational Policy Analysis," *Education Policy* 30, no. 1 (2016), 117.

53. Campbell et al., "Exploring the Nanoworld," 17.

54. Campbell et al., 17–18.

55. Campbell et al., 6.

56. Partnership for 21st Century Learning, "Framework for 21st Century Learning," Battelle for Kids, 2019, http://static.battelleforkids.org/documents/p21/P21_Framework_Brief.pdf; Lachney, "Building the Lego Classroom," 174.

57. Beth Warren et al., "Rethinking Diversity in Learning Science: The Logic of Everyday Sensemaking," *Journal of Research in Science Teaching* 38, no. 5 (2001), 539.

58. On the arbitrary nature of these demarcations, see essays in Letts and Fifield, *STEM of Desire*.

59. NGSS Lead States, "Appendix H: The Nature of Science," in *Next Generation Science Standards*.

60. Jones et al., *Nanoscale Science*, 69.

61. See Barbara Townley, *Reframing Human Resource Management: Power, Ethics and the Subject at Work* (London: Sage, 1994); David A. Hollinger, "Inquiry and Uplift: Late Nineteenth-Century American Academics and the Moral Efficacy of Scientific Practice," in *The Authority of Experts: Studies in History and Theory*, ed. Thomas L. Haskell (Bloomington: University of Indiana Press, 1984), 142–156; and many of the essays in Kaiser's *Pedagogy and the Practice of Science*, which delve deeply into the contingent and morally freighted character of scientific replication and certainty. The most helpful history of science literature from the perspective of structural discrimination has picked up science's emphasis on the identities, broadly conceived, of "good" research actors

at the level of experimenter, technician, or witness, as in Schaffer, "Accurate Measurement," and Steven Shapin, "The Invisible Technician," *American Scientist* 77, no. 6 (1989): 554–563.

62. Jeff Nordine and Ruben Torres, "Enhancing Science Kits with the Driving Question Board," *Science and Children* 50, no. 8 (2013): 57–61.

63. Michel Foucault, *Discipline and Punish*, trans. Alan Sheridan (New York: Vintage Books, 1977), 136; Amy E. Slaton, Erin A. Cech, and Donna M. Riley, "Yearning, Learning, Earning: The Gritty Ontologies of American Engineering Education," in Letts and Fifield, *STEM of Desire*, 319–340; Slaton, *Reinforced Concrete*; and Slaton, *Race, Rigor, and Selectivity*. Recent work on intellectual disability is central here; see note 20.

64. O'Connor, Peck, and Cafarella, "Struggling for Legitimacy," 177; Hoeg and Bencze, "Values Underpinning STEM Education."

65. Campbell et al., "Exploring the Nanoworld," 5.

66. Taylor, "Knowledge Citizens," 2.

67. National Research Council, *Inquiry*, xiii; Loucks-Horsley et al., *Elementary School Science*, 18.

68. Examples include Warwick Anderson, "Introduction: Postcolonial Technoscience," *Social Studies of Science* 32, no. 5 (2002): 643–658; and Ann Laura Stoler, ed., *Haunted by Empire* (Durham, NC: Duke University Press, 2006).

69. See Taylor, "Knowledge Citizens."

70. Gert Biesta, *The Beautiful Risk of Education* (Boulder, CO: Paradigm Press, 2014), 1, 34.

71. Projit Bihari Mukharji, "Occulted Materialities," in "Thinking with the World: Histories of Science and Technology from the 'Out There,'" edited by Gabriela Soto Laveaga and Pablo F. Gómez, special issue, *History and Technology* 34, no. 1 (2018), 33–34.

72. NGSS Lead States, "Appendix H: The Nature of Science."

73. See Eduardo Viveiros de Castro, "Perspectival Anthropology and the Method of Controlled Equivocation," *Tipiti: Journal of the Society for the Anthropology of Lowland South America* 2, no. 1 (2004): 3–22.

74. Jones et al., *Nanoscale Science*, 42–43.

75. Jones et al., 42–43.

76. Mukharji, "Occulted Materialities," 35–36.

77. See Taylor, "Knowledge Citizens"; Cech et al., "Epistemological Dominance"; and Slaton, *Race, Rigor, and Selectivity*.

78. Megan Bang and Douglas Medin, "Cultural Processes in Science Education: Supporting the Navigation of Multiple Epistemologies," *Science Education* 94, no. 6 (2010): 1008–1026.

79. Secules et al., "Zooming Out," 60; see Cech et al., "Epistemological Dominance."

80. Patel, "Reaching beyond Democracy," 117.

81. In critiquing the pipeline model, Carone problematizes economic disadvantage as arbitrary exclusion (see "Fixing Value"); O'Connor, Peck, and Cafarella articulate, following Leigh Star, how those deemed peripheral actors secure the centrality of those in authority ("Struggling for Legitimacy," 168–169). Research literature on science education in America is vast, and in addition to unreflective projects aiming to support meritocratic visions, it includes important critical work that helps explain such unjust social instrumentalities/ For excellent overviews, see Secules et al., "Zooming Out," and Donna M. Riley et al., "Feminisms in Engineering Education: Transformative Possibilities," *NWSA Journal* 21, no. 2 (2009): 21–40. Here I want to draw in particular on ideas of epistemic imperialism and epistemic injustice to displace rhetoric of "inclusion" and "diversity" in research and historical studies of technical education that routinely hide how epistemic operations produce human difference. On epistemic imperialism, see Dan Goodley and Griet Roets, "The (Be)comings and Goings of 'Developmental Disabilities': The Cultural Politics of 'Impairment,'" *Discourse Studies in the Cultural Politics of Education* 29, no. 2 (2008): 239–255, and on epistemic injustice, see Taylor, "Knowledge Citizens."

82. Taylor, "Knowledge Citizens," 2–3; Warren et al., "Rethinking Diversity," 546; O'Connor, Peck, and Cafarella, "Struggling for Legitimacy," 181. Warwick and Kaiser problematize Kuhn's point that the learning of physics was not based in "understanding, but practice." That disarticulation of "mastery of craft skill" and "insight into meaning" is one of the stubborn framings of science pedagogy that this essay also challenges. Andrew Warwick and David Kaiser, "Conclusion: Kuhn, Foucault, and the Power of Pedagogy," in Kaiser, *Pedagogy and the Practice of Science*, 395.

83. John Law and Marianne Elisabeth Lien, "Slippery: Field Notes in Empirical Ontology," *Social Studies of Science* 43, no. 3 (2013): 363–378.

84. See in particular Goodley and Roets on "poststructuralist thinking" in disabilities scholarship ("(Be)comings and Goings," 242–243). Also, Steve Fifield and Will Letts, "[Re]considering Queer Theories and Science Education," *Cultural Studies of Science Education* 9, no. 2 (2014): 393–407.

85. See Paolo Freire, *Pedagogy of the Oppressed* (1970; London: Continuum, 2007); Amy E. Slaton, "Merit Makes the Difference" (paper presented at "Technologies in Use" Conference, Max Planck Institute for the History of Science, Berlin, April 5, 2018); and Ferguson, *Reorder of Things*, 223.

86. I am deeply indebted to Jesse Smith for his articulation of this distinction.

87. Fenwick, "Re-thinking the 'Thing,'" 110–112.

88. Warren et al., "Rethinking Diversity," 546; Susan Gabel, "Some Conceptual Problems with Critical Pedagogy," *Curriculum Inquiry* 32, no. 2 (2002), 184; Goodley and Roets, "(Be)comings and Goings," 243; David J. Connor and Jan W. Valle, "A Socio-cultural Reframing of Science and Dis/Ability in Education: Past Problems, Current Concerns, and Future Possibilities," *Cultural Studies of Science Education* 10, no. 4 (2015): 1103–1122.

89. See La paperson, *A Third University Is Possible* (Minneapolis: University of Minnesota Press, 2017); and Slaton, Cech, and Riley, "Yearning, Learning, Earning." The historicization of inventive

deficit in STEM settings as a racialized concept is very well supported by André Brock, *Distributed Blackness: African American Cybercultures* (New York: New York University Press, 2020).

Bibliography

Ahmed, Sara. *On Being Included: Racism and Diversity in Institutional Life*. Durham, NC: Duke University Press, 2012.

Anderson, Warwick. "Introduction: Postcolonial Technoscience." *Social Studies of Science* 32, no. 5/6 (2002): 643–658.

Bainbridge, William Sims, ed. *Societal Implications of Nanoscience and Nanotechnology*. Boston: Kluwer Academic, 2001.

Bang, Megan, and Douglas Medin. "Cultural Processes in Science Education: Supporting the Navigation of Multiple Epistemologies." *Science Education* 94, no. 6 (2010): 1008–1026.

Barad, Karen. *Meeting the Universe Halfway: Quantum Physics and the Entanglement of Matter and Meaning*. Durham, NC: Duke University Press, 2007.

Biesta, Gert. *The Beautiful Risk of Education*. Boulder, CO: Paradigm Press, 2014.

Biesta, Gert. "Why 'What Works' Still Won't Work: From Evidence-Based Education to Value-Based Education." *Studies in Philosophy and Education* 29, no. 5 (2010): 491–503.

Bowker, Geoffrey C. *Science on the Run: Information Management and Industrial Geophysics at Schlumberger, 1920–1940*. Cambridge, MA: MIT Press, 1994.

Bowles, Steven, and Herbert Gintis. *Schooling in Capitalist America: Educational Reform and the Contradictions of Economic Life*. Chicago: Haymarket Books, 1976.

Brint, Steven, and Jerome Karabel. *The Diverted Dream: Community Colleges and the Promise of Educational Opportunity in America, 1900–1985*. Oxford: Oxford University Press, 1989.

Brock, André. *Distributed Blackness: African American Cybercultures*. New York: New York University Press, 2020.

Carlone, Heidi B., Julie Haun-Frank, and Sue C. Kimmel. "Tempered Radicals: Elementary Teachers' Narratives of Teaching Science within and against Prevailing Meanings of Schooling." *Cultural Studies of Science Education* 5, no. 4 (2010): 941–965.

Carone, Justin. "Fixing Value: History, Ethnography, and Material Ontologies of Deservingness in a Philadelphia Repair Shop." *History and Technology* 33, no. 4 (2017): 367–395.

Carson, John. *The Measure of Merit: Talents, Intelligence, and Inequality in the French and American Republics, 1750–1940*. Princeton, NJ: Princeton University Press, 2007.

Cech, Erin A., Anneke Metz, Jessi L. Smith, and Karen DeVries. "Epistemological Dominance and Social Inequality: Experiences of Native American Science, Engineering, and Health Students." *Science, Technology & Human Values* 42, no. 5 (2017): 743–774.

Connor, David J., and Jan W. Valle. "A Socio-cultural Reframing of Science and Dis/Ability in Education: Past Problems, Current Concerns, and Future Possibilities." *Cultural Studies of Science Education* 10, no. 4 (2015): 1103–1122.

Darder, Antonia. Foreword to *Neoliberalizing Education Reform*, edited by Keith M. Sturges, ix–xvii. Rotterdam: Sense Publishers, 2015.

Drew, David E. *STEM the Tide: Reforming Science, Technology, Engineering and Math Education in America*. Baltimore: Johns Hopkins University Press, 2012.

Dumas, Michael A. "Against the Dark: Antiblackness in Education Policy and Discourse." *Theory into Practice* 55, no. 1 (2016): 11–19.

Ebeling, Mary F. E. "Mediating Uncertainty: Communicating the Financial Risks of Nanotechnologies," *Science Communication* 29, no. 3 (2008): 335–361.

Ebeling, Mary F. E., and Amy E. Slaton. "Promise Her Anything: Education for Work in the U.S. 'Nanoeconomy.'" Unpublished manuscript, 2018. Microsoft Word file.

Fenwick, Tara. "Re-thinking the 'Thing': Sociomaterial Approaches to Understanding and Researching Learning in Work." In "Selected Papers from the 6th International Conference on Researching Work and Learning, Denmark, June 2009," edited by Henning Salling Olesen. Special issue, *Journal of Workplace Learning* 22, no. 1/2 (2010): 104–116.

Ferguson, Roderick A. *The Reorder of Things: The University and Its Pedagogies of Minority Difference*. Minneapolis, MN: University of Minnesota Press, 2012.

Fifield, Steve, and Will Letts. "[Re]considering Queer Theories and Science Education." *Cultural Studies of Science Education* 9, no. 2 (2014): 393–407.

Foucault, Michel. *Discipline and Punish: The Birth of the Prison*. Translated by Alan Sheridan. New York: Vintage Books, 1977.

Freire, Paolo. "The Banking Concept of Education." In *Thinking about Schools: A Foundations of Education Reader*, edited by Eleanor Blair Hilty, 117–128. Boulder, CO: Westview Press, 2011.

Freire, Paolo. *Pedagogy of the Oppressed*. 1970. Reprint, London: Continuum, 2007.

Gabel, Susan. "Some Conceptual Problems with Critical Pedagogy." *Curriculum Inquiry* 32, no. 2 (2002): 177–201.

Goodley, Dan, and Griet Roets. "The (Be)comings and Goings of 'Developmental Disabilities': The Cultural Politics of 'Impairment.'" *Discourse Studies in the Cultural Politics of Education* 29, no. 2 (2008): 239–255.

Gould, Stephen Jay. *The Mismeasure of Man*. New York: Norton, 1986.

Gunckel, Kristin L., and Sara Tolbert. "The Imperative to Move toward a Dimension of Care in Engineering Education." In "A Critical Examination of the Next Generation Science Standards," edited by Troy D. Sadler and David E Brown. Special issue, *Journal of Research in Science Teaching* 55, no. 7 (2018): 938–961.

Hacking, Ian. "Making Up People." In *Reconstructing Individualism: Autonomy, Individuality, and the Self in Western Thought*, edited by Thomas C. Heller, Morton Sosna, and David E. Wellbery, 222–236. Stanford, CA: Stanford University Press, 1986.

Hess, Frederick M., and Michael J. Petrelli. *No Child Left Behind: Primer*. New York: Peter Lang, 2006.

Hoeg, Darren G., and John Lawrence Bencze. "Values Underpinning STEM Education in the USA: An Analysis of the Next Generation Science Standards." *Science Education* 101, no. 2 (2017): 278–301.

Hollin, Gregory. "Failing, Hacking, Passing: Autism, Entanglement and the Ethics of Transformation." *BioSocieties* 12, no. 4 (2017): 611–633.

Hollinger, David A. "Inquiry and Uplift: Late Nineteenth-Century American Academics and the Moral Efficacy of Scientific Practice." In *The Authority of Experts: Studies in History and Theory*, edited by Thomas L. Haskell, 142–156. Bloomington: University of Indiana Press, 1984.

HoSang, Daniel Martinez, Oneka LaBennett, and Laura Pulido, eds. *Racial Formation in the Twenty-First Century*. Berkeley: University of California Press, 2012.

Jones, Jacqueline. *American Work: Four Centuries of Black and White Labor*. New York: Norton, 1998.

Jones, M. Gail, Michael R. Falvo, Amy R. Taylor, and Bethany P. Broadwell. *Nanoscale Science: Activities for Grades 6–12*. Arlington, VA: NSTA Press, 2007.

Kaiser, David. "Introduction: Moving Pedagogy from the Periphery to the Center." In Kaiser, *Pedagogy and the Practice of Science*, 1–8.

Kaiser, David, ed. *Pedagogy and the Practice of Science: Historical and Contemporary Perspectives*. Cambridge, MA: MIT Press, 2005.

Kao, Yvonne S., Anthony Cina, and J. Aura Gimm. "Inside the Black Box." *Science Teacher* 73, no. 9 (2006): 46–49.

Kirchgasler, Kathryn L. "Dangers of 'Making Diversity Visible': Historicizing Metrics of Science Achievement in U.S. Education Policy." In *Handbook of Education Policy Studies*, edited by G. Fan and T. S. Popkewitz, 331–348. Singapore: Springer, 2020.

Kirchgasler, Kathryn L. "Strange Precipitate." In *STEM of Desire: Queer Theories and Science Education*, edited by Will Letts and Steve Fifield, 191–208. Leiden: Brill, 2019.

Lachney, Michael. "Building the LEGO Classroom." In *Lego Studies: Examining the Building Blocks of a Transmedial Phenomenon*, edited by Mark J. P. Wolf, 166–186. New York: Routledge, 2014.

La paperson [pseud.]. *A Third University Is Possible*. Minneapolis: University of Minnesota Press, 2017.

Law, John, and Marianne Elisabeth Lien. "Slippery: Field Notes in Empirical Ontology." *Social Studies of Science* 43, no. 3 (2013): 363–378.

Letts, Will, and Steve Fifield, eds. *STEM of Desire: Queer Theories and Science Education*. Leiden: Brill, 2019.

Light Feather, Judith, and Miguel F. Aznar. *Nanoscience Education, Workforce Training, and K–12 Resources*. Boca Raton, FL: CRC Press, 2011.

Loucks-Horsley, Susan, Roxanne Kapitan, Maura D. Carlson, Paul J. Kuerbis, Richard C. Clark, G. Marge Melle, Thomas P. Sachse, and Emma Walton. *Elementary School Science for the '90s*. Andover, MA: NSTA Press, 1990.

Miller, Jonson. *Engineering Manhood: Race and the Antebellum Virginia Military Institute*. Ann Arbor, MI: Lever Press, 2020.

Mody, Cyrus C. M. *Instrumental Community: Probe Microscopy and the Path to Nanotechnology*. Cambridge, MA: MIT Press, 2011.

Mowery, David C., Richard R. Nelson, Bhaven N. Sampat, and Arvids A. Ziedonis. *Ivory Tower and Industrial Innovation: University-Industry Technology Transfer before and after the Bayh-Dole Act*. Stanford, CA: Stanford University Press, 2004.

Mukharji, Projit Bihari. "Occulted Materialities." In "Thinking with the World: Histories of Science and Technology from the 'Out There,'" edited by Gabriela Soto Laveaga and Pablo F. Gómez. Special issue, *History and Technology* 34, no. 1 (2018): 31–40.

National Academy of Sciences, National Academy of Engineering, and National Institute of Medicine of the National Academies. *Rising above the Gathering Storm: Energizing and Employing America for a Brighter Future*. Washington, DC: National Academies Press, 2007.

National Research Council. *A Framework for K–12 Science Education: Practices, Crosscutting Concepts, and Core Ideas*. Washington, DC: National Academies Press, 2012.

National Research Council. *Inquiry and the National Science Education Standards: A Guide for Teaching and Learning*. Washington, DC: National Academies Press, 2000.

Newfield, Christopher. *Ivy and Industry: Business and the Making of the American University, 1880–1980*. Durham, NC: Duke University Press, 2003.

Nordine, Jeff, and Ruben Torres. "Enhancing Science Kits with the Driving Question Board." *Science and Children* 50, no. 8 (2013): 57–61.

O'Connor, Kevin, Frederick A. Peck, and Julie Cafarella. "Struggling for Legitimacy: Trajectories of Membership and Naturalization in the Sorting Out of Engineering Students." *Mind, Culture, and Activity* 22, no. 2 (2015): 168–183.

Olsesko, Kathryn M. "Science Pedagogy as a Category of Historical Analysis: Past, Present, and Future." In "Textbooks in the Scientific Periphery," edited by Antonio Garcia-Belmar, José Ramón Bertomeu-Sánchez, Manolis Patiniotis, and Anders Lundgren. Special issue, *Science & Education* 15, no. 7/8 (2006): 863–880.

O'Shea, Clara. "Learning How Kinds Matter: A Posthuman Rethinking Ian Hacking's Concepts of Kinds, Dynamic Nominalism, and the Looping Effect." In *Proceedings of the 11th International Conference on Networked Learning 2018*, edited by Milan Bajic, Nina Bonderup Dohn, Maarten de Laat, Petar Jandric, and Thomas Ryberg, 203–209. Zagreb: Zagreb University of Applied Science, 2018.

Partridge, Damani J., and Matthew Chin. "Interrogating the Histories and Futures of 'Diversity': Transnational Perspectives." *Public Culture* 31, no. 2 (2019): 197–214.

Patel, Leigh. "Reaching beyond Democracy in Educational Policy Analysis." *Education Policy* 30, no. 1 (2016): 114–127.

Peller, Gary. *Critical Race Consciousness*. Boulder, CO: Paradigm Publishers, 2011.

Phillips, Christopher J. *The New Math: A Political History*. Chicago: University of Chicago Press, 2015.

Pickering, Andrew. "New Ontologies." In *The Mangle in Practice*, edited by Andrew Pickering and Keith Guzik, 1–16. Durham, NC: Duke University Press, 2007.

Riley, Donna, Alice L. Pawley, Jessica Tucker, and George D. Catalano. "Feminisms in Engineering Education: Transformative Possibilities." *NWSA Journal* 21, no. 2 (2009): 21–40.

Roco, Mihail C., Chad A. Mirkin, and Mark C. Hersam. *Nanotechnology Research Directions for Societal Needs in 2020: Retrospective and Outlook*. Berlin: Springer, 2011.

Rose, Mike. *The Mind at Work: Valuing the Intelligence of the American Worker*. New York: Penguin, 2004.

Rudolph, John L. "Epistemology for the Masses: The Origins of 'The Scientific Method' in American Schools." *History of Education Quarterly* 45, no. 3 (2005): 341–376.

Sacks, David O., and Peter A. Thiel. *The Diversity Myth: "Multiculturalism" and the Politics of Intolerance at Stanford*. Oakland, CA: Independent Institute, 1995.

Saraiva, Tiago. *Fascist Pigs: Technoscientific Organisms and the History of Fascism*. Cambridge, MA: MIT Press, 2016.

Schaffer, Simon. "Accurate Measurement Is an English Science." In *Values of Precision*, edited by M. Norton Wise, 135–171. Princeton, NJ: Princeton University Press, 1995.

Schwartz, Christina V., Cynthia Passmore, and Brian J. Reiser, eds. *Helping Students Make Sense of the World Using Next Generation Science and Engineering Practices*. Arlington, VA: NSTA Press, 2017.

Secules, Stephen, Ayush Gupta, Andrew Elby, and Chandra Turpen. "Zooming Out from the Struggling Individual Student: An Account of the Cultural Construction of Engineering Ability in an Undergraduate Programming Class." *Journal of Engineering Education* 107, no. 1 (2018): 56–86.

Shapin, Steven. "The Invisible Technician." *American Scientist* 77, no. 6 (1989): 554–563.

Singh, Nikhil Pal. *Black Is a Country: Race and the Unfinished Struggle for Democracy*. Cambridge, MA: Harvard University Press, 2004.

Slaton, Amy E. "Merit Makes the Difference." Paper presented at "Technologies in Use" Conference, Max Planck Institute for the History of Science, Berlin, April 5, 2018.

Slaton, Amy E. *Race, Rigor, and Selectivity in U.S. Engineering*. Cambridge, MA: Harvard University Press, 2010.

Slaton, Amy E. *Reinforced Concrete and the Modernization of American Building, 1900–1930*. Baltimore: Johns Hopkins University Press, 2001.

Slaton, Amy E., Erin A. Cech, and Donna M. Riley. "Yearning, Learning, Earning: The Gritty Ontologies of American Engineering Education." In Letts and Fifield, *STEM of Desire*, 319–340.

Stoler, Ann Laura, ed. *Haunted by Empire*. Durham, NC: Duke University Press, 2006.

Symonds, William C., Robert Schwartz, and Ronald F. Ferguson. *Pathways to Prosperity: Meeting the Challenge of Preparing Young Americans for the 21st Century*. Cambridge, MA: Pathways to Prosperity Project, Harvard University Graduate School of Education, 2011.

Taylor, Ashley. "Knowledge Citizens? Intellectual Disability and the Production of Social Meanings within Educational Research." *Harvard Educational Review* 88, no. 1 (2018): 1–25.

Townley, Barbara. *Reframing Human Resource Management: Power, Ethics and the Subject at Work*. London: Sage, 1994.

Traweek, Sharon. *Beamtimes and Lifetimes: The World of High Energy Physicists*. Cambridge, MA: Harvard University Press, 1988.

Turner, Kenneth, Emma Tevaarwerk, Nathan Unterman, Marcel Grdinic, Jason Campbell, Venkat Chandrasekhar, and R. P. H. Chang. "Seeing the Unseen: The Scanning Probe Microscope and Nanoscale Measurement." *Science Teacher* 73, no. 9 (2006): 58–61.

Viveiros de Castro, Eduardo. "Perspectival Anthropology and the Method of Controlled Equivocation." *Tipití: Journal of the Society for the Anthropology of Lowland South America* 2, no. 1 (2004): 3–22.

Walker, Tim. "Despite Progress, the 'Charade' of High-Stakes Testing Persists." *NEA Today*, April 4, 2018. https://www.nea.org/advocating-for-change/new-from-nea/despite-progress-charade-high-stakes-testing-persists.

Warren, Beth, Cynthia Ballenger, Mark Ogonowski, Ann S. Rosebery, and Josiane Hudicourt-Barnes. "Rethinking Diversity in Learning Science: The Logic of Everyday Sensemaking." *Journal of Research in Science Teaching* 38, no. 5 (2001): 529–552.

Warwick, Andrew, and David Kaiser. "Conclusion: Kuhn, Foucault, and the Power of Pedagogy." In Kaiser, *Pedagogy and the Practice of Science*, 393–409.

Whalley, Peter, and Stephen R. Barley. "Technical Work in the Division of Labor: Stalking the Wily Anomaly." In *Between Science and Craft: Technical Work in U.S. Settings*, edited by Stephen R. Barley and Julian E. Orr, 23–52. Ithaca, NY: Cornell University Press, 1997.

Welton, Michael R., ed. *In Defense of the Lifeworld: Critical Perspectives on Adult Learning*. Albany: State University of New York Press, 1995.

Young, Michael, and Geoff Whitty, eds. *Society, State and Schooling: Reading on the Possibilities of Radical Education*. Ringmer, UK: Falmer Press, 1977.

III THE THREE DOMAINS: SOCIETY, ECONOMY, EPISTEMOLOGY

7
AN AESTHETIC OF KNOWLEDGE: RELATIONS AND THE DOCUMENTATION OF TRADITIONAL KNOWLEDGE IN PAPUA NEW GUINEA

James Leach

INTRODUCTION: KNOWLEDGE FORMS, RELATIONALITY, AND AESTHETICS

The questions this chapter addresses revolve around how one retains the relationality of knowledge. Questions about relationality are of particular relevance when it comes to the subject of this chapter: *traditional knowledge* (TK), a form of knowledge that consistently poses epistemological as well as political and ethical challenges to modern, scientific, and academic understandings.[1] As we will see below, the term *TK* is itself problematic in various ways. But exploring the difficulties it poses in comparison to "modernist" ways of knowing will prove fruitful for understanding a relational kind of knowing that differs from how science or IP owns knowledge.[2]

The value and importance of traditional knowledge is widely acknowledged, as is its significance for cultural identity, sustainable ecological practices, medicinal and health-related uses, and uses in development, to name just a few. The terminology and definition are much debated—the World Intellectual Property Organization (WIPO) asserts that "traditional knowledge can be found in a wide variety of contexts, including: agricultural, scientific, technical, ecological and medicinal knowledge as well as biodiversity-related knowledge."[3] The UNESCO Convention on the Protection and Promotion of the Diversity of Cultural Expressions, agreed in Paris on October 20, 2005, recognized "the importance of traditional knowledge as a source of intangible and material wealth, and in particular the knowledge systems of indigenous peoples, and its positive contribution to sustainable development, as well as the need for its adequate protection and promotion."[4] These statements, which speak of utility, economics, and fragility, point to anxiety over loss and to untapped value.

Whatever traditional knowledge is, allying traditional knowledge with development and wealth creation may be a good strategy to convey its importance to a skeptical audience that includes many who persist in considering indigenous peoples a throwback to a lost past. However, an approach to traditional knowledge that echoes a colonial assimilation of resources to contribute to developing an economy addresses the problem of how to approach traditional knowledge in a depressingly familiar manner.[5] This quotation illustrates that the problem of traditional knowledge lies not with the diverse ritual and practical activities of indigenous people, but with the narrow, usually instrumental, and often exploitative, rendition of indigenous knowledge by academics, development agencies, pharmaceutical companies, and so forth. There remain serious and unresolved questions, then, as to what the right way is to apprehend or engage with such knowledge.[6]

Consider the assumption that because traditional knowledge is "traditional," it is inherently situated as a timeless and ancient lore. That is seldom true. Traditional knowledge is always contemporary with other forms of knowledge. The innovative nature of indigenous people's practices is often misunderstood, as it seldom aims at developing the next product. Or consider the notion that traditional knowledge can apparently only be redeemed from its convoluted and complex expression—in stories, myths, and rituals—by comparison to, and assimilation with, the dominant knowledge forms of colonizers and developers.[7]

In response, this chapter is concerned with how one might think about and represent a knowledge form in which relations come first, where "knowledge" as an object in its own right, or as a truth claim, is not at issue so much as how people manage knowledge as a resource within, and for, referencing, establishing, or transforming relationships. I emphasize relationality because in the examples that I describe, relations are a central aspect, and they appear in ways that are unfamiliar in a modernist conception. *Relationality* refers to the form that knowledge takes, to what I am calling the *aesthetic* of its form.

To this end, the chapter describes how a particular group of people, "traditional knowledge holders" from the Rai Coast of Madang Province in Papua New Guinea (PNG), shape their social relations of knowledge ownership—that is, understand the value of the circulation and use of knowledge. Priority, secrecy, and license are all part of the picture, refracted though some assumptions in this region that are rather different from those that give rise to conceptions of the utility or economic exploitability of traditional knowledge. This is reflected in particular when it comes to considering the value, or the ownership, of this knowledge through intellectual property (IP) law.

Nowhere in the world is "outside" the purview of IP in the early twenty-first century. The promotion of IP as a means of protection feeds on fears about appropriation; it promotes state and bureaucratic control of resources that are imagined to be exploitable, while the mechanisms of appropriation and propertization that actually utilize

IP shape, and provide the context to, interactions within which traditional knowledge plays a part. This is as true for the area that I will draw on in this chapter as for anywhere else, whether or not people there understand or can leverage the legal process in any particular case.[8] Yet IP has not, and does not, provide either the problem or the solution for a negotiation of what knowledge is, why it is valuable and to whom, and who has the power to control or benefit from its revelation and circulation.[9] Rather than seeking to critique IP directly, or using it as a frame for analysis, in this chapter I follow through my assertion about the "relationality" of knowledge, in this instance to ask a different question that might be phrased as, "How can one own a relationship?"

The question of owning a relationship with reference to the material that follows arises in direct response to the two themes advanced by this volume's conveners: What is knowable, and what is ownable? In light of both historic and contemporary ethnographies of the Rai Coast, approaching these questions involves a recognition that "knowledge," as Rai Coast people understand it, requires some form of relation—both for it to be acquired, and for it to manifest. Relationships themselves are coveted, and what is "owned" about knowledge might be said to be the relationships it constitutes. Rai Coast objectifications of knowledge—of which documents, records, and writings are new examples—can be seen as more or less successful moments and experiments in forming or transforming relations. This casts a series of inflections on "knowledge" that shape the practice of documentation and require different approaches to circulation and protection than those framed by IP regimes.

This need for alternatives connects this specific discussion to a wide scholarship on diverse strategies for and interests in knowledge, informing our understanding of IP as a specific and historically situated practice.[10] Influential work in the history of science has documented many examples of the relationships forged or necessitated *around* claims over knowledge, both before and outside of intellectual property regimes.[11] In this chapter, the emphasis is slightly different, as restriction is geared toward protecting access to the potential of relationships themselves, not to any knowledge "object." Toon van Meijl has recently demonstrated "compellingly that scientific metaphors of knowledge are unhelpful for understanding indigenous knowledge practices. Māori ways of knowing cannot simply be collected."[12]

In this chapter I use the term *aesthetics* to mean how something must appear in order to be recognized as a particular thing. So, for example, "mathematics" involves abstract numeration, or "a person" is a human animal with agreed moral, social, or legal status. Anthropologists have often described alternatives to these assumed forms, detailing conditions for recognizing personhood in different societies and pointing to the different criteria used to recognize combinations of agency, autonomy, origin, and other factors,[13] or indeed, describing different ways of counting or reckoning that can denote "mathematics" in certain spatial or kinship organizations.[14] I draw on the anthropological practice of describing unfamiliar forms for things to illustrate the particularity of our own conceptions, and utilize this practice to interrogate something that scholars often take for granted: "knowledge." This presents certain and specific challenges, including the resistance of pervasive "modernist" assumptions about what constitutes knowledge.[15] I will come to this later in the chapter, where I contrast the modernist conception of knowledge as being reliant on nature as a register for effect with a Melanesian form of knowledge that is reliant on effecting or transforming relations to other people.

The main idea in using the term *aesthetics* is that the way that "knowledge" registers in different historically situated contexts is a percept as well as a concept, and that its form also implies particular forms of connection or disconnection to people and to things. These perceptions then become the basis for different forms of claiming ownership. One of this chapter's aims is to establish the possibility of recognizing different kinds of knowledge aesthetics and to illustrate this idea by depicting a specific contrast between a contemporary Melanesian form and an anthropological, modernist-academic aesthetic of knowledge. This choice of comparison is consequent on the case study about traditional knowledge on the Rai Coast of Madang, PNG. I anticipate that the case study, and the comparison, will reveal more general principles about the connection of knowledge aesthetics and different forms of knowledge ownership.

The subject matter is a contemporary experiment to develop a documentation process for what the indigenous people from the Madang province call *kastom*. *Kastom* and its relationship to knowledge is an important concept in this story. Strathern and Hirsch offer the following useful definition of *kastom*:

> The Tok Pisin (neo-Melanesian) concept of *kastom* [is one] by which people indicate what makes them distinctive. Whether or not it is appropriately translated as "tradition" is a moot point. *Kastom* refers to practices flowing across the generations which (like reproductive power) are to be found in habits, conduct and well-being definitive of the present; in Bolton's words, *kastom* is not conserved but enacted, and may have a transactable or communicational value in relation to outsiders.[16]

The term *traditional knowledge* is a problematic phrase, especially when used as a synonym for *kastom*. In this essay, however, its use is a conscious choice. The choice arises directly from the impetus to associate the documentation project on which this chapter focuses with a wider interest in the field of traditional knowledge, indigenous knowledge (IK), and traditional ecological knowledge (TEK). Using traditional knowledge in the title of the documentation project, whether analytically appropriate or not, links what Nekgini speakers are doing to a wider world with whom they can generate recognition and connection. Through the designation, they have already received some support. The analysis of this chapter illuminates that old category (TK) because it demonstrates the impossibility of the TK categorization pointing to something that can be simply "added" to other kinds of "knowledge."

This analysis begins by turning to the early literature on the region that emphasizes the relationships in which knowledge figures. In the next section, I aim to show how relationships motivate action in which knowledge is manifest, and how knowledge takes the form of something that connects people. The ethnographic discussion that follows lays the ground for an outline of these recent experiments in the documentation of knowledge on the Rai Coast that highlight the relationality of both possession and appropriation and leads to a careful consideration of how different "aesthetics of knowledge" (i.e., what is recognized as knowledge, and how that is ownable) might be approached in ways that are "responsive" to its form. The conclusion then turns to what might best be described as methodology: how to engage—reflexively and ethically—in the production and exchange of knowledge in which relationships and ownership are always key components. Making knowledge unfamiliar in this way perhaps begs the question whether *knowledge* is an appropriate term at all, or if we need to understand practices of relational knowing within a wider conceptual framework of kn/own/ables—not only to conceptually grasp but also to responsibly practice an ethnographic project concerned with "traditional" ways of knowing such as *kastom*.

Other scholars have addressed questions of whether *knowledge* is an appropriate term in such contexts. Responding to a reader who thinks this kind of emphasis on knowledge is misguided in such small-scale Pacific societies, for instance, Lindstrom goes so far as to write, "Islanders know they live in information societies. They realize the power of talk. They recognize both the value and the danger of knowledge."[17] This chapter takes this observation seriously, while arguing that we must shift our concept of knowledge in order to respond to the issues of power and value that it raises.

KNOWLEDGE AND OWNERSHIP: COVETING RELATIONSHIPS

The most significant twentieth-century ethnographer of the Rai Coast, Peter Lawrence, concisely describes what he calls "the cosmic order."[18] Referring to the precolonial past as well as the time that he was writing in (the 1950s and 1960s) Lawrence emphasizes that all the processes and practices that shaped human life in the region were given to people by deities and ancestors. People's relations to those entities were vital, as only those with a relationship to specific deities had important knowledge, and they "accepted myths as the sole and unquestionable source of all important truth."[19] Sacred knowledge was paramount over everyday knowledge, and "the hard core of knowledge was the mastery of esoteric formula, only passed on during and after initiation."[20] Lawrence opines that religion is essentially a "technology" for living on the Rai Coast. That is, "religion" was the source of all valuable knowledge. Religious activity consisted of managing one's relations to ancestors and deities, and what those relations made possible. Without an equal distribution of knowledge—deities populated the area with different valuables and capacities; myths were distributed geographically, as were groups of people—the most valuable and significant possession anyone or any group could have were these myths and the knowledge they encoded. Therefore, "knowledge" of how to undertake the everyday and the ritual tasks necessary to life was a matter of such connections. Knowledge connected people to places, to deities, and to other people. "Rights to deities had to be established by genealogy or purchase. Otherwise they were invariably withheld from outsiders who, it was believed, would exploit them to their own advantage and so impoverish the original owners."[21] Characterizing Rai Coast thought with an "essential materialism and anthropocentrism," Lawrence asserts that the relations people had with each other, and those they had with their deities and ancestors, were based on similar principles: "What counted was that each party to a relationship should be forced to 'think on' . . . the other by the fulfilment of specific obligations—as in kinship and exchange commitments."[22] Lawrence concludes that "where there was no exchange of goods and services, there could be no sense of relationship, mutual obligation and value."[23]

As a more recent ethnographer in this area, I can attest to the fact that there is a vast everyday, narrative, artistic, and esoteric knowledge on the Rai Coast. The things we think of as practical (gardening techniques, healing plants) and things we think of as esoteric or mythic (narratives about ancestral activities, modes of divination or magic) do not fall into these neat categories there, despite Lawrence's mid-twentieth-century language of description. The practical—how to plant food crops in your garden, how to make them grow—are specified in the myths. What nurtures the main staple crops of taro and yam[24] *are* the spirits of the place, the ancestors, so looking after and attending to these is just as practical a matter as how the earth is prepared for the tubers. In fact, it *is* how the earth is prepared for tubers. It is not only the case that the "social" and "cultural" are always inherent in any practical activity in this way, they are also *the reason* for undertaking it. People do not grow food in a garden just to eat (subsist), as that would mean isolating themselves from the ongoing cycles of reciprocal work and exchange that result in finding potential marriage partners for their children, supporters in times of hunger, or protection from hostile others. Exchange makes the social principles of interdependence and co-constitution explicit, and its elaboration is as "practical" as it is "religious." Although Lawrence emphasizes certain distinctions between sacred and secular knowledge, and between pragmatism and materialism, it is clear that in this careful ethnographer's summation, knowledge is both about and proof of, as well as dependent on, relations to specific other entities—be they deities or the people already in productive relations with those deities.

I turn now to my own more recent ethnography to provide a sense of how an act or process is turned into recognition—how Reite people demonstrate knowing and how they make a claim over the outcome. This involves using ethnographic examples to make three key points of analysis. These observations are, first, that knowledge is apparent in the effects it has on other people or spirits; second, that knowing something means performing it—that is, demonstrating the capacity to have the effect; and third, that these capacities connect people in specific relations of obligation. To know is to perform, and performance makes relations visible. This overview sets the scene for a discussion of the contemporary impetus toward making documents from knowledge and provides more ground for understanding how and why relationships in which knowledge plays a central role are coveted.

I refer here specifically to Reite, a collection of hamlets of Nekgini-speaking people located about ten kilometers inland of the Rai Coast. Nekgini is a small language group of around 1,500 individuals who live by horticulture and hunting in a rainforest environment. Reite people cultivate taro and yam, supplemented by native vegetables, and more recently introduced crops. Cash cropping in the area began in the 1970s and has been sporadically practiced. They had—and continue to have—a strong sense of their distinctiveness as people who consciously choose to live according to ancestral practices, their *kastom*. Their economic activities as a whole are geared toward processes of social reproduction. Gardening and cash cropping are channeled into kinship exchange cycles through which people marry, raise children, go through initiation, and attend to illness and death. The Rai Coast traditionally had no institutions of inherited hierarchy, wealth, or political authority. Thus, in every generation, people compete for influence, using their ability to gather supporters and obligate others to themselves, usually based on being able to produce or attract amounts of exchangeable wealth which is fed into reproductive, kin-based exchange cycles. Possessive individualism[25] does not characterize personhood in the area,[26] since people conceptualize themselves as both being comprised of and forming part of other people's creative and generative efforts.

Pressure from an expanding population due to the availability of simple new technologies since the 1970s[27] has been exacerbated in the last fifteen years by the arrival of extractive industries—mining and logging—in what was, until then, a very isolated region. The rainforest is being depleted at an alarming rate as a result of these recent internal and external forces. Reite is connected to the local urban center of Madang through marketing cash crops and through limited educational, employment, and healthcare possibilities. Most people there have both a sense of inevitable social change and a deep anxiety about how to maintain their autonomy and distinctiveness while benefiting from "development."[28]

Nekgini speakers plant taro and yam following the strict and often complex processes specified in their taro myth.[29] These include the use of various other plants, planting patterns, the order of crops, specific gender and kin roles for the gardeners and the taro plants themselves, and secret names, as well as tunes in which those names are sung. An important part of a garden's gestation is what is known in Nekgini as the *wating*. Literally translated, *wating* is a "garden's shoot" or "garden's point of growth." Taro grown within this special area encourages the other taro in the garden. The *wating* is specified in a key taro myth, and it in turn attracts the mythical mother of taro, who is cajoled or coerced into caring for this garden's tubers at the explicit expense of other peoples' gardens. Prior to harvest, the man who owns the garden will "block the road" to and from his garden by planting another series of plants in the same style, which keeps the taro mother from leaving until all the taro has been gathered. It is fair to say that Reite people are competitive in their gardening.

Each hamlet group, or those associated with it through descent or trade, has a different form of *wating* and different names and tunes used during the procedure. Thus, each place is associated with what Lawrence refers to as deities, either because they have a direct connection to the characters in a myth who had divulged a particular style of *wating*, or because this style has been passed on to them at the quintessential moment of (male) knowledge transfer during initiation.

In the context of my anthropological research, people often told me about the specificity of their style of *wating* and emphasized the limitations on its use. I quote at length here from one garden magician:

> If someone else uses our style of *wating* we can charge them. We would ask, "Where did you get the knowledge to do this? It was not one of us who gave you this knowledge." We will say this, but we will be thinking that someone must have given him the knowledge of it. The strength of each style is in the *paru* [secret name or spell] that accompanies it. To use a style the person must know what the spell is and what it refers to, and it is for this that we would charge someone we found using our particular style. He may say, "I just decorated my garden in this way, and now you want to charge me" [i.e., this is not fair/I didn't use the *paru*]. But we would reply, "You are lying." A man cannot make up such a thing in his own head, he must have got this knowledge from someone else, even though we never gave permission for this transfer. We would therefore charge him. Once he has paid us, he can use our style, and even pass it on to whichever sister's son he chooses.[30]

Some key characteristics of what can be known, and what can be owned, are apparent from this short description and statement. One is that knowledge is performative. People show knowledge by doing or making something. Another is that it is restricted to those who have a right, through kinship or payment, to use it. I say "use it," as that is the crucial element. People who are aware of but do not use the form of a *wating* would not be seen to know.

Then there is what lies behind the appearance. Making a form appear in Nekgini action is a claim to, or a demonstration of, an underlying connection to power.[31] In the case of charging for the appropriation of a *wating*, it is not the style itself that is the issue, but the fact that there is knowledge that gives the form power. This in turn has to have come from other people in the past, and it takes the form of a relation to a deity. Hence, having knowledge is always seen as being part of a relationship in which obligation and reciprocation are crucial.

The hamlets of Reite are grouped into clusters, reflecting local principles of social organization and kinship. In one such group of hamlets, a successful trade store was operated by a young, unmarried man during the 1990s. The store was attached to the house where he lived with other unmarried young men, as is the usual practice in the region. These "boys' houses" are dwellings and are not the ritual homes of the Nekgini speakers' male spirit cult. Cult houses—*passae*—are off-limits to noninitiates, whereas boys' houses are not usually so. Yet suddenly, kinswomen were warned against visiting the trade store because of danger to their health from the proximity of the spirit cult in the house adjoining it. Now the presence of the male cult is a very public matter, albeit a matter of "consensual secrecy"[32] as to its objects and methods. In this case though, everyone was confounded. When had the male cult arrived there? Who had constituted it? Why was its presence not apparent in any other ways? As it turned out, a young Reite man who had been traveling in another region of Papua New Guinea was staying in the storekeeper's house. On his travels, he had become close to a certain group of islanders connected by language and *kastom* to other Rai Coast language groups who initiate their young men. This young man had participated in their initiation ritual. The surgical operation that forms the core element of this initiation had not been successful, so he had come home and the young men living in the storekeeper's house had redone the surgery at his request. Talk of the presence of the spirit cult was, in reality, a euphemism for his seclusion while recovering from the operation.

Two major problems were immediately apparent. Despite the consequences—which we will come to in a moment—this young man's maternal uncle was so incensed by these events that he complained vociferously in public. In doing so, he made it inevitable that the people who have the right to conduct this form of initiation on the Rai Coast would also hear about it. His public revelation turned what many had hoped would remain a village-level complaint by the uncle over his nephew's initiation into an intervillage and interlanguage group dispute with potentially grave consequences for many people. Without overburdening the reader with context and detail, there are different forms of initiation on the Rai Coast, and certain villages and language groups have the right, either through revelation by a deity or through purchase, to perform the particular surgical operations in question. These initiation sequences are held to be powerful and dangerous, so they are jealously and fiercely guarded.[33] It is no exaggeration to say that Rai Coast people consider their misuse a matter of life and death. I used Gourlay's term *consensual secrecy* above, as it aptly conveys the fact that people choose not to know, or *choose not to perform*, things that they may well know *about*.

People know about other people's garden designs. They know about other people's initiations. But they do not use them, partly out of pride, partly out of fear of reprisals. As I have discussed at some length elsewhere, people will not even tell the narrative of a myth they do not own themselves, however many times they have heard others recount it.[34] *To act on knowledge is to claim inclusion in the relationships of its origin*. Such action is understood as future-oriented; that is, actions based on these powerful forms of knowledge generate and sustain relationships. As the economy there is geared toward the production of people—through labor flowing into kin-based exchange cycles—it is the recognition of relationships that defines a person's worth and wealth. This perhaps explains Lawrence's emphasis on pragmatism and materialism. Relationships are seen, they are made visible, in flows of wealth between people.

There is also a point to be made here about whether knowledge *as an object or thing in itself* is the matter at hand. The young man and his assistants did know what to do. It is not the knowledge in this sense that is at issue, but rather the validation of the performance. The man who appropriates a *wating* is accused of appropriating the relationship to the deity that is encapsulated by the connection between style, *paru*, and reference. The anger demonstrated by the uncle arose because it was his *right* to initiate this boy. He had *worked* to produce the opportunity to be named and paid to pass knowledge on in the context of an initiation sequence. When quizzed over why he had undertaken the initiation elsewhere, the young man complained about all the hard work involved in following the process at home.

So, this young man had made two thefts. He had stolen the opportunity his initiation provided from his mother's brothers, and he had stolen the right to perform a particular surgical operation from the owners of the initiation. *Work* is the cover-all term in Nekgini for anything that involves organizing elements and people to achieve growth or transformation in others' bodies. Knowledge, then, not only is a relationship but also resides in people's bodies *as capacities that have been given by others*. By bypassing his uncle, this young man had short-circuited, as it were, the healthy and appropriate flow of exchange items and knowledge that amounts to a life substance moving between kin. Tony Crook, discussing knowledge forms in Bolivip, a village in the Star Mountains of PNG, writes that "'knowledge' appear[s] as an activity of the body, and a circulating, nurturant bodily substance."[35]

From this perspective, knowledge is owned as certain aspects of a person's capacities in relation to others. Knowledge arises in places and is shared but not communal. What an individual knows is all about where they are from, and their capacities for action—from growing tubers successfully, to transforming youths into adults via initiations—all of which come from somewhere and someone else. This knowledge both connects and differentiates people. The economy is one in which people's efforts are directed not toward subsistence but toward the production of food and wealth that forms part of kin-based exchange cycles. In their understanding, you grow others by feeding them—requiring knowledge—and you transform them at key life-cycle moments by sharing knowledge and capacities *for which you retain recognition*. Knowledge is effective in the way the social world is shaped and made to appear through practical everyday activity, including ritual and exchange. Knowledge is not apparent in its effect on nature, on something external to the human world,[36] but in its effects on other people, on their bodies and thoughts, on their capacities and orientations.

The aesthetic of knowledge here does not work with knowledge objects, with bits of information as discrete units, but with connection. Knowledge is performed as a relation and it requires a relation. It is fundamental to both parties. In this sense, knowledge *is* the relation. Knowledge is ownable just as a relationship is ownable.

DOCUMENTING REITE *KASTOM*

My association with Reite began in the early 1990s when I was enthusiastically welcomed there to undertake anthropological fieldwork.[37] The enthusiasm was indicative. They had their own reasons for engaging me. During negotiations around my arrival and presence, Reite people stated clearly in publicly staged events that I was being welcomed as a student who was trained to write about *kastom* and history. Reite people told me that they wanted their *kastom* written down for future generations, and as a means to achieve recognition from the wider world. Strathern and Hirsch's definition of *kastom* cited above is helpful in comprehending this impetus.[38] They emphasize that *kastom* is not conserved but enacted, that it is "like reproductive power" and that it may "have communicational or transactable value." *Kastom*: its value, ownership, and potential is at the heart of my relationship to people in Reite. It is—if not always explicitly—what we explore together.

The personal and self-referential aspect of what follows is thus not only an accident of my own personal involvement as an ethnographer. As will be shown below, it is a crucial consequence of engaging with *kastom* as a relational way of knowing. It has consequences for ethnography as a method. This began, for instance, with the fact that what I wanted to know in order to write a good ethnography—the way I saw myself fulfilling our agreement that I was there to document *kastom*—was not always what Reite people thought was relevant to tell me. Lindstrom puts it well: "The hidden task that any anthropologist faces—no matter what his or her research interest—is to figure out the rules and conditions that govern talking and access to knowledge in a society. Landing on an island and asking, 'tell me about your economy and religion' is not enough."[39] Reite people clearly understood they were in a relationship with me involving knowledge. This brought my knowledge aesthetic into contact with theirs. In many instances, I was expected to *recognize* key myths and *respond* to their revelation. The information they offered was often partial. Investigations of genealogy always seemed to turn to demonstrations of connections to particular myths or stories rather than a comprehensive map of who was related to whom. What they thought was important to tell me followed from what they thought the purpose of the exercise was. A comprehensive sociological survey did not really figure in that purpose for them. None of this is surprising, as different expectations, hopes, and understandings are grist to the mill of anthropological endeavor. It was clear that they understood the relationship with me, and texts, as a potential vehicle for making Reite *kastom* into a form that might have more directly beneficial outcomes—be it practical or material for instance—than those they had previously achieved through outsiders ("white people"), where "white people" (*ol wetman*) is the local term used generally for all foreigners. My willingness to engage with this project was interpreted as evidence of a prior connection to them—it was assumed I was already related. This illustrates a particular cast on, and temporality for, knowledge according to their aesthetic. As Lawrence implies, there is no *new* knowledge.[40] Knowledge exists as particular relationships with particular others—deities and ancestors—and it always appears in the context of shaping or effecting other people's responses, growth, or capacities. It follows that there are no *new* people.

Assumptions around knowledge and its value and ownership pervaded the enterprise of ethnographic fieldwork. For one thing, and in addition to the above material setting out some of the contours of a Reite "knowledge form," the power of writing things down has been impressed on these people by all their contacts with powerful outsiders who do just that. Their question was, *what are the relationships that writing brings into being, or can produce?*[41] Much of their desire for documentation has arisen in the context of their conviction of the power of their "knowledge," and the perception that it has not been properly recognized.

THE AESTHETICS OF KNOWLEDGE AND THE PROBLEM OF MYTH

From the outset, explicitly to protect me from accusations of stealing or appropriating knowledge, Reite people decided that I would not be *given*—that is, I would not write down or even hear—the *paru*, the secret magical names that make procedures effective. They decided that it was important for me to record mythic narratives, but not mythic character's names. Likewise, I should record rituals in gardens and initiations, but the specific and very personal spoken formulae that activate the procedures were not revealed. In fact, fears of appropriation were always expressed as fears of the appropriation of *power*.

I want to draw attention here briefly to two aspects of different knowledge aesthetics, which relate to the form things must appear in if they are to be counted as knowledge. As I have made clear in this chapter, for Reite, that form is all about the connection between one thing and another, and about the specific relationships that make any action or process effective. Thus, it was considered acceptable to give me narratives of the origins of taro, as I could not exploit that knowledge without the associated *paru*.

I was not at all concerned by this decision. In fact, I was grateful for it, since it had been made to protect me from accusations of stealing knowledge—which were hovering in the background anyway. But at a deeper level, and without any intention of conveying petulance, magical names were not actually what I was there for. Techniques for how to plant a garden or the social and technical process of carving an artifact are much more easily translated into anthropological knowledge than esoteric formulae. Similarly, social organization, details about who exchanges what with whom, and other particulars provide the necessary sociological information from which to craft an ethnographic description. They look like knowledge—how to do things, what is done, what is believed, and so on—and are convertible into ethnography. I was faced with the problem that our different aesthetics of knowledge meant that Reite people were mainly concerned with magic.

This brings us to a stage in the argument where we might label magic as "relations to powerful others." Their fear was not for the loss of an object—a name—but the appropriation of a relationship in which they would not benefit. As Lawrence notes, "Rights to deities . . . were invariably withheld from outsiders who . . . would exploit them to their own advantage."[42] Villagers do not share their relation to the taro goddess. Keeping her in your garden comes at the expense of someone else. Having knowledge is not the point—knowing how to activate the relationship is. And this brings us to a crucial distinction between a Reite aesthetic of knowledge and the more familiar academic aesthetic. We can accept the notion that knowledge is personal, but the idea that something practical and effective in one place might be ineffective in another *because of who practices it* begins to challenge our sense of knowledge altogether.

This was brought home to me forcefully in thinking through a successful moment in our documentation endeavors, the publication of the coauthored book *Reite Plants*. This book was the result of a long collaboration between Reite elder Porer Nombo and myself. Porer approached me early on in my time in Reite and asked if I could make a book about the plants he used for healing and ritual to pass onto his grandchildren. This culminated in the publication by Australian National University Press of a color volume containing descriptions of more than one hundred plants and their uses. Happily, this publisher's policy is to enable free distribution of the text by PDF download, and there have been many downloads. Porer and Reite have received recognition—although no direct payment—and *Reite Plants* has stimulated other relationships, including those with funders and groups supporting the TK Reite Notebooks project described below.

Yet I, and some of the book's audience, perceived problems with its presentation[43]— that is, with the aesthetic of knowledge into which our text was fitted. Papua New Guinean readers observed that there were no spells or magical formulae accompanying the text. I noticed that Porer's extensive knowledge of plants was arranged in the book in a way that made (partial) sense in his terms but would not make sense to most academic readers. Put simply, where the PNG readers found *too little* magic in the book, from the ethnobotany perspective it looks as if there is *too much* magic. Undertaking the exercise made me realize certain inadequacies in using the form of an ethnobotanical textbook to represent what Porer *knows* about plants. To summarize a complex problem, although it looks as if plants are the subject of a book that we published together, for Porer, plants do not work in isolation or without practitioners who know about their provenance and the ritual and magic that make them effective. Writing down what Porer knows *about plants* makes it appear that the plants can be isolated from people and processes. He uses plants based not on their chemical properties alone, but on mythic narratives, magical associations, and relations with spirits and ancestors. The title of a 1999 editorial in the leading scientific journal, *Nature*, demonstrates the point emphatically: "Caution: Traditional Knowledge. Principles of Merit Need to Be Spelt Out in Distinguishing Valuable Knowledge from Myth."[44]

Reite Plants hovered between two different aesthetics of knowledge, and in a mirror image to the concerns of the editors of *Nature*, I became aware of the danger that traditional knowledge, in our rendering, may look like an inadequate version of science. Plants work, but do the practitioners really know why or how? The assumption I desperately want to avoid fostering is that use is a matter of "superstition and belief,"[45] and therefore that, for this knowledge to have *any value*, it must be stripped of its relational and contextual factors. This is the crux of the *Nature* article: that knowledge must be distinguished from myth.

As we know, in many instances, indigenous and traditional knowledge holders are left in the position of providing clues that corporations and laboratories can follow up, to their advantage. That is, making the move from a piece of traditional knowledge to an intellectual property claim rarely benefits indigenous villagers such as Nekgini speakers.[46] Cori Hayden recounts that pharmaceutical companies actively avoid collecting plants from any place where they might have to acknowledge or share benefit with the original users.[47] In other words, an aesthetic of knowledge that isolates information from stories and myths has the potential to also isolate it from its holders. "Whether or not others appropriate that tradition *as* tradition (their own particular tradition, generalized national tradition) will depend on context. The reverse may be taken as even more problematic, that is, when tradition is *erased* in so far as what is being taken is being valued for quite different properties than those it originally encapsulated and thus not for connection to (anyone's) ancestral values at all."[48]

In these circumstances we asked ourselves (Reite villagers and relevant others): What would it look like if knowledge of *kastom* were recorded, not according to an academic aesthetic of knowledge, but at a meeting point between academic and Nekgini?

A strong motivation for undertaking the experiment was the enthusiasm with which *Reite Plants* was greeted by other Papua New Guineans, even those who understood that something vital was missing. Many expressed a desire to produce a similar kind of record. Despite their fabled ubiquity in PNG, anthropologists are quite a rare resource, and not many have the skills or the time and inclination to undertake the painstaking process of ethnobotanical or biological documentation. We wondered whether it would be possible to devise a simple, cheap, and accessible system with which village-based people could make, and keep, their own records.

The idea of people making records for themselves, where they would control both the content and the circulation, also promised a potential solution for certain worries, including the fear that outsiders might exploit the content for their own gain. So, we began a process in which, again, the meaning, value, and form of knowledge was under investigation and negotiation. The TK Reite Notebooks project (TKRN), funded by the Christensen Fund, instigated an experiment into the codesign of a self-documentation process that was responsive to this context.

We began by drawing on an individual who had a specific set of skills and approaches that had already been introduced to a couple of Reite people. This was Giles Lane, a London-based artist from an arts organization called Proboscis,[49] which has developed a system for "public authoring."[50] Porer Nombo and Pinbin Sisau met Giles and began exploring his public authoring system during a visit to the British Museum in 2009.[51] Public authoring is based on a paper-folding technique that allows commonly used paper formats to be hand-modified into self-binding booklets. These booklets require simple tools yet are designed to become hybrid entities existing both physically and digitally. PDF templates are created that are the basis for the booklets. When printed out, the sheets are cut and folded. The templates can be designed with, for example, different rubrics, questions, and information, and of course in different languages. They are then available for people to fill in any way they see fit—for instance, they choose the number of pages, any prompt or subheading, written or drawn records, and so on. More or less guidance or direction is available when designing specific templates. Once complete, the booklets can be unfolded and scanned, offering a potentially more durable digital copy of the original. Digital files can be printed and refolded to provide a facsimile. The scans can be stored, combined with others to generate a series or set, and shared through digital media formats if desired. Giles offered a way forward with his booklets, enhanced by his experience in working with people to design ways of using them, and we began what we thought of as an extended codesign process of the templates and a process with protocols for their use. This involved intense discussions around what templates for specific booklets that were designed to be useful to villagers would look like, and we had many public meetings in Reite villages to gauge interest, receive concerns, hear oppositions, and solicit advice.

The TKRN project was never conceived as outsiders helping villagers to preserve traditional knowledge. We approached the endeavor with the notion of an exchange and common exploration of what a Reite documentation of their *kastom* could look like. Whatever the process and outcomes, these were always going to be about more or less well-comprehended—on both sides—meeting points. As I now describe some of what happened during our engagements, it is important to keep in mind that this was approached as a common project and, necessarily, as an exchange. That meant each side offering what they had, while thinking about how that might be made responsive to other possibilities.

From the outset, we recognized the importance of embedding clear and unambiguous information about the project, its aims as well as its limits, and thus conscious and informed agreement by participants. I will use the term *authors* to designate these participants, even though "authorship"—as I come to below—is not an entirely adequate category here. Author-participants are asked to confirm on the front cover that they have understood that TKRN is providing materials with which they can, if they so choose, make a personal document about something. They are asked to indicate the scope of sharing for the particular booklet that is being produced. The options range from being completely private to having them scanned and returned to them as a digital copy, to various other restrictions on circulation—for instance, limited to just family, village, women, or men. Each time we worked with a different group of people, there were further articulations of what people might be concerned about, and what control they would like to exercise over the documents they produce. These discussions have led to evolving iterations and have been an important mode of engagement with the idea of documenting *kastom* in the first place. Though far from perfect, the requirement to consider circulation ensures that participants are definitely made aware of the possibility of appropriation and, in response, the potential for absolute control—and, indeed, the totally voluntary nature of using the templates.

We came to call this method *engaged consent*. In contrast to the procedures that I had to follow in my universities for this project to go ahead—gaining permission from ethics committees, guaranteeing ascertaining the participants' free, prior, and informed consent—our form of engaged consent is not simply a signing off that allows me or other researchers to use *their data*. It is instead a moment where a booklet's writer considers their intentions and interest in the outcome. The emphasis is firmly on articulating the wishes of those filling in the booklets, rather than on asking them to agree to an already established framework that absolves the documenter of responsibility. It also makes the writer the *author* of their version of what is produced.

Alongside this engaged consent section on the front cover of each booklet is a space for a photograph of the writer and their name, place, date, and booklet title, if they choose to give one. The photograph helps to make evident a personal connection. By attaching the photographs—in most instances, immediately—and maintaining the digitized records as facsimiles, the process keeps writers in view. Whatever they choose to record appears in their own handwriting,[52] and in a booklet that was handmade by them. Obviously, it is never able to fully address the complex interleaving of knowledge and person, but this format does make a move in that direction. In addition, Giles and I worked with many Reite people to phrase prompts, questions, or guidance notes for the use of the booklets, placing a strong emphasis on recording from whom the story or process came and where it originated. We also reminded writers of the importance of recording only things that would not cause disputes. We then tried to find appropriate analogies with local protocols in which knowledge is passed on in particular relationships.

For those people who choose to do so, it is possible to make the scanned files of their booklets widely available through digital channels. There is no encouragement to do this in the method or protocol itself. Somewhat to my surprise, Reite people have collectively agreed, after repeated discussion and questioning, that their booklets should be made freely accessible online. There is now a simple website on which they post their booklets.[53] It is true that, once these stories and practices are in the public domain, anybody can use, mix, remix, or otherwise appropriate them. As we will come to below, Reite people have devised their own system for managing this risk on their own terms.

The fact that people independently decide on the topic and scope of their documents makes for some interesting outcomes, and discussions. Many people recorded similar things. Taro and yam figure prominently, for example, with levels of detail ranging from complex and intricate to very minimal, and differently phrased accounts of taro and its origins, gardening techniques, and so forth. Rather than being concerned about repetition or duplication, we realized that these individual records allow for a diversity of representation on any one topic, and for different aspects of the same thing to be recorded. Although this was unplanned, it also mitigates the emergence of overly canonical or authoritative versions that come under the control of one generation or group and reflects the multiple connections and types of relationship that constitute knowledge, since it also makes it possible for younger people to document *kastom*. No one has made any complaint about these duplications or different versions. Thus, each booklet is just one actualization that, in fact, is only part of an actualization anyway. Writing things down does amount to a kind of fixing, but it does not guarantee that any reader is necessarily going to practice it in that same way.

Some absolutely beautiful documents are being produced. Many are detailed, careful, and superbly artistic. The spontaneous introduction of beautiful drawings into many of the booklets is significant. Of course, artistry is a key aspect of any process involving knowledge in Reite, as aesthetic effect is crucial. As I have emphasized, knowledge is a relation to or manifestation of a particular kind of power here.

Most, if not all booklets—however carefully produced—are incomplete in some way. Booklets are more often than not *indicative* of a story or process rather than a complete rendering of it. Several people have approached me to evaluate the booklets they had produced, and several have also asked if they were *complete*, clearly displaying an understanding that the form they are content with and those they imagine I require are different. Yet, even those people who display a clear understanding of this disparity and are the most vocal advocates of the booklet project have not used it to make a comprehensive record of knowledge that is in danger of being "lost." Most records are of things that are quite well known. There has been no systematic effort to use them to record the knowledge of frail old people, for example, nor, seemingly, to prioritize rare esoteric knowledge. Perhaps this reflects the fact that there is *no sense of an existential need to document knowledge as such*. The desire for documentation does not come from the same aesthetic of knowledge as that of the academy.

As Crook argues, the existential need to document knowledge arises from a different aesthetic, one in which there are objects or units of information that will disappear if not transmitted or recorded.[54] The immediate lesson to be learned from this is that knowledge is operating in a different manner, where performance and enactment is key. Reite people are performing versions for each other and to each other. So, the documentation is about capacity and relationality, not a catalog of objects that, lined up together, could ever be called encyclopedic or comprehensive.

Although—or perhaps because—the booklet form was intended as a way of facilitating Reite documentation processes for their own ends, it has been revealing to see how closely I am implicated in their production. People did not engage in making booklets unless I was physically there in the village. In some ways, this confirmed that documentation is tied into expectations and interest in a relationship with me. More widely, that knowledge and its performance require a relationship to motivate action. Questions that concerned Reite people about the form and content have been more about an aesthetic of effectiveness in these terms than about completeness or coverage per se. And even more interestingly, incompleteness was absolutely deliberate in many cases, and explicitly motivated. As Annelise Riles observes while discussing the modern ubiquity of documents as forms of knowledge, "There is nothing inherently passive or automatic about actors' responses to documents. . . . The agency of the form and the form-filler are not neatly circumscribed."[55]

VALUE AND CIRCULATION

Some Reite people remain suspicious of the whole endeavor, maintaining a narrative of external benefit, which would be realized by me, and possibly my local supporters. The tactic of giving people control over content and circulation was only partially successful. In trying to take myself out of the frame, as it were, as author, steward, or beneficiary, I might have been wholly missing the point. Hence, we are still struggling at the meeting point between knowledge aesthetics. For example, some younger Reite people have noticed that they have to pay for data on their mobile phones, and they associate the charge for data with a charge for access to the booklets. They are correct, of course. While university academics in wealthy countries are used to assuming that access to information is free, in reality it is not. The fact that the income from selling data does not come to me, or anyone associated with the booklets, is both true and, in a sense, irrelevant. Someone is benefiting from their knowledge and making a gain in which they are not included. I am constantly reminded that documenting knowledge *for its own value* is not part of the aesthetic here.

This brings us to a question that was discussed at length in Reite, and that I have been asked much more forcefully by academics and activists in various public presentations in the United States and in Europe. Is it not naïve to make traditional knowledge available online? Surely this will make it far too easy to appropriate? The answer to this is multilayered. For one, controls on circulation of the booklets' content were both formal—the embedded definition of how public the writer wanted them to be—and, more importantly, informal. That is, Reite people instinctively use the booklets to provide information about a process or myth but omit the esoteric or secret aspects necessary for the process to have any effect. They leave things out. Those words are of no practical use or interest to outsiders, but they are the key element for Reite. Of course, one point that tallies with my argument so far is that *paru could not be* embedded in the notebooks because *they require specific transmission*. Taboos and restrictions are personal and specific, and the transfer of power relies on an ongoing relationship of obligation to the donor.

Like the Papua New Guineans who perceived that something was missing in *Reite Plants*, we might ask, What use is the documentation if the key aspects are excluded? There was a clear answer to this from writers in Reite, who explained that the value is in creating a new potential route to sustain their *kastom*. This can be realized locally when children, nephews and nieces, and grandchildren are engaged with their elders in making the booklets. They are a new and intriguing context for transmission and exchange. Or the potential may be regional for them. If the booklets are made public, other Papua New Guineans will see and be stimulated to engage with Reite in exchanges and relations around *kastom*. The booklets demonstrate Reite people's expertise. They are seen as a stimulus for transmission that involves a personal relationship with the knowledge holder. It is here that the booklets take their inspiration, in terms of process, from Reite protocols. They are there as a reason for further relationship making. For this they need to be available, to circulate. Reite people have knowledge of their own knowledge in the sense of knowing the purpose and process of what they do and observing its effects and whether these meet the ethical and practical purposes for which they exist. Knowledge becomes an exchange item or boundary object between them and others.

The issue for Reite people is not that other people might copy their booklets or their stories, but that other people might utilize them for some purpose that brings a benefit from which they are excluded. Their solution is to make available some things that they are happy to see act as a link to others, things that will draw others into a relationship with them. What they choose to circulate is knowledge, acting in a different way from its practical application in everyday life. In the booklets, knowledge takes a form in which a different value can be realized. So, rather than stop people from copying or circulating the booklets, many individuals have adopted a different strategy altogether: encouraging people to print out and view the booklets. In return, they assume that they will achieve recognition.

Reite people's obvious inability to pursue IP claims against any appropriators is part of the inequality of global capitalism. Sillitoe argues that "the idea that IPRs [intellectual property rights] offer some kind of general protection to local science by assigning rights and excluding others, should not be misconstrued as protecting knowledge from extinction under the relentless onslaught of the economic and social forces of capitalism. Regarding IPRs as a solution to the protection of local knowledge is thus to misdiagnose the problem."[56] IPRs require states to enforce them, and, as Antons explains in his summary of a lot of earlier work, this tends to put local practices into the frame of knowledge as an economic resource, where techniques, designs, and even plants come to be controlled by urban elites and bureaucracies.[57] The registration of IP in traditional knowledge misconstrues—one might argue, willfully misconstrues—both the forms of knowledge and the ownership practices of traditional knowledge holders.[58] Antons, for example, cautions strongly against the move from perceiving local traditional knowledge as forms of cultural heritage to determining them as intellectual property.[59] In this he echoes Hirsch and Strathern.[60] While WIPO's 2016 statement on the value of traditional knowledge—as "dynamic and evolving"—moves toward a more accurate recognition than their previous renderings, its language inexorably draws attention to "innovation" and "scientific value" and inevitably points toward the potentially propertizable aspects of traditional knowledge.[61]

The TK Reite Notebooks project is opening up different possibilities and forcing us to think about appropriation in its relationality, rather than as an act in and of itself. We have designed the booklets to retain as much context—that is, connection and relationality—as possible. Yet, the Reite documenters were way ahead of us in this regard. There *is* information in the booklets, and context is there in terms of the potential to identify and trace the source of a practice or story. But inevitably, the booklet is not the knowledge in a text, but a potential for other relationships. The actual relations that connect these things to the real world have to be lived; this is a different kind of knowledge that means something different to different readers. What Reite people record is there to remind them and their children of the particular kinds of stories, practices, ceremonies, designs, and relations to history that they will be able to embody as practice. People who are not there can only really view it as a representation of something. They can behold it, not practice it. While clearly not following the logic of intellectual property, the documents are *also* an a priori claim, a mark of knowledge and understanding.

What, then, does making something available do in this case? These Papua New Guinean villagers understand the potential of knowledge to create relationships. For that to be possible, knowledge is attached to people in ways that recast vulnerability through revealing and hiding aspects of itself. They shift attention away from knowledge *objects* toward the relationships involved or created by transmission.

APPROPRIATE DOCUMENTATION?

Eugene Hunn advocates that we should "write" ethnobiological anthropology in a way that allows a wide audience to "appreciate traditional ecological knowledge."[62] He argues that a level of expertise on the part of the ethnographer is an important starting point. Translating "their" "cultural knowledge"[63] of "the natural world"[64] requires a strong understanding of that natural world on the part of the translator. Hunn says that a skilled and knowledgeable ethnobiologist has the opportunity to meet indigenous people with enough understanding for a mutually interesting exchange. This exchange has the potential to represent the wonders of the natural world *and* the deep and important understandings of traditional knowledge holders. His model example is the much-admired collaborative work of Saem Majnep and Ralph Bulmer. Majnep was from the Kaironk valley in the Schraeder mountain range in Papua New Guinea, and Bulmer was an anthropologist and expert botanist and ornithologist. Their classic coauthored book *Birds of My Kalam Country* laces Majnep's descriptions of flora and fauna with contextualizing descriptions containing Latin names and scientific identifications where possible, alongside botanical and ornithological commentary from Bulmer.[65] Hunn cites this as an example of "expert meets expert," where their common ground—the natural world and a close and technical appreciation of it—provided the opportunity to create a compelling text that "writes culture" in a way that transcends the concerns over reflexivity or master narratives that have plagued social and cultural anthropologists since the end of the last century.[66]

While I can only agree with him on the value of traditional knowledge, and of course on the centrality of the classic—and brilliant—work of Majnep and Bulmer, I believe that Hunn's position does not take account of the times when a traditional knowledge holder's expertise is phrased in ways that confound scientific expertise—or, even worse, that seem to be at odds with scientific understandings and explanations. Take one example of many: pages 38 to 40 of *Birds of My Kalam Country* describe a healing ritual undertaken by Majnep's forebears. In this, pigs are sacrificed, and the patient is rubbed with slimy substances from specific plants and vines "to make the sick person's skin all slippery, so that the sickness cannot get a hold on him."[67] Other leaves and plants are used that "drive out the sickness and the evil that is causing it" before the whole group collectively plants cordyline shrubs: "With the planting of the cordylines the sorcery that has been causing sickness is planted also. They are a sign that everything is now straight with this group, for planting them causes all the *kawnan* ('souls' or 'shadows') of the members of the group to return and stay safely at that place."[68]

I have quoted only certain points here to draw attention to Majnep's emphasis on what we might term magic, and to the explanation of the use of certain plants through this narrative route.

Annelise Riles takes a different approach to Hunn, but one within the same ethos: the engagement and mutual recognition between anthropologists and those among whom they work. She calls for a focus on "ethnographic response"[69] in a time when anthropologists can no longer imagine that they will find culture or knowledge as objects out there in the world to be discovered.[70] Riles asserts, "If anthropologists ever believed that facts were there to be 'collected' in the 'field' rather than produced collaboratively in the ethnographic encounter, they have abandoned any such pretense."[71] In a striking parallel to Hunn—striking because of their different starting points—she describes the "mutual empathy"[72] necessary for successful ethnographic exchanges. Riles considers this to involve an explicit tolerance of the different aesthetics that apply to knowledge, and a willingness to try and accommodate or imagine that they may be talking about very different things, even if using the same language(s).

I have already drawn on Crook's argument about knowledge in Bolivip, PNG. He proposes that in any knowledge exchange there, it is a social relation that is at issue, and not any object of knowing. Crook argues that this has created a huge challenge for anthropologists to comprehend: "Part of the trouble is having all-too-familiar metaphors of knowledge falling easily to hand. . . . For example, knowledge is perceived in building-block-like 'domains' . . . lined up."[73] He makes a strong critique of anthropology itself for always assuming there is *something* to be discovered or uncovered, an elusive and evasive core to other people's lives, which *looks like something we call knowledge*. Something, indeed, that can be discovered, cataloged, and documented as key pieces of information found ready-made "out there" in the field—in other words, the "knowledge objects" that make up a culture.

Instead, Crook focuses on a series of relationships in which people transact care for understanding; "these exchanges involving knowledge would be understood as the mutual support of persons who become encompassed together as if they were one person."[74] According to this, what looks to an outsider like "secrecy"[75] and the constant loss of bits of knowledge as generations pass away are in fact nothing like secrets at all. The value placed on knowledge in Bolivip, and the shifting restrictions on its circulation, are dependent on the relational power of knowledge that flows between people, helping to constitute their bodies and persons at key moments. Each Bolivip statement or performance is anchored in an approach to knowing that makes relations between things and people the focus and object of effort, not the discovery or transmission of any item of information. Crook writes that, by undertaking field research in Bolivip, "knowledge implicates the anthropologist and involves their person! Important knowledge here is not knowledge information or data, but a form constitutive of persons which draws upon the bodily resources and substances of one to grow and bolster the other."[76]

In his recent book detailing complex and extraordinary environmental and social adaptation in the societies of the Massim—a region consisting of the islands that lie to the east of the PNG mainland—Fred Damon makes analogous observations.[77] He vividly demonstrates the impossibility of approaching knowledge as a series of fixed points. His attention to local variations, and to the different information received over an extended enquiry, offers a picture of a variegated and adaptive socio-ecological practice in which what is true in one place and for one person is *not true* in another. He asserts that even though the chapters in his book record in detail what people know, they do not provide a coherent set of things that are known. Again, despite the information on plants, planting, and the effects of plants and soils on each other, none of the knowledge captured there will be effective outside the particularity of its situation. So, anyone looking for a simple formula that could be applied to managing such environments will be disappointed; but that is the point. By revealing partiality, distribution, contrast, and alternatives, Damon is eventually able to describe two things clearly: details of the variable practices, trees, ecology, landscape management, and other things, *and* the overall shape of a knowledge system—which is also a social and technical system—in which knowledge about trees and gardens, fallow fields and orchards, are tied together in different ways by different places. Knowledge *of* plants is also of places and terrains, of soils, and the conditions of growth. Knowledge moves toward usage, and usage is always a matter of social positioning, time, place, and need. Need is conditioned by who an individual is in relation to others.[78]

CONCLUSION

The TKRN project is an attempt to think, along with Reite people, about their stated desire to preserve their *kastom* for future generations and to be responsive to my relationship with them and its foundations. Riles writes that "ethnographic response is part art and part technique, part invention and part convention, part the ethnographer's own work and part the effect of allowing others to work upon the ethnographer. It is theoretically informed but not theoretically determined."[79] She proposes that "anthropologists [should] begin to think of what we share with our subjects as a source of the very conceptual distance that makes analytic progress possible."[80] Here the contrast between Hunn's and Riles's approaches is apparent. However, they share a source of conceptual distance that will enable an appreciation of knowledge forms in different guises.

In this chapter, I have emphasized relationality because relations have proved to be a central aspect of knowledge on the Rai Coast, and because they appear there in ways that are unfamiliar in a modernist conception of knowledge. *Relationality* refers to the form that knowledge takes, to the aesthetic of its form. The criteria for recognition—the aesthetics of the knowledge form—include a reference to where and how knowledge is constituted and exists, and that its effects are apparent on relations between people, and between people and things. Restriction is geared toward protecting access to the potential of relationships themselves, not to any knowledge object. This chapter has intended to establish the possibility of recognizing different knowledge aesthetics such as this, and to illustrate the idea by use of a specific contrast between a Rai Coast form and an anthropological, modernist academic aesthetic of knowledge. I have assumed that there is a connection between the modernist context of academic knowledge production and intellectual property regimes that have arisen in the same historical and social context. Notions of intellectual property being a form of ownership over knowledge, and a modernist form of knowledge, are both part of the same frame. That is a frame in which knowledge can be, and regularly is, separated from its relations of production and use and has to be reconnected to owners as an abstracted, knowledge object. Reite people, on the other hand, maintain that ownership is *built into* knowledge because it is always in, and of, a relationship.

ACKNOWLEDGMENTS

The Christensen Fund (US) and the Australian Research Council (FT120100262) funded elements of the TKRN project. Catherine Sparks's imaginative and supportive engagement was crucial. I am grateful to colleagues at the University of Western Australia and the Centre for Research and Documentation in the Pacific (CREDO UMR 7308, Aix-Marseille University) for their comments on aspects of the project. Giles Lane has been an insightful colleague and collaborator, and I gratefully acknowledge his centrality to the development of many of the ideas presented here. Fleur Rodgers, Kriss Ravetto, Marilyn Strathern, Laurent Dousset, Banak Gamui, and Yat Paol join a long list of people to whom TKRN is indebted in various ways, and to that list I now add the current volume's editors and the participants of the "Ownership of Knowledge" workshops at MPIWG.

Notes

1. Paul Sillitoe, "Local Science vs Global Science: An Overview," in *Local Science vs Global Science: Approaches to Indigenous Knowledge in International Development*, ed. Paul Sillitoe (Oxford: Berghahn Books, 2007), 1–27; Toon van Meijl, "Doing Indigenous Epistemology: Internal Debates about Inside Knowledge in Māori Society," *Current Anthropology* 60, no. 2 (2019): 155–156.

2. The chapter refers to speakers of the Nekgini language, residents of Reite village, on the north coast of Papua New Guinea (PNG). These villagers are multilingual, also utilizing one of the official languages of PNG called *Neo-Melanesian*.

3. "Traditional Knowledge," WIPO World Intellectual Property Organization (website), accessed October 31, 2019, https://www.wipo.int/tk/en/tk.

4. UNESCO, *The 2005 Convention on the Protection and Promotion of the Diversity of Cultural Expressions*, Paris, October 20, 2005, https://en.unesco.org/creativity/sites/creativity/files/passeport-convention2005-web2.pdf, 2.

5. Epeli Hau'ofa, "Our Sea of Islands," in *A New Oceania: Rediscovering Our Sea of Islands*, ed. Eric Waddell, Vijay Naidu, and Epeli Hau'ofa (Suva, Fiji: University of the South Pacific School of Social and Economic Development, in association with Beake House, 1993), 2–16.

6. Subramani, "The Oceanic Imagery," *Contemporary Pacific* 13, no. 1 (2001): 149–162; James Leach and Richard Davis, "Recognising and Translating Knowledge: Navigating the Political, Epistemological, Legal and Ontological," *Anthropological Forum: A Journal of Social Anthropology and Comparative Sociology*, 22, no. 3 (2012): 209–223.

7. See Porer Nombo and James Leach, *Reite Plants: An Ethnobotanical Study in Tok Pisin and English* (Canberra: Australian National University Press, 2010), 169.

8. An important point in any discussion of IP and *traditional knowledge* holders.

9. As Strathern and Hirsch pithily put it, the concepts of "'indigeneity,' 'heritage,' 'traditional knowledge' and such get filtered through the discourse of an international community that must trade in its own conceptual currency (NGO forums, UNESCO, WIPO). But what may be important politically may be less than useful analytically." Marilyn Strathern and Eric Hirsch, introduction to *Transactions and Creations: Property Debates and the Stimulus of Melanesia*, ed. Eric Hirsch and Marylin Strathern (Oxford: Berghahn Books, 2004), 5.

10. See Mario Biagioli, "From Ciphers to Confidentiality: Secrecy, Openness and Priority in Science," *British Society for the History of Science* 45, no. 2 (2012): 1–21.

11. For example, how one can reveal an invention or discovery while protecting against its appropriation, e.g., Biagioli.

12. Van Meijl, "Doing Indigenous Epistemology," 165.

13. In other words, interrogating the criteria used to evaluate whether a being or entity is a "person" in the sense of socially recognized as such. For examples and elaborations, see chapters in Michael Carrithers, Steven Collins, and Steven Lukes, eds., *The Category of the Person: Anthropology, Philosophy, History* (Cambridge: Cambridge University Press, 1985).

14. For example, Jadran Mimica, *Intimations of Infinity: The Mythopoeia of the Iqwaye Counting System and Number* (Oxford: Berg, 1988). From a description of the counting system of Iqwaye people of Papua New Guinea, the author develops an interpretation of the Iqwaye kinship system and cosmology, culminating in a powerful critique of Western assumptions about the development of rational thought.

15. Tony Crook, *Anthropological Knowledge, Secrecy and Bolivip, Papua New Guinea: Exchanging Skin* (Oxford: Oxford University Press, 2007), 10.

16. Strathern and Hirsch, introduction to *Transactions and Creations*, 5.

17. Lamont Lindstrom, *Knowledge and Power in a South Pacific Society* (Washington, DC: Smithsonian Institution Press, 1990), xiv.

18. Peter Lawrence, *Road Belong Cargo: A Study of the Cargo Movement in the Southern Madang District, Papua New Guinea* (Melbourne: Melbourne University Press, 1964), 9.

19. Lawrence, 30.

20. Lawrence, 31.

21. Lawrence, 30.

22. Lawrence, 28.

23. Lawrence, 29.

24. The starchy vegetables *Colocasia esculenta* var. *antiquarium* and *Dioscorea* sp., respectively.

25. C. B. MacPherson, *The Political Theory of Possessive Individualism: Hobbes to Locke* (Oxford: Clarendon Press, 1962). MacPherson details the development of the notion of an individual and a unitary self, distinct from other people, and who is the logical and natural possessor of the things they labor to produce, in European statecraft.

26. See James Leach, "'Twenty Toea Has No Power Now': Property, Customary Tenure, and Pressure on Land near the Ramu Nickel Mine, Madang, Papua New Guinea," *Pacific Studies* 34, no. 2/3 (2011): 295–322.

27. Such as steel tools that allow for larger food gardens, and penicillin that has reduced infant mortality.

28. See Leach, "Twenty Toea," 305–306.

29. Procedures outlined by their taro "deity." "Story" here is a translation of the neo-Melanesian *stori*, a synonym for myth, but also a translation of the Nekgini term *patuki*. *Patuki* refers to not just a narrative but also the characters/deities of the myth and, crucially, is used locally as a term meaning knowledge in general.

30. Anyone making such a payment would divide it into an obligatory amount and an extra amount, and demand to be given the relevant *paru*. Payment would be made in the currency of kin exchange: ancestral valuables and pork. Today, cash and trade store goods are part of such a payment, but cannot be used to replace pigs or other valuables.

31. Simon Harrison's 1990 ethnography, *Stealing People's Names*, about the Sepik River (PNG), is illustrative here. Harrison describes public ritualized battles in which groups of men vie to show, in whispered duels, that they know the true names of ancestors, and thus have rights over territory and ritual power. He details a similar connection between secret knowledge and everyday accomplishments. Simon Harrison, *Stealing People's Names: History and Politics in a Sepik River Cosmology* (Cambridge: Cambridge University Press, 1990). Similarly, Lawrence notes, "As elsewhere in

Melanesia, restricted knowledge (technological and otherwise) is part of the requisite capital that supports the production of material goods." *Road Belong Cargo*, 29–30. Meanwhile, Lindstrom adds that "people attempt purposely to control those procedures that order the production, circulation, and consumption of knowledge statements." Lindstrom, *Knowledge and Power*, 10.

32. K. A. Gourlay, *Sound Producing Instruments in Traditional Societies: A Study of Esoteric Instruments and Their Role in Male/Female Relations* (Port Moresby: Australian National University, 1975), 126.

33. I explore this in relation to "brands" and "branding" in James Leach, "'*We* Will Make a Man Out of You': Taro 'Brands' and Initiation 'Styles' on the Rai Coast of Papua New Guinea," *UC Davis Law Review* 47, no. 2 (2013): 633–656.

34. James Leach, "Situated Connections: Rights and Intellectual Resources in a Rai Coast Society," *Social Anthropology* 8, no. 2 (2000): 163–179.

35. Crook, *Anthropological Knowledge*, 23–24.

36. James Leach, "Leaving the Magic Out: Knowledge and Effect in Different Places," *Anthropological Forum* 22, no. 3 (2012): 251–270.

37. I ask the reader to bear with the personal and self-referential aspect of what follows. It is unfortunate but necessary, as I am unavoidably involved in the documentation of Reite *kastom*.

38. Strathern and Hirsch, introduction to *Transactions and Creations*, 5.

39. Lindstrom, *Knowledge and Power*, xi.

40. Lawrence, *Road Belong Cargo*, 17.

41. On this point, see Don Kulick and Christopher Stroud, "Christianity, Cargo and Ideas of Self: Patterns of Literacy in a Papua New Guinean Village," *Man* 25, no. 2 (1990): 286–304; and Roy Wagner, *The Invention of Culture* (Chicago: Chicago University Press, 1975), 31–33.

42. Lawrence, *Road Belong Cargo*, 30.

43. See Leach, "Leaving the Magic Out."

44. "Caution: Traditional Knowledge. Principles of Merit Need to Be Spelt Out in Distinguishing Valuable Knowledge from Myth," *Nature* 401, no. 6754 (1999): 623, highlighted by Sillitoe, "Local vs Global Science," 5.

45. "Caution: Traditional Knowledge," 623.

46. See, e.g., Silitoe "Local vs. Global Science," 15–16.

47. Cori Hayden, *When Nature Goes Public: The Making and Unmaking of Bioprospecting in Mexico* (Princeton, NJ: Princeton University Press, 2003), 5–7.

48. Strathern and Hirsch, introduction to *Transactions and Creations*, 2; emphasis added.

49. "Giles Lane," Proboscis, accessed July 15, 2018, http://proboscis.org.uk/about/people/giles-lane/.

50. Giles Lane, "Indigenous Public Authoring in Papua New Guinea," Proboscis, October 2, 2013, http://proboscis.org.uk/5309/indigenous-public-authoring-in-papua-new-guinea/.

51. Porer Nombo and Pinbin Sisau, "*Mi sori long ol*: Seeing the Ancestors in the Collection," in *Melanesia: Art and Encounter*, ed. Lissant Bolton et al. (London: British Museum Press, 2013), 92–95.

52. Many people in Reite are illiterate, and in several of these cases, their children or grandchildren have assisted them in the writing of stories or processes.

53. "TK Reite Notebooks," Reite TKRN Library, accessed November 27, 2020, https://reitetkrnlibrary.wordpress.com/. For a time, Reite people created and uploaded booklets, managed their own website, and taught others how to use the TKRN process.

54. Crook, *Anthropological Knowledge*, 10–11.

55. Annelise Riles, "Introduction: In Response," in *Documents: Artifacts of Modern Knowledge*, ed. Annelise Riles (Ann Arbor: University of Michigan Press, 2006), 23.

56. Sillitoe, "Local vs Global Science," 15–16.

57. Christoph Antons, "Biodiversity, Intangible Cultural Heritage and Intellectual Property," in *The Routledge Handbook on Biodiversity and the Law*, ed. Charles R. McManis and Burton Ong (London: Routledge, 2018), 313–325.

58. See, for example, Lorraine Aragon and James Leach, "Arts and Owners: Intellectual Property Law and the Politics of Scale in Indonesian Arts," *American Ethnologist* 35, no. 4 (2008): 607–631.

59. Christoph Antons, "What Is 'Traditional Cultural Expression'?—International Definitions and Their Application in Developing Asia," *WIPO Journal* 1 (2009): 103–116; Christoph Antons, "Epistemic Communities and the 'People without History': The Contribution of Intellectual Property Law to the 'Safeguarding' of Intangible Cultural Heritage," in *Diversity in Intellectual Property: Identities, Interests and Intersections*, ed. Irene Calboli and Srividhya Ragavan (New York: Cambridge University Press, 2015), 453–471.

60. Strathern and Hirsch, introduction to *Transactions and Creations*.

61. *The Protection of Traditional Knowledge, Genetic Resources and Expressions of Folklore Act 2016*, WIPO 2016, accessed November 27, 2020, https://www.wipo.int/edocs/lexdocs/laws/en/zm/zm056en.pdf.

62. Eugene Hunn, "Meeting of Minds: How Do We Share Our Appreciation of Traditional Environmental Knowledge?," in "Ethnobiology and the Science of Humankind," ed. Roy Allen, special issue, *Journal of the Royal Anthropological Institute* 12, no. 1 (2006): 143.

63. Hunn, 143.

64. Hunn, 156.

65. Ian Saem Majnep and Ralph Bulmer, *Birds of My Kalam Country* (Auckland: Auckland University Press, 1977). Their book was also an inspiration for Porer Nombo and myself in our collaboration on *Reite Plants* (see above).

66. Hunn, "Meeting of Minds," 154–158.

67. Majnep and Bulmer, *Birds*, 39.

68. Majnep and Bulmer, 39–40.

69. Riles, "Introduction: In Response," 4.

70. Riles, 2; see also van Meijl, "Doing Indigenous Epistemology," 155–156.

71. Riles, "Introduction: In Response," 2.

72. Riles, 24–28.

73. Crook, *Anthropological Knowledge*, 10–11.

74. Crook, 11.

75. Fredrik Barth, *Ritual and Knowledge among the Baktaman of New Guinea* (New Haven, CT: Yale University Press, 1975).

76. Crook, *Anthropological Knowledge*, 11.

77. Fred Damon, *Trees, Knots and Outriggers: Environmental Knowledge in the Northeast Kula Ring* (Oxford: Berghahn Books, 2017).

78. Damon, 249, 270–272.

79. Riles, "Introduction: In Response," 5.

80. Annelise Riles, "[Deadlines]: Removing the Brackets on Politics in Bureaucratic and Anthropological Analysis," in Riles, *Documents*, 89.

Bibliography

Antons, Christoph. "Biodiversity, Intangible Cultural Heritage and Intellectual Property." In *The Routledge Handbook on Biodiversity and the Law*, edited by Charles R. McManis and Burton Ong, 313–325. London: Routledge, 2018.

Antons, Christoph. "Epistemic Communities and the 'People without History': The Contribution of Intellectual Property Law to the 'Safeguarding' of Intangible Cultural Heritage." In *Diversity in Intellectual Property: Identities, Interests and Intersections*, edited by Irene Calboli and Srividhya Ragavan, 453–471. New York: Cambridge University Press, 2015.

Antons, Christoph. "What Is 'Traditional Cultural Expression'? International Definitions and Their Application in Developing Asia." *WIPO Journal* 1 (2009): 103–116.

Aragon, Lorraine, and James Leach. "Arts and Owners: Intellectual Property Law and the Politics of Scale in Indonesian Arts." *American Ethnologist* 35, no. 4 (2008): 607–631.

Barth, Fredrik. *Ritual and Knowledge among the Baktaman of New Guinea*. New Haven, CT: Yale University Press, 1975.

Biagioli, Mario. "From Ciphers to Confidentiality: Secrecy, Openness and Priority in Science." *British Society for the History of Science* 45, no. 2 (2012): 1–21.

Bolton, Lissant. *Unfolding the Moon: Enacting Women's "Kastom" in Vanuatu*. Honolulu: University of Hawai'i Press, 2003.

Carrithers, Michael, Steven Collins, and Steven Lukes, eds. *The Category of the Person: Anthropology, Philosophy, History*. Cambridge: Cambridge University Press, 1985.

Crook, Tony. *Anthropological Knowledge, Secrecy and Bolivip, Papua New Guinea: Exchanging Skin*. Oxford: Oxford University Press, 2007.

Damon, Fred. K. *Trees, Knots and Outriggers: Environmental Knowledge in the Northeast Kula Ring*. Oxford: Berghahn Books, 2017.

Gourlay, K. A. *Sound Producing Instruments in Traditional Societies: A Study of Esoteric Instruments and Their Role in Male/Female Relations*. Port Moresby: Australian National University, 1975.

Harrison, Simon. *Stealing People's Names: History and Politics in a Sepik River Cosmology*. Cambridge: Cambridge University Press, 1990.

Hau'ofa, Epeli. "Our Sea of Islands." In *A New Oceania: Rediscovering Our Sea of Islands*, edited by Eric Waddell, Vijay Naidu, and Epeli Hau'ofa, 2–16. Suva, Fiji: University of the South Pacific School of Social and Economic Development, in association with Beake House, 1993.

Hayden, Cori. *When Nature Goes Public: The Making and Unmaking of Bioprospecting in Mexico*. Princeton, NJ: Princeton University Press, 2003.

Hunn, Eugene. "Meeting of Minds: How Do We Share Our Appreciation of Traditional Environmental Knowledge?" In "Ethnobiology and the Science of Humankind," edited by Roy Allen. Special issue, *Journal of the Royal Anthropological Institute* 12, no. 1 (2006): 143–160.

Kulick, Don, and Christopher Stroud. "Christianity, Cargo and Ideas of Self: Patterns of Literacy in a Papua New Guinean Village." *Man* 25, no. 2 (1990): 286–304.

Lane, Giles. "Indigenous Public Authoring in Papua New Guinea." Proboscis, October 2, 2013, http://proboscis.org.uk/5309/indigenous-public-authoring-in-papua-new-guinea/.

Lawrence, Peter. *Road Belong Cargo: A Study of the Cargo Movement in the Southern Madang District, Papua New Guinea*. Melbourne: Melbourne University Press, 1964.

Leach, James. "Leaving the Magic Out: Knowledge and Effect in Different Places." *Anthropological Forum* 22, no. 3 (2012): 251–270.

Leach, James. "Situated Connections: Rights and Intellectual Resources in a Rai Coast Society." *Social Anthropology* 8, no. 2 (2000): 163–179.

Leach, James. "'Twenty Toea Has No Power Now': Property, Customary Tenure, and Pressure on Land near the Ramu Nickel Mine, Madang, Papua New Guinea." *Pacific Studies* 34, no. 2/3 (2011): 295–322.

Leach, James. "'*We* Will Make a Man Out of You': Taro 'Brands' and Initiation 'Styles' on the Rai Coast of Papua New Guinea." *UC Davis Law Review* 47, no. 2 (2013): 633–656.

Leach, James, and Richard Davis. "Recognising and Translating Knowledge: Navigating the Political, Epistemological, Legal and Ontological." *Anthropological Forum: A Journal of Social Anthropology and Comparative Sociology* 22, no. 3 (2012): 209–223.

Lindstrom, Lamont. *Knowledge and Power in a South Pacific Society*. Washington, DC: Smithsonian Institution Press, 1990.

MacPherson, C. B. *The Political Theory of Possessive Individualism: Hobbes to Locke*. Oxford: Clarendon Press, 1962.

Majnep, Ian Saem, and Ralph Bulmer. *Birds of My Kalam Country*. Auckland: Auckland University Press, 1977.

Mimica, Jadran. *Intimations of Infinity: The Mythopoeia of the Iqwaye Counting System and Number*. Oxford: Berg, 1988.

Nombo, Porer, and James Leach. *Reite Plants: An Ethnobotanical Study in Tok Pisin and English*. Canberra: Australian National University Press, 2010.

Nombo, Porer, and Pinbin Sisau. "*Mi sori long ol*: Seeing the Ancestors in the Collection." In *Melanesia: Art and Encounter*, edited by Lissant Bolton, Nicholas Thomas, Elizabeth Bonshek, Julie Adams, and Ben Burt, 92–95. London: British Museum Press, 2013.

Riles, Annelise. "[Deadlines]: Removing the Brackets on Politics in Bureaucratic and Anthropological Analysis." In *Documents: Artifacts of Modern Knowledge*, edited by Annelise Riles, 71–93. Ann Arbor: University of Michigan Press, 2006.

Riles, Annelise. "Introduction: In Response." In *Documents: Artifacts of Modern Knowledge*, edited by Annelise Riles, 1–39. Ann Arbor: University of Michigan Press, 2006.

Sillitoe, Paul. "Local Science vs Global Science: An Overview." In *Local Science vs Global Science: Approaches to Indigenous Knowledge in International Development*, edited by Paul Sillitoe, 1–27. Oxford: Berghahn Books, 2007.

Strathern, Marilyn, and Eric Hirsch. Introduction to *Transactions and Creations: Property Debates and the Stimulus of Melanesia*, edited by Eric Hirsch and Marylin Strathern, 1–20. Oxford: Berghahn Books, 2004.

Subramani. "The Oceanic Imagery." *Contemporary Pacific* 13, no. 1 (2001): 149–162.

United Nations Educational, Scientific and Cultural Organization (UNESCO). *The 2005 Convention on the Protection and Promotion of the Diversity of Cultural Expressions*. Paris: UNESCO, October 20, 2005. https://en.unesco.org/creativity/sites/creativity/files/passeport-convention2005-web2.pdf.

Van Meijl, Toon. "Doing Indigenous Epistemology: Internal Debates about Inside Knowledge in Māori Society." *Current Anthropology* 60, no. 2 (2019): 155–174.

Wagner, Roy. *The Invention of Culture*. Chicago: Chicago University Press, 1975.

8

NAMES FOR WORK: CRAFTS, BUREAUCRACY, AND LAW IN YUAN AND MING CHINA (THIRTEENTH–SEVENTEENTH CENTURY)

Dagmar Schäfer

"Bans on manufacturing (*jinzhi* 禁制) refer to the private production of things, such as weaponry and utensils produced privately without good cause."[1] This entry is from the *Guidebook for Clerks* published in 1301 by a local clerk (*li* 吏) called Xu Yuanrui 徐元瑞 living under Yuan (1271–1368) rule. In this Chinese-language primer for legal and administrative practices, bans make up one of five categories—among a total of eighty-four (and 1,405 clauses)—that directly address craft knowledge and thus tackle how state and individual were able to own knowledge through owning bodies and the products of their work. The other four categories tackle the relation between land and silk taxes (*qianliang zaozuo* 錢糧造作), household registers (*huji* 戶籍), material provisions (*wuke* 五科), and tax services, as well as their distribution (*zhenglian chafa* 征斂差發).[2] Historians have therefore taken Xu's *Guidebook* as quintessentially representative of the nature of craft knowledge and ownership in premodern China in three aspects: first, the imperial state as the center point; second, laws addressed "access to" craftsmen and not "ownership of" their craft "work"—nor their "knowledge"; and third, clerks who mediated as managers between craftsmen and the state were actually the knowers. Its author, Xu Yuanrui, represents a major stakeholder in this game by identifying himself as a member of the *Ru* 儒 literati.[3] This historical group distinguished themselves at the most basic level by their ability to read and write, and they served Chinese imperial rulers in this function to order society and state.

As I suggest in this chapter, all three of the above-mentioned aspects are deeply flawed and biased, as they are resting on Eurocentric and modernist assumptions about how knowledge is approached or owned, and how it relates to law. I have chosen to focus on Xu's work because it features the major practice by which the literati governed craft knowledge and its ownership in premodern China, and which still dominates our historical view of it today. In seven of his eight chapters, Xu explained legal and administrative processes by means of a process he called "rectifying names" (*zhengming*

正名). This meant he defined words in the style of a dictionary. Classical *Ru* (Confucian) literature connected this scholarly practice with a claim to the authority of organizing society and state. Over the course of imperial Chinese history, as historian Yang Shitie notes, "determining meanings became the contemporary method to settle affairs and understand things."[4] Such affairs and things, as the *Guidebook* illustrates, included the possibility of, or limits on, craftsmen owning their knowledge and skills, and of rulers and elites appropriating craftsmen's bodies, their labor, and the fruits thereof, for their purposes.

The *Guidebook* reflects a particular moment in the practice of rectifying names that made practitioners' abilities and talents visible in new ways. It also enabled clerks to manipulate ownership over craft knowledge in new ways, because clerks were tasked with naming abilities so that they could be owned by the state. The imperial house of the Yuan established a state-owned manufacturing network for textile, porcelain, and lacquer production in order to profit from crafts, and it honed its bureaucracy to secure access to the whole spectrum of craftwork through a complex of tax, trade, and transport. The *Guidebook* showcases the central role of bureaucracy. While it addresses the contents of codices (*lü* 律), it more prominently features administrative conduct (*ge* 格) and rules (*fa* 法), thus reflecting the mindset of a dynastic house that favored the ad hoc generation of regulative measures to rule its people and lands.[5] It also showcases a major innovation of the Yuan Dynasty who were adamant that craftsmen be recruited and registered systematically in a hereditary system of household categories.

This moment of visibility had lasting consequences, as politicians and elites of the subsequent Ming and Qing dynasties would continue to employ and build on such registers and the abilities they cataloged to make use of crafts. To provide points of reference for the *longue durée* view, in the final section I compare this case study with another from the early Ming period, in which a *Ru* literatus categorized craft practices to fix craft knowledge and ownership. I offer this comparison as a chance to critically reflect on the perceived continuity of a world that placed crafts and craftsmen under the regime of an intellectually engaged scholarly elite. This continuity is caused by the fact that throughout the subsequent seven centuries, until the republican era, historians have seen the *Ru* as acting continuously in three major roles. First, *Ru* functioned as de facto clerks who ordered and classified bodies as "work" and materials as "goods" that could be appropriated for the benefit of society (that is, to serve elite needs) and the state. Second, as part of the social or political elite, *Ru* negotiated with the imperial house in order to specify the ownership of craft materials and products by identifying products as either "art" or conspicuous consumption, or by assessing such work within moral terms, warning emperors and elites when their desires for such goods threatened social peace and political rule. Third, as the intellectual class, *Ru* defined which tasks counted as labor and crafts, and therefore, as knowledge or not. In short, then, *Ru* wielded words to determine who could know and own crafts in domains of epistemology, society, and the economy. But as will become clear later in this chapter, the traditional tripartite view of the function played by *Ru* has failed to grasp how these domains interrelate, while emphasizing the continuity of the *Ru*'s roles has led to a historiographic segregation in which researchers approach crafts and property in premodern China either as unrelated or as a resolved case of disinherited craftsmen set in opposition to an imperial state and literati elite to whom everything belonged.

In this chapter I critically engage with the notion of continuity and scrutinize it for fissures and cracks. The fact that the *Guidebook* is an artifact of the Yuan makes it an excellent case study inasmuch as this era was, in fact, not unambiguously "Chinese." On the one hand, the Yuan rulers were Mongolian and assembled people from different regions, with varied practical and linguistic capacities and cultural backgrounds. But on the other hand, the crafts of this era were predominantly documented in and as Chinese. As philologist Victor Mair has emphasized, this is because "in the East Asian heartlands, dynasties came and went, and were headed by non-Sinitic people as often as not. What persisted was the bureaucracy and the command of the sinographic/sinophone script that constituted its very essence."[6] The *Ru*'s wielding of the power of words during the Yuan has enhanced the impression of historical continuity, even though historians are well aware that the Yuan mistrusted the *Ru* because of their association with the preceding Song dynasty and awarded influential office positions (*guan* 官) mainly to Mongols and their allies. Xu thus served as a minor clerk for rulers who appreciated and "needed artisans to provide them with the rich array of products and services available in the sedentary zone of their domain,"[7] as they expanded their territories through Asia. This raises the question: How much actual power over crafts did words give clerks?

Xu's headings constitute the categories under which I trace notions of knowing and owning crafts and explore the roles that legal and bureaucratic framings played. The subheading to each section is a definition quoted from the *Guidebook*, with the key term in bold, followed by my gloss in parentheses. While previous research has elucidated the *Guidebook* as part of a linguistic effort of communicating with a multilingual elite, my focus is on the practice of explaining Chinese words by other words or by references to classical Chinese literature as a way to affix realities by affixing meanings.[8] As I will trace throughout the chapter, Xu identified crafts variously as tax, labor, and work, while highlighting specific relations. He did not name crafts to learn about, appropriate, or own such knowledge, though. Instead, Xu is a prime example of an effort to claim the practice of rectifying names—that is, of matching words to realities and objectifying meanings—as a skill and as a means to make himself relevant. The Yuan elites mistrusted exactly this skill of the *Ru*, however, and their use of words as a way to regulate social relationships and achieve (or improve) social status and power. We can thus infer that the actual power of naming as a way to know and own was limited.

There is no doubt, though, that the *Ru* of the Yuan era substantially affected how crafts became knowable and ownable to the state, in their own time and throughout time. Their efforts made the work and expertise of practitioners visible in public and to the ruling class in a way that, two centuries later, caused the scholar official and minister of rites, Qiu Jun, great concern. Qiu used the same scholarly weapon/skill/practice, the rectification of names—this time, within a context of ritual and ethics—to rein in the power of crafts and hinder the emperor from paying too much attention to the products of such work. A subversive reading could take this to mean that craftsmen *owned* their knowledge in performance in society and economy and clerks were once more in a weak position (and wanted to reassert control).

As different as their approaches may seem, Xu during the Yuan and Qiu during the Ming both operated in a world that knew and acknowledged many valid ways of owning knowledge—even though not all of the actors may have validated all forms equally. In this world, the two scholars executed naming as a collective meaning-making practice in the classical sense of Émile Durkheim in a given organizational setting[9]—while also defining and embracing it purposefully as a *skill* to manipulate knowledge ownership across different domains. Reflecting on a given craft culture, Xu had to rely on political enforcement and laws to empower his bureaucratic sense-making, whereas Qiu Jun operationalized ritual regulations (*lifa* 禮法) and morals as a frame of reference and authority in silencing practitioners' expertise and defining standards for the production of utensils (*qi* 器). Both efforts relied on premises of rights and duties, objects and work; and both defined the family unit (*jia* 家) and its validity as a taxable entity (*hu* 戶), the "economic family," which formed a patri-corporation.[10] In this sense we could say that Chinese actors defy a Durkheimian logic of sense-making and thus a sociology of organizations in that these actors *operationalized* classification as their practice and a worthy skill.[11] By giving names to the tasks a family or individual owned and had to transfer to the state as tax, *Ru* were not able to own craft skills, but they were able to actively influence how craft skills became visible to the state and therefore historically knowable and ownable.

造作謂: 董督工程, 確其物料也

"MANUFACTURING: means overseeing work processes, authenticating their materials." (Or: the control of bodies as labor, and material as property, in order to control knowledge)

When seen through its artifacts, Yuan China appears to have been an innovative, extremely productive, and highly diversified manufacturing site. Zhou Liulang in Siming (Ningbo) as well as Zhang Cheng and Yang Mao in Xitang, for instance, verifiably carved red lacquer wares when commissioned by the imperial court. Zhu Bishan and Peng Baojun smelted and wrought silver wares in the county of Jiaxing; Zhang Linnan molded ceramics in Shangxi; Yu Shengkan carved fine wood sculptures in Hangzhou; and the Sun clan wielded sturdy armaments in Datong—all equally putting their skills into the service of the state.[12] Whereas we can only imagine their homes and lives, the names of these individuals—to which many could be added—can be verified, since practitioners of diverse crafts during the Yuan consistently inscribed their names on the products of their handiwork.

From a contemporary Yuan literati view, the carving of craftsmen's names (*wule gongming* 物勒工名) was an established practice by which the state calculated taxes and traced quality concerns. This practice can be linked back to a method of controlling bodies as labor and material as property, in order to make craft knowledge accessible to, and useful for, the state.[13] As a clerk, Xu had to select and "oversee" (*du* 督) such craftsmen's work. In his namings, Xu emphasized the managerial role of *Ru*, stressing thereby his relevance and indispensability for the ruling house, rather than his power over or superior status to craftsmen. Overseeing meant fairly allocating the burden of corvée and balancing imperial demands with commoners' needs; goods had to be distributed without causing shortages, materials collected based on administrative rules, and people spared crippling taxes.[14] Clerks were perfectly suited for this intermediary role between artisan and court, and Xu further substantiated their timeless authority as arbiters through his historical exemplifications. As Xu explained, countless accounts of historical "clerks" of both high and low ranking, with yielding or fierce mindsets, showed that the position of a clerk was central to any kind of rulership: "In the Qin and Han dynasties clerks functioned as generals and ministers. They were established to hold the reins of government. They held key positions and were no weaklings."[15] Referencing an earlier Song dynasty work, Xu recounted stories of individuals who exerted power and influence skillfully and ethically, or bluntly and by forceful acts of will, in order to demonstrate that the breadth of their proficiency included diplomacy and pragmatism, as well as the simple enforcement of imperial rules.[16] In short, the *Guidebook* makes a strong case for giving power to clerks, during a period when, as historians largely concur, *Ru* of Chinese or local origin (*hanren* 漢人, *nanren* 南人) had to serve under officials (*guan* 官) who had been awarded such influential posts because of their Mongol pedigree (*Yuanshi baiguan* 元世百官).[17] Clerkship thus offered a chance to regain higher social status along with greater political power, and wealth.

There is an obvious correlation between the availability of sources and historical analysis: the more sources, the more historical attention. Given that contemporary *Ru* scholars pontificated about their importance, or the necessary distinctions between themselves and those clerks who operated as lowly translators or mere scribes, historical analysis has concentrated on exploring the nuances of *Ru* power and posts—which once again highlights the role of *Ru*. It is thus important to note that the *Guidebook* makes no such distinction. Rather, Xu intentionally addressed the post generically, thus rejecting the idea that clerks were auxiliary figures in relation to such officials. For Xu, clerkship was the new pathway to power and wealth for the *Ru*.

The sum of Xu's interventions suggests that during the Yuan, the *Ru* were faced with rulers who not only had abolished the civil service exams as a possible route to political and social power, but also generally approached authority and ability based on values quite different to those of their predecessors, as the early disciple of Kublai Khan Wang E 王鶚 (1190–1273) noted in 1267: "Since the civil service exam has been abolished, literati can no longer enter the ranks of officialdom and they either practice writing [literally, carve words on bamboo] while acting as government clerks (*lixu* 吏胥) or organize corvée labor *or* act as craftsmen or traders producing, selling, or buying."[18] That the *Ru* were useful for the Yuan only when they applied their ability to read and write as a technical skill—rather than an intellectual agenda, as propagated in *Ru* books—is a point that Xu's meaning-making emphasizes in two regards: first, by highlighting the practical combination of talents that a clerk embodied; and second, by showing that the scholarly practice of ordering by meaning-making was a way to use the *Ru* to implement the state's control over craftspeople's talents. In his preface, Xu grounds his arguments for the importance of *Ru* by emphasizing that rulership needed order and clarity: "I have heard that ordering must be prioritized for good governance; attempts to order have to clarify laws/rules (*fa*)."[19] For such ordering, he claimed, men who could read and write were key.

Four of the six skills that Xu identified as being crucial for the profession of clerks addressed clearly practical tasks that were related to administration, namely, "clear calligraphy, knowing the laws, good debating, and calculation skills." Another skill concerned ethics, as clerks had to show "proper behavior." Apart from all those issues, Xu insisted, clerks needed to know "how to interpret the books of the *Ru*."[20] The books of the *Ru* represented an ideological entitlement to a role in governance that the Yuan may not have appreciated at all; Xu therefore carefully explained that their trained skill was necessary to enable clerks to act as intermediaries on behalf of rulers.

That Xu mentions the books of *Ru* at all, but also takes care to promote interpretation of them primarily as a practical or even technical skill (and not an intellectual stance), is historically telling. It is the kind of fissure that is frequently identified by historiography that acknowledges a loss of political power but stresses the continuity of skill sets that the *Ru* wishes to offer the imperial state. Xu, at least, was treading a narrow path between being perceived as an asset or a threat to his Yuan rulers. Copious but also somewhat arbitrary references to classical pre-imperial literature following no clear pattern explicated and generated genealogies of meaning-making, while also testifying to the political authority and social power that such skills could engender. Xu quoted the "Discussion of Writing and Explanation of Words" (*Shuowen jiezi* 說文解字) to substantiate his assertion that *Ru* were "flexible," depicting the Han dynasty poet, musician, and politician Sima Xiangru 司馬相如 (197–117 BCE) as the epitome of *Ru*. Of all the historical figures Xu could have chosen, Sima Xiangru was particularly appreciated during the Yuan for his relation to the literary genre of a rhapsody (*fu* 賦) that had been a key examination topic in the Song era, and his promotion of Confucianism as a method of governance.[21] Xu explicitly pointed to Sima as a role model who had "wielded the art of the Dao (*daoshu* 道術)," which equally refers to the art of governing (*zhiguo zhi shu* 治國之術) and to devising convincing lines of rhetoric.[22] Xu furthermore solidified the role of meaning-making by associating the style of his *Guidebook* with the literary philosophy of "The Literary Mind and Carving of Dragons" (*Wenxin diaolong* 文心雕龍, ca. sixth century), which had "explained names and titles in an orderly sense (*shiming yi zhang yi* 釋名以章義)"[23] and had similarly been written during an era of foreign rule.

Showcasing the usefulness of his skills also guided Xu in his choice of judiciary themes, and in laying out a vocabulary of official documents. He took up terms essential for understanding social ordering, addressing topics such as disease, kinship and social relations, and military concerns.[24] Via careful headings, Xu signposted the *Ru* as being the imperial disciples for the control of social order and the key stakeholder group that would keep vital abilities accessible to the state: "The eminent and the humble; when status is high, they are respected and called eminent, the menial are without place and called humble." Clerk Xu then explained that *Ru* "organize names [of high and low] in the household registration system. People of plain origin and commoners are high; shopkeepers, prostitutes, and private slaves of bureaucrats are low."[25] The clarification of meanings and belongings was also key for clerks to make sure that craftsmen paid "tax provisions and without exception did their duty as manservants" for the state.[26]

It is in passages like these that Xu is at his most convincing, yet it is also in such passages that we find one more of the fissures and cracks revealing that all was not well. In discussing social hierarchy, Xu added ominously: "Sometimes the gentleman resides among the humble."[27] This signposting reflects the fact that the *Ru*'s obligation to organize and categorize objects, practices, and subjects for further use by the state, and thus their skill to perform words, gave them power. Yet it also indicates that sociopolitical reality was defined by the very household categories that had to be clarified by clerks. Xu listed *Ru* as one of seventeen household categories (*hu* 戶) that served as the basis to calculate taxes during the Yuan, followed by representative "schools" such as Buddhist or Zen monks, nuns, Daoists, or male and female clergy.[28] Religious and intellectual-philosophical groups or factions thus clearly mattered to the Yuan, as Xu reflected by noting them first before continuing to list scholars/gentlemen and farmers and then "workers" (*gong* 工), which he grouped with merchants (*shang* 商) who established monetary relations and sold goods. These were followed by doctors and healers, and by traders and store holders, who mainly sought profit through the redistribution and transport of wares, and by socially defined groups such as elders without family to support them.

That Xu lists the *Ru* first is telling. But does this mean they were higher ranked or socially and politically more influential than workers or doctors? One reason to argue that any playing up of *Ru* signifies only Xu's private opinion is that other official or historiographic listings are ordered differently.[29] Although I have not come across a list that put craftsmen first, many such lists expanded the categories and groups and gave detailed descriptions of craft expertise and work tasks.[30] The fact that clerks spent considerable time "naming" crafts invites a close look at their actual relation to practitioner groups and further scrutiny of the kind of power that words had over the artisans who owned their knowledge in performance and through the use of their bodies or tools.

作巧成器曰：工

"One who produces ingenuity and creates utensils is called WORK/CRAFTSMEN." (Or: one who performs knowledge and achieves status and wealth.)

There is little doubt that the reality of being a craftsman was defined by an imperial desire to control craftsmen experts, and that craftsmen were far from being free and able to own their skills and products. Certainly, if we understand ownership of knowledge and skills as the freedom of craftsmen to express their creativity, then it is also correct that the legal canon strictly regulated sensitive crafts such as armaments, weaponry, and textile production—the latter of which provided an important source of income and status. The *Statutes of the Yuan Dynasty* (*Yuan dianzhang* 元典章, published ca. 1322, henceforth *Statutes*), for instance, introduced sumptuary laws that restricted the use of patterns.[31] Like all other commoners, craftsmen were penalized for making "improper claims, threatening administrative clerks, or giving excuses, and [they] should not enter the bureau or hold up the work process. This is to be punished by cutting off limbs."[32] Laws kept craftsmen in place.

Historical scholarship has portrayed the roles of clerks and craftsmen in Yuan China in black and white terms, with them as *Ru* being the morally responsible protagonists who attempted to contain a willful ruling class of Mongol elites, who disdained them, and at the same time treated craftsmen as pawns in more or less sensible policies and power struggles. In contrast, the literature of the time actually reflects the ambiguities of social status and the realities that marked every person's life. Some of these ambiguities can be explained by the nature of the sources that historians have relied on to inquire about artisans and their roles during the Yuan. History is written by Chinese actors or, at least, in Chinese. Thus, it is Chinese voices, such as that of the late Song official Xu Ting 徐霆 (active ca. thirteenth century), that dominate our understanding of his period. Xu declared: "I have investigated it, the Tartars were originally quite uncivilized. They had nothing such as the affairs of the hundred works/crafts (*baigong* 百工)."[33] We will return to the "hundred works/crafts" as an indicator of diversity in more detail later. Here it is mainly important to note its unique association with "Chinese" (imperial) culture and its standards. These standards prove to be of overbearing relevance. In analyses based on a comparison of material culture, Liu Liya and Chen Peng exemplify a type of scholarship that explains the Mongol elite's appreciation of Chinese crafts as being a question of quality; they further argue that the Mongols' own artisanship was not well developed and they had no highly qualified artisans. Following this argument, we can say that one of the reasons the Mongols conquered the XiXia, Liao, Jin, and finally Song territories was to access their artisanal riches and exquisite wares.[34]

Historians have pinpointed the fallacies that lie in following sources mainly produced by Chinese actors, without considering the texts and fragmented landscape of a material culture that has survived through a process of natural selection (as some products are more perishable than others) in research on the social, epistemic, economic, and political roles of clerks and craftsmen. Xu Yuanrui's assessment, like that of his predecessor Xu, may have willfully ignored, for instance, all arts relevant to a Mongol lifestyle—such as tannery, felt and leather processing, distillery techniques, and tent production.[35] Clerks neither understood nor wrote about such crafts, which were carried out by Mongols themselves. Such crafts suffer from a double bias in historiography because the Yuan rulers identified and "owned" them as part of their identity—and consequently, Yuan rulers did not integrate these crafts into Chinese state governance, address them via laws, or make them economically relevant. An additional, important point regarding ownership is that not all cultures emphasize preservation. With use, objects decay and the very artifacts that perhaps could have attested to the Yuan's excellence are erased/have perished.

What has survived until today makes a valuable point about a distinctively imperial and elite Yuan-era approach to knowledge and its ownership in regard to the court's growing appreciation of certain individuals and their practical and aesthetic skills.[36] This appreciation stands in contrast to contemporary *Ru* who rarely acknowledged anything other than Chinese arts and crafts, while they mainly pitied laborers and craftsmen and called on emperors not to exhaust their labor force. Appeals like this often had little effect, as *Ru* themselves bemoaned and as a critical reading of sources also suggests. Historical scholarship, once again, has mainly toed the line and repeated such literati-*Ru* accounts.

Craftsmen were powerful actors in Yuan courts. *Ru*, by comparison were in a precarious position. When serving the emperor as scribes, *Ru* not only had to identify specialized talents among groups of hundreds and thousands of commoners or people seized as war booty, they also had to sift out those who were pretending to have craft skills in order to avoid prosecution or harm—and some of those "pretenders" were clearly from their own group (i.e., the literati).[37] After the civil service examination had been discarded, Yuan literati began craving a different household status, because craftsmen taken as spoils of war were spared from death and desired by the emperors for their skills.

As we will see later on, the Yuan offered more than one incentive that spurred clerks to attain the status of craftsmen, and not the other way around. With regard to the clerks' management tasks in manufacturing as well as in collecting various forms of levies (*zhenglian chafa* 征斂差發) and procuring and hiring artisans, the *Guidebook*'s nomenclature indicates a state that controlled the role of clerks while thoughtfully caring about the artisans' life, work, and products. A clerk's duty was to handle construction projects: assigning work outside quota restrictions set by the imperial state (*hengzao* 橫造), segmenting tasks among different crafts (*sanzhi* 散支), and taking over project organization and measurements (*cuozhi* 措置).

In organizing and ordering work, clerks would neatly differentiate between different forms of recruitment, allowing us to conclude that the state carefully protected crafts rather than "willfully" appropriating or exploiting any possible skill, as scourned contemporary Yuan literati and, later, Ming scholars regularly implied. Clerks would assign or lay off recruits (*guicuo* 規措), finely distinguishing between corvée raised for military garrisons or forced labor (*yao* 徭) and work done as a levy (*yi* 役), and determining whether such work was to be forced on "one body" (*yishen* 一身= *yao*) or an entire household had to be conscripted. An "agreement of hiring" (*hegu* 和雇) points to an open labor market with free hiring policies, whereas in other fields, expert labor was exchanged for grain (*hezhong* 和中). An "agreement of sale" (*hemai* 和買) meant that "both sides exchange money for products," which usually included the state provisioning necessary materials or funds in advance; this was so habitually practiced in textile manufacture over several different dynasties that it earned its own label of "agreement for weaving" (*hezhi* 和織).³⁸ The *Statutes* distinguished "proper craftsmen" (*zhengjiang* 正匠) who could be "turned" (*fa* 撥), "taken out" (*chou* 抽), "included" (*kuo* 括), or "recruited" (*zhao* 招), as appropriate, by changing their designation in the household register from "people" (*min* 民) to "craftsmen" (*jiang* 匠).³⁹ As the proliferation of nomenclature suggests, there was more than one way to appropriate—and thus also more than one way for craftsmen to own their knowledge and benefit from it.

Many of these terms—and practices—existed before and after the Yuan dynasty. It is their assembled listing here that invites a second look at the Yuan as a particularly exploitative state. From a chronological point of view, Xu compiled the *Guidebook* at a point in time when methods of recruiting artisans had been consolidated. Violent ransacking and other forms of compulsory employment were increasingly being replaced by a sophisticated and complex approach to recruitment through levy and hiring. Though this seems like two very different methods from a legal or social point of view, practices of enslavement, sparing expert lives, and institutionalizing intergenerational inheritance and training all have one thing in common: they situate the human body as the major reference point for knowledge circulation/transmission. Another literatus of the Song-Yuan transition, Hu Qiyu 胡祇遹 (1227–1293), drove home this point by noting that whenever *Ru* as servants of the state selected workers and assigned tasks, they performed "the art of selection to obtain the people [i.e., their bodies and their arts/skills]" (*ze ren de ren zhi shu* 擇人得人之術).⁴⁰ With this emphasis in mind, the Yuan made the household register its major tool to control practitioners' skills and keep them available to the state at all times.

戶籍：生齒之總

"HOUSEHOLD REGISTER: it gives the total of the population." (Or: organizes the performance of knowledge and owns it as society.)

Census records exist for almost every period of imperial China, and they have been almost continuously analyzed by historians of China to reveal financial relations between subject and state, as well as social and political order. Although the Yuan continued many established ways of using this instrument, few historians would deny that the Yuan system was simply copied from its predecessor. For instance, the Yuan-era vocabulary provided by Xu mirrored the Jurchen Jin (1125–1234) system described by historian Hok-Lam Chan. Chan described the labor system of this dynasty that ruled the northern part of China in conjunction with the Song as "riddled with racial and ethnic inequalities and discriminatory haphazard practices."⁴¹ Household categories of the Yuan, however, were far from arbitrary; they specified practices and social skills that were either in demand or in need of control by the state. Standard categories in Yuan records are *Ru*, scholars, farmers, *yinyang* diviners, doctors, and Buddhist monks, and social roles such as elderly people and households with just one member. Apart from generic lists in scholarly literature, on the administrative and local level, we can also find an increasing number of categories that detail ironsmiths, tanners, weavers, reelers, and carpenters, as well as clerks, monks, hunters, and beggars.[42]

The concern about skill is also reflected in basic tax distinctions between military and civilian tasks, and between commoner and artisan households.[43] Historical scholarship of the twentieth century regularly implied that such differentiation reflected, or was even a tool of, social and political ordering.[44] As one of the first of these historians, Gao Rongsheng in the 1990s cast doubt on historical ideas of any categorical dividing lines that made *Ru* the socially or politically high (or higher) group and craftsmen the low (or lower) group during the Yuan.[45] Into the twenty-first century, legal historians have critically engaged with this question from the viewpoint of contemporary legal debates. For instance, Liu Liya and Chen Peng asserted convincingly that even though the situation of craftsmen in the early Yuan days must have been hard and political control over them became increasingly strong, they cannot be sweepingly characterized as slaves or expelled convicts lacking any rights—not least because in the *Yiwei* (乙未) household registration system, expelled convicts and slaves were comparatively low in mumber and clearly distinguished from "workers" (*gong* 工) and "craftsmen" (*jiang* 匠).[46] Xu's *Guidebook* substantiates this point by making no distinction between craftsmen or *Ru*—or any other household group. In another passage, Xu offers a refined and unique catalog of terms for different groups of servants: those performing menial duties in regard to supervisors (*siyi* 私役), or manservants (*shenyi* 身役). Another clue suggesting that craftsmen were an important rather than a suppressed social group is that legal and administrative regulations were put in place to ensure that practices and skills would survive the test of time.

紹業謂: 承繼其產業

"CONTINUING AN OCCUPATION: means inheriting a producing business." (Or: owning use and ordering society.)

The key topics to which legal historians refer to discuss property in China concern inheritance. Historian Brian E. McKnight pinpoints "transmitting assets across generations . . . and a vertical handing down . . . coupled with an abiding concern for the continuation of lineage."[47] The research focus has been on "tangible" assets such as land and dowries, while the transmission of skills and practices among craftsmen has rarely been examined, and even then, mainly as a secondary issue or collaterally within lineage concerns. *Ru* ideology enforced the family socially and epistemologically as the mandatory unit in which a father would hand his skills down to his son, and a mother to her daughters. While "tensions between government laws, Confucian ideology, social practice and ethnic norms"[48] are apparent during all periods, the Yuan code established some interesting nuances in determining, in the case of artisans (*jiang*), that "in all male and female offspring of the various craft households, the males have to practice labor and the females have to practice needlework. It is forbidden to avoid service or eliminate their status."[49] Two issues are noteworthy. First, in relation to household registers, the Yuan instrumentalized intergenerational succession explicitly for the transfer of skills. Second, this demand addressed the workforce holistically, including both males and females, the latter in particular in the context of the textile industry.

While the topic of gender is too rich to be discussed sufficiently in this chapter, the explicit reference foregrounds the wholesale approach of the Yuan toward skills, which we see reflected in the rules of intermarriage as well. Women had traditionally played a central role in silk production, as Francesca Bray has noted,[50] and often organized large-scale weaving workshops in their households, with dozens of female workers weaving, reeling, and embroidering silk under the preceding Song (976–1279; presumably also between 1125 and 1234 under Jurchen Jin rule in the northern territories, which the Mongols conquered first). Local gazetteers substantiate that officials relied on women as the real experts in the setting-up and running of state-owned weaving and dyeing offices in the Southern Song territories,[51] even though the nature of written language at the time concealed how central such female forces actually were, since the working units were household (*hu* 戶), taxpayer (*ding* 丁), or body (*shen* 身), meaning that gender was not specified.

The key authority files for tracking tasks and skills over time were the local mousetail registers[52] at the village level, and the key feature that secured reliability in the intergenerational lines of transmission was the practice of leaving a page blank. With the mousetail register, accounting regulations aimed to secure a truthful record that reflected quantitative and qualitative changes in the workforce. A mandatory blank page after each household listing had a specific purpose: whenever a taxpayer died, or someone reached adulthood and thus became required to pay taxes, or when the number of workers and farmers increased or declined, the village head was to report this to the clerk, who was "ordered to annotate under each household issues such as tribute service (*chaifa* 差發), in silver or tax provisions, corvée (*fuyi* 夫役), cart horses, production, requisitions, and military service, scrutinizing the book so that each addition and subtraction to a household is made based on personal observation of physical strength."[53] The blank-page policy signifies a high level of control, as well as a pragmatic acknowledgement that a hereditary system needed to recognize dynamics of population growth and changes in the skill sets of people over time.

Reading further into Xu's administrative dictionary, we find two indications—one about eyewitnessing and the other about cross-checking—that invite questions with regard to who was actually controlling whom. While some regulations—and corresponding meaning-makings by Xu—confirm the clerks' supervisory function, others are mainly concerned with the accuracy of the clerks' work, which was done by "comparing calculations" (*bijiao* 比較), "comparing for matches" (*bidui* 比對), "surveying on the spot" (*jianta* 檢踏), or "inspecting amounts" (*jianliang* 檢量).[54] These tasks reflect imperial regulations that required clerks to

> determine the household status on eyewitnessing the actual taxpayers, actual production, and suitable resources, and then produce the registers. Furthermore, [they are to] meticulously annotate following the original signature chronologically as to which category it subordinates; reaffirm the type of category and what category of tax service they have verified as an eyewitness on the day of recruitment, and then which of the household registers applies—civil register, postal station register, or craftsmen register. [They are also to] catalogue each and every detail meticulously, finishing the production of registers on all three accounts [that is, original household, service, and household category] so as to set an example.[55]

The *Tongzhi tiaoge* 通制條格 (Legislative Articles from the *Comprehensive Regulations*), a fragmentarily preserved collection of law cases and edicts, equally stresses the clerks' duty of keeping the records up to date and making sure

> not to appropriate craftsmen for special use from the various households who were attributed to either military or civil registers in the *renzi* year without having the supervising office change/amend the records with a clarifying statement. Whether they are named and enter the bureaus for manufacture on personal observation or are allocated funds for production in their private households, they are to rely on the registers to confine them to services.[56]

The various levels of recording and the additional insistence on seeing things firsthand protected and controlled both sides—the craftsmen who had to deliver tax labor and the executors of the related bureaucratic measures, the clerks.

In 1301, three decades after the statewide introduction of such imperial regulations, Xu's *Guidebook* echoed their importance, in particular with regard to timeliness and precision of reporting, as household registers were useful instruments for recruiting labor and collecting tax only when clerks performed their work faithfully. Xu furthermore noted the clerks' crucial ability to keep track of changes over time: "Increase in household numbers: refers to a maximum of clan members. Civil registers increase as new taxpayers appear or grow old; annotate and comment income and outcome; ascertain the facts as to whether there are escapees or some reluctant people who can be redirected to return to their craft."⁵⁷ It was the clerk's duty to ensure that households produced and reproduced knowledge important for the state.

This discrepancy between historical narration and the original Yuan historical viewpoint is the crux of the matter. Historians have highlighted how literati pitied the poor craftsmen who are exploited by the state. However, administrative books, of which the *Guidebook* is representative, focus on penalizing bad behavior or substandard performance in clerks—not craftsmen. It was the clerk's responsibility to promote and to police—that is, to select those with specific talents after mass recruitment or registration efforts, and to identify those who pretended to have certain abilities to avoid tax payments. In 1271 and again in 1280, for instance, orders had to be given to reassign those with no actual skills to the civilian households.⁵⁸ Similarly, in the sixth lunar month of 1290, an official ordered the "release of 341 Baoding Chutong laborers into civilian households."⁵⁹ An often quoted example for the exploitative nature of Yuan rule states that a good one-third of Jiangnan's three hundred thousand civilian households were struck from the craftsman registers, "after every kind of craftsman had been selected and determined."⁶⁰ In many of these cases, individuals hoping to become categorized as a craftsman tried to cheat the system to attain a change of rank.⁶¹

Clerks had to thwart these efforts because all kinds of commoners, including *Ru*, were trying to be classified as craftsmen—not only to escape beheading as representative of the former *Ru* elite, but to alleviate their tax burden or to become involved in a trade that was in high demand and regard during the Yuan. *Ru* trying to be registered as craftsmen were the target of court "officials and investigators who will survey the foundations of each household, the members' physical strength, and their hands to ascertain whether all of these fit with the categories noted in the registers."⁶² Regardless of whether someone was registered as official, commoner, or craftsman, all were reassigned to their original household category if they lacked the wherewithal (*jiacai* 家財).⁶³ In a eulogy, for instance, Luo Wenjie 羅文節, an administrative assistant in Fuzhou (*Fuzhou panguan* 撫州判官), is praised for having offended influential figures because he relentlessly—and rightly—prosecuted all who attempted to evade service, both scholars and craftsmen.⁶⁴ The Yuan dynasty's great concern for artisans found expression in institutional structures as well, and with the "superintendency" (*tiling suo* 提領所), a special department in the Ministry of Works was established to deal with litigation from craftsmen.⁶⁵

It was not just that the status of craftsman protected *Ru* from political prosecution. The status of artisan was attractive during the Yuan economically, too, as a considerable part of the household registration laws (*hukou tiaohua* 戶口條畫) made sure that craftsmen were taken care of and well fed whenever they were called on to perform labor for a state bureau.[66] Mongol laws made sure that Chinese clerks did not run wild. In his *Guidebook*, for instance, Xu noted that the "beating up of craftsmen by several officials resulting in the craftsmen's death has to be equally penalized by execution."[67] Between clerks and craftsmen, duty went both ways: craftsmen had to deliver service and perform their skills and clerks had to treat craftsmen well—and remain ethical in their behavior in order to survive.

良吏 謂政尚寬和，人懷其惠者，如晉吳隱之等

"BEING A VIRTUOUS CLERK means: governing with fairness, tolerance, and tenderness. Commoners cherish his benevolence as exemplified by Wu Yinzhi (d. 413) of the Jin Era." (Or: performing words to own skill in order to achieve status and wealth.)

Because *Ru* as clerks were under as much scrutiny as artisans (and may have led even riskier lives), they had to work with great care. Xu invoked the prefect of Guangzhou in South China, Wu Yinzhi, who had fought against corruption and reestablished order under a foreign ruling house during the fifth century as an example. At that time, elite corruption and administrative misbehavior was the greatest threat to craftsmen's lives. A considerable number of accounting mechanisms were set in place to protect not only the state but also craftsmen from greed. For instance, provisions for artisans were stored in the recruiting office, and the Ministry of Finance was usually in charge of funding them—to keep local officials in check. Whenever the court issued an additional order, it usually provided the funding required for materials and labor, as well. The details of these were quite complicated and changed frequently, but in general a "craftsman received a provision as the work process was investigated with each production."[68]

After 1273, as the Yuan pushed further south into Song territories, additional regulations were imposed to make sure craftsmen would still be compensated even if they fell ill and no additional labor for production was available, so that the workforce was not depleted unnecessarily. Or, in cases when a household was too small to support itself, it was decreed that an allowance should be provided by the storehouses.[69] In 1283 clerks were asked to consider an artisan's household size when recruiting them to the labor force, by, for instance, calling one person up for service but paying enough to support three people.[70] By around 1287 at the latest, these payments were being calculated based on the number of mouths to be fed in each household. This system was changed again in 1314, when some craftsmen bought, or were given, a small patch of land mostly for subsistence farming, on which they had to pay taxes (*nashui* 納稅).[71] Apart from military households, all craftsmen households were allowed to produce and sell their products or offer their services on the market.

Such institutionalization facilitated further exploitation, which historians have amply discussed, seeing such claims substantiated in the criticism of contemporary literati about the impoverishment of craftsmen or the fact that some craftsmen tried to escape service. However, when we read opinions against regulations that penalized corruption and officials who did not take proper care of artisans, we might also conclude that literati were highlighting the state's responsibility and craftsmen's poverty not primarily for humanitarian reasons, but to protect themselves. The state, for instance, held clerks responsible when craftsmen escaped service, which they did regularly. After all, the *Guidebook* emphasizes what clerks should do—namely, "check every month that all items are being produced according to the regulations. The standard procedure cannot be disregarded."[72] Xu cataloged instances of fraud or theft generically, while the state also made sure that its officials could not embezzle goods or compel craftsmen to produce items for them personally, issuing a "prohibition against ordering craftsmen to commence private production (*shadow possession* 影占)" to protect the very assets of the state. This section in the *Statutes* was tellingly titled "Harassed and Troubled Craftsmen" (*saorao gongjiang* 騷擾工匠).[73]

From the viewpoint of penalty laws, it was clearly in the *Ru*'s own interests to champion craftsmen's needs—to make sure the office looked after the artisans and that they "took care of their tools, insisted that weavers' households carried out repairs to their looms, and ensured that everything else was taken care of, that their dwellings were protected from wind and rain, that there was sufficient firewood, window sheets (made of paper), lamp oil for night work, and paper for reports"[74]—since the clerks would also suffer if the artisans were derelict in their duties. Occasionally, officials even bought agricultural implements for craftsmen to enable them to work.[75] Within the lines of state responsibility, the clerks had to become advocates for the craftsmen, complaining on their behalf to a higher power if necessary, to shield themselves from greater harm.

One reason why later generations have judged Yuan approaches to craftsmen as exploitative lies in the rigidity of the rules concerning keeping the accounts and records in the registers up to date. But the first addressee for any lack of control could only be the clerk who had to make sure that each household properly delivered its inherited levy over generations. Whenever clerks or state officials had to fill the ranks in state workshops, they relied on local lists that correctly designated abilities and skills.[76] According to the section "Prelude Record on Varied Crafts" in the "Great Statutes to Statecraft" (*Jingshi dadian* 經世大典) from 1304,

> households for state manufacture were first determined in mid-summer, for which an [administrative] procedure already existed. All the workers under heaven were gathered and all masters' departments were classified so that the procedures and measurements could

be investigated and provisions assigned. By restoring the households, it is possible to have them concentrate on their craftsmanship.[77]

By such means, local officials under the reign of Tugh Temür in the year 1304 allocated 240 tanners to a bespoke production of 587 pieces of armor.[78] The source also explicitly notes that the tanners were exempt from annual tax service so that they could concentrate on producing exemplary pieces for the court. Not exploitation or greed, but purpose and order ruled.

It thus seems that during the Yuan, *Ru* clerks wielded their power to make *meanings* by performing words in texts mainly to defend their own position. Or we could say that they claimed such performance of words as their knowledge and the very reason for their usefulness to the new ruling elite. It fits this picture that the imperial house treated *Ru* in the same way as craftsmen or traders, that is, as technicians, making no exception for them in terms of either household categories or social status. We can see such pragmatism in the imperial attitude of the Mongols toward social status, substantiated by the fact that in the early days or Yuan rule, workers (*gong*) and traders (*shang*) were allowed to participate in the civil service examination—unlike during the Song reign, which had restricted this route to literarily trained men. No source verifies that craftsmen successfully passed the exam—nor is there any indication that anyone attempted to or wished to participate in it, before its abolition in 1267. By comparison, in combing through biographies and the official historiography, we find several artisans such as bow makers, weavers, and masons who obtained esteemed posts in officialdom because of their excellent craftsmanship.[79] Under Shundi (1341–1368), the lacquer craftsman Mr. Wang from Pingjiang

> tried to manufacture a boat out of a cowhide and adorned it on the inside and outside with varnish/lacquer; after taking a rest, he produced numerous decorative joints and transported them to the capital. Floating on the Luan River, it could hold 20 people. . . . He received a decree to produce a collapsible armillary sphere, which was easy to store. His brilliant ideas met and surpassed any expectation; thereafter the order was issued to designate him as a state craftsman.[80]

Such cases show not only that craftsmen could rise in status based on their practical skills and innovative ideas rather than through exams or scholarship. Or to put it another way, craftsmen were in charge and clerks were subservient to them, and in fact, this was often the case.

As a final clue, we can consider contemporary notions of *Ru* toward their specific duty—namely, selecting skills and being selected for their skills. After the Yuan conquest, Hu Qiyu invoked the need of good tools for good craftsmanship, reinterpreting a passage of the *Classic of Rites*. That he therein criticized the Yuan approach to expertise—of both craftsmen and clerks—becomes apparent only when one sees that Hu played with the double meaning of the Chinese term for "workers for the state" (*gong*) to address all crafts in the broad or literati arts in the narrow sense:

> *Gong* who desire to be good at their affairs must first make use of/profit from their tools; if their tools are not effective, affairs will not succeed. Even if you have skilled workers, there is nothing for them to execute their ingenuity with.[81]

Hu continued by exemplifying the way of Tang Taizong (598–649), who had

> showed this with the bow. The archer said: this is not a fine bow. The wooden core is not exact, the limbs are all arranged irregularly. Hence when you shoot, the arrow will not fly straight. When the very essence that Taizong uses to shoot an arrow is not good, then the mark will not be hit.[82]

Xu alluded to the archers' bow as well when explaining that "chaffing and felting (*jiqiu* 箕裘) means to carry on the business/trade of the ancestors,"[83] making a connection, like Hu, to classic debates of governance from the pre-imperial period. One of these, the "Artificers' Record" (*Kaogong ji* 考工記), was part of the "Rituals of Statecraft" (*Zhouli* 周禮) and explained the structure of the so-called Winter Offices organized by the Ministry of Works under Zhou rule, noting that simple skills had to be mastered before approaching the complicated ones mastered by the fathers (i.e., seniors of the trades): "The son of an ingenious archer had to learn chaffing; the son of a fine smith [which can also mean ruler] had to engage in fur-making/clothing."[84] Xu was effectively pointing to a time when power lay in the hands of practitioners. Major classical texts trace the core influential hundred clans (*baijia* 百家) of antiquity back to core crafts such as carpentry, smithing, or weaving; only later, under subsequent imperial rulers, did scholars start to fill this role of "the hundred workers" (*baigong* 百工) and fill the ranks of the "hundred offices" (*baiguan* 百官).

It was this debate between the relative merits of scholarly and artisanal knowledge for political rule to which Ming scholars and politicians would allude two centuries later, in 1489, in negotiating the relationship of craftsman and clerk in their time. By then, the ruling house of the Ming dynasty had returned the power of statecraft to the *Ru*, as historians have seen verified in figures such as Qiu Jun 邱濬 (1421–1495), who reached the peak of his career as a highly decorated minister of rites. Unlike the Yuan, the Ming installed a codex that Qiu Jun acknowledged. Analyzing Qiu's discussion of the contents of codices, the legal historian Huang Yin describes Qiu as more interested in structural issues than in specifics, noting mainly that he doubted the effectiveness of penalties.[85] In such contexts, Qiu never addressed crafts. However, when we examine the words Qiu used and his approach to rectifying names, we can see that Qiu was extremely concerned about such themes, discussing crafts and their regulation thoroughly in relation to ritual norms and economy.

It is when one follows changes across political periodization in terms of dynasties that the question of continuity must be addressed. In contrast to Xu, who mainly quoted high antiquity and literature up until the Tang (leaving out Song scholars), two centuries later, Qiu carefully quoted *Ru* books, including other Yuan thinkers who upheld prior Song traditions. In such literature, crafts were important mainly when it came to the quality of ritual utensils (these were taken as the highest standard) and to how the imperial house was legitimized to rule only as long as it cared for the people and followed, among other proscriptions, the seasonality of work in order to not exhaust resources. On a practical level of governance practices, Qiu's entire debate was directed toward the fact that Yuan practices lingered on, probably most persistently when it came to crafts, as the Ming also relied on household registration to recruit workers for the construction of huge palace complexes and for the production of intricate lacquer boxes and textile wares.[86] The Ming thus profited from the lists that had been drawn up in each locality by the Yuan to recruit tax levies for the state, which ran into the thousands in terms of households in rich places such as Jinling and were equally available for metropolitan counties such as Jiangning and Shangyuan, or even rural counties such as Tanshouzhou and Tanyangzhou.[87] Even though this system was constantly adapted to meet new needs as it persisted over the subsequent centuries and into Qiu's time, one continuity was the partial association of household registers with practical tasks. The final section of this chapter sketches how the efforts of the Yuan to identify and name crafts for their appropriation by the state were thought of and received during the succeeding Ming dynasty, which also wished to profit from craftsmen's skills—by similar means, but clearly on its own terms.

工而謂之百，不止一工也

"WORK was named by a hundred, as there is not just one kind." (Or: disowning skills by ignoring words.)

When Qiu Jun wrote down this comment on craftwork in 1489, he could look back on his steady upward career under three Ming rulers. His *Supplement to the Great Learning* was meant to guide the offspring of his ruling house through the arts of statecraft, which for Qiu included the management of crafts. Having conquered the weakened Yuan by 1369, the Ming emperor Zhu Yuanzhang 朱元璋 had been able to build up his empire despite the destruction wrought by warfare, not least because he was able to make use of craftsmen recruited by the Yuan from all parts of Asia. He relied in particular on access to a cosmopolitan group of experts from the inner Asian lands, South Asia, and the Near

East, who could dye, weave, and carve, build, and tan; smelt metals, mine, and mint; and produce ceramics, weaponry, and lacquer works. This was, in short, a veritable army of masters in the practical arts.

Zhu continued the principle of household tax inheritance but added multiple innovations, among them a quota system that allowed craftsmen to rotate in and out of service. In an early effort to balance the demands of the state against craftsmen's needs to replenish their private coffers for sustenance and life, the Ming state institutionalized two forms of "shift craftsmen" (*lunban jiang* 輪班匠), who rotated in and out of service over cycles of three or five years, and "resident craftsmen" (*zhuzuo jiang* 住坐匠), who delivered the taxes for their household in permanent residential positions. Over the course of the Ming period, such tax assignments were negotiated among family members. Some cases culminated, as historian Thomas Nimick has shown, in legal disputes over how working for the state was to be compensated by other family members.[88]

My focus in this chapter on the practice of name rectification and *Ru* literati meaning-making reveals a change in the role law played in the state's access to craftsmen's labor and knowledge during the Ming. Two trends are apparent. First, classic ritual texts provided the guiding framework for craft production—a return to a Song-era practice.[89] Second, legal measures no longer concerned access to crafts but rather almost exclusively the penalizing of craftsmen who did not perform well: "For the production of prohibited goods, bludgeon 100 times; when the goods are not up to standard, flog 40 to 50 times."[90] Whereas clarifying nomenclature and rectifying names and status remained an important legal practice—similar to rhetoric techniques of analogous argumentation, as legal historian Chen Xinyu has recently suggested—neither legal codices nor bureaucratic practices relied on such methods to manage crafts.[91] The category of "workers" or "craftsmen" is not even mentioned in the "Ming Penal Code" (*Da Ming Lü* 大明律), and the "Collected Statutes of the Great Ming" (*Da Ming huidian* 大明會典) only cursorily touches on them as one of the four social groups—scholars, farmers, workers, and merchants. Many of the technical terms and processes that Xu identified around contracting craft work or recruiting levy, such as "agreements about weaving," are equally absent.[92] Instead, official historiography suggests that Ming *Ru* officials began enforcing administrative regulations on craftsmen originally implemented by the Yuan in an ad hoc manner and mostly, it seems, on a local level.

It was this world, where the management of crafts relied on laws mainly as a penalizing tool, into which Qiu Jun was born and where he grew up as a fatherless child on the southern island of Hainan. He passed the provincial civil service exam at the age of twenty-three, gained his first ranked position in 1466 and thereafter served at a court and under emperors who favored eunuchs, skilled artisans, and artists.[93] He stood side by side with artisans such as the craftsman Yao Wang 姚旺, who had entered the very institution that provided the court with luxury wares such as fine silks and jewelry, the Courtyard for Cultured Thoughts (*wensi yuan* 文思院) in 1464;[94] the carpenter-architect Kuai Xiang 蒯祥, who reached the exalted position of vice minister of work (*gongbu zuo shilang* 工部左侍郎, rank 2a, then 1) in 1465;[95] and Kang Yongshao 康永韶, a magistrate from the countryside of Fujian who was awarded the salary of an astronomy supervisor simply because he had helped to heal the Chenghua emperor when he had fallen ill.

By the early fifteenth century, as emperors no longer felt obliged to follow the restrictions set by their ancestors, the number of craftsmen attempting to evade service—by either fleeing or deliberately delivering substandard work—had risen to the thousands. Imposing legal penalties on artisans proved of little effect whenever the state asked for more than its share. Thus, in his first year of rule, 1436, Emperor Yingzong allowed "southern craftsmen" (*nan jiang* 南匠)—a term that originally denoted the geographic origin of such craftsmen but had come to designate high expertise more generally—to substitute their corvée with a tax payment in silver (*zhengyin* 徵引) so that the government could hire suitable craftsmen in Beijing "for the convenience of both sides."[96] In the twenty-first year of Chenghua (1485), about two years before Qiu Jun published his *Supplement*, the Ministry of Works allowed craftsmen from Zhejiang, Jiangxi, and other places to pay tax as a substitute for corvée labor,[97] which marks the first evident signs of a shift to monetary payments that historians of the 1970s such as Peng Zeyi see culminating in the "first sprouts" (*mengya* 萌芽) of capitalist structures, a commodity economy, and a free labor market.[98]

It was against this background that Qiu debated "naming" as a mechanism to regulate craft work—and for empowering or disempowering it by the state, promoting not laws but ritual and morality as the correct domain of exchange. Qiu had realized that craftsmen could exert enormous power because their exquisite and sophisticated wares stimulated emperors' desire for acquisition, which strained the state's coffers and caused the emperor to ignore literati advice. It was an essential part of Qiu's strategy to caution the emperor and his peers against individual skillfulness and the ingenuity of the technical arts (*jiyi* 技藝), which, to Qiu, was primarily a question of how the state and emperor recognized the different categories of craft know-how—that is, the hundred crafts.

As Edward Farmer has noted, the Ming continued the Yuan system of household registers to the extent of "smaller specialized categories such as artisans (*jiang*), and physicians (*yi*)."[99] As a matter of fact, the "Yellow Register" tax system (*lijia huangceng* 里甲黃冊), established in 1381, was based on Yuan records for the assignment of each tax household category. Until the end of the Ming dynasty, the household category could not be changed without the consent or signed release by an official.

In 1487, though, Qiu wanted to see the diversification of names limited—at least when it came to how such names made expertise visible to the emperors of the Ming. How many crafts actually were registered or whether the "hundred" indicated a specific number is debatable. Qiu's initial quote points to a sophisticated culture of crafts during the early to mid-Ming—and probably to a proliferation of nomenclature as well, as there were indeed *many* "works/crafts," as the quote at the beginning of this paragraph indicates—and according to Qiu, there was *no need* for emperors or officials to further itemize them. He reminded his colleagues that the grand Song thinker Zhu Xi 朱熹 (1130–1200) had already noted that "when asked who was able to smoothly regulate the affair of bestowing the hundred works, the Yellow Emperor [pointed out that it was] passed down through the generations and named an office. This is a skillful thought."[100] Qiu thus explicitly *denied* crafts their names. Or we could say he rectified names, and by refusing to name all crafts, he disowned the artisans—for without a name, there was no political and intellectual recognition of the work and the person/body who performed it.

This interpretation assumes that Qiu deliberately left out the Yuan period when making meanings, preferring to quote the *Rites of Zhou*, as "the Grand Steward recruited all people within nine assignments.[101] The fifth, called the hundred workers/crafts, processes and transforms the eight materials."[102] Though literati may have argued about the institutionalization of crafts, as Qiu emphasized, any evaluation of such skills had always followed basic rules of decency:

> Heaven has seasons, earth has *qi*, in materials there is beauty and in work there is artistry. Decency depends on the combination of these four factors. If materials are beautiful and work is artistic but indecent, then it is because it is out of its proper season and has not achieved the earthly *qi*.[103]

With this reference, Qiu attempted to resituate crafts into an agricultural state model that recruited farmers who also could serve as skilled workers for public needs, but who did not produce luxury goods that threatened a morally upright and frugal emperor. Farmers had to secure grain first—in contrast to Xu and the emperors of the Yuan, who had employed and utilized crafts for a good life and associated skills with political authority. In due course, Qiu reverted to past incidents when the desire of elites—and in particular, the continuous emphasis of emperors on the hundred crafts over time—had wrought considerable confusion over political structures and social order: specifically, people had come to mistake the workers of the state—that is, officials—to mean craftsmen, and vice versa. For instance, in some eras, designations of social hierarchy (*shangxia* 上下) had been conflated with occupations and businesses/trades/jobs (*zhiye* 職業). Even early Song literati such as Wang Zhaoyu 王昭禹 (fl. 1080), as Qiu complained, had mistaken classic references that "do not particularly address an 'official's' duties.

Producers of records set up the affairs of the hundred workers [hundred crafts].[104] And because officials held on to the principle that communal efforts ease affairs,"[105] and there were more crafts than officials, "each [of the five offices] took over six of the businesses/trades/jobs." The five offices, he further explained, "address the five materials processed by the hundred trades and also the possible usefulness of differentiating the people's tools and utensils."[106] Clerks were useful because they were able to classify, order, and summarize the otherwise multifarious crafts.

In tax records of the textile trade of this period, nomenclature was sophisticated. While reelers (*luosi* 絡絲), silk walkers (*daxian* 打線), and dyers (*ransi* 染絲) had always been identified, officials now frequently differentiated different weavers on the basis of loom type, such as "waist loom weavers" (*yaoji* 腰機) as opposed to those who worked on a drawloom (*tihua ji* 提花機).[107] Against this background, Qiu Jun concluded that "men of wit had especially clustered tasks with others of their kind (*lei* 類)." Qiu thus wanted to constrain the power of crafts by restricting their naming, as chaos was more likely to ensue if names increased beyond one hundred—at least on the level of court and central state debates.

From a *longue durée* point of view, Qiu's efforts seem like an attempt to limit the consequences of the Yuan's systematic state involvement in crafts, which had made craft expertise visible in new ways. Qiu's countermeasure was to standardize and restrict names and meanings. Fast-forwarding, we can see these efforts bearing fruit by the end of the Ming. The editors of the "Local Gazetteer of Jiangxi" (*Jiaxing fuzhi* 嘉興府志) in the year 1600 give seventy-two as the standard number for a generic list of registered craft occupations. While the tax and levy sections in local gazetteers—a genre that was published regularly over the entire territory of the Ming and would amount to ten thousand titles by the end of the Qing—include a considerable variety of terms, but few lists exceed one hundred.

Over the course of the centuries, the efforts of men such as Xu—a man who had embraced the rectification of names as a skill and a way to substantiate the *Ru*'s relevance and usefulness for the imperial state—thus had two significant effects. First, the lists that these efforts generated became important reference points for the continued recruitment of skilled labor by the state, to the point that intergenerational continuity was enforced because the state wanted to secure access to such knowledge and the economic benefits that came from performing such skills and producing craft wares. Second, the need to carefully record and register such skills locally over generations in order to ensure such access by the state led to a geographical mapping of such skills—as local and central state officials and local elites consulted such archived registers over the centuries to further recruitment drives. When we consider how local gazetteers in

China by the end of the Ming (the early seventeenth century) had started to contribute to local identity, generating and enforcing it,[108] we can see that the real power of the *Ru* and their practices lay in the *longue durée*, gaining new power with the modern emphasis on laws as a framework for knowledge property. This emphasis assigned new power to the Yuan identification of crafts when it began to identify such references to crafts as "traditional knowledge," or when it related the Yuan household tax categories in local gazetteers and/or genealogical local sources to modern legal and economic entities such as the Chinese "old brands" (*Lao zihao* 老字號).[109] These are the most obvious signs of how modern ideas interfere with historical concepts and practices of regulating the ownership of knowledge.

察知也

"SCRUTINY is knowing." (Or: ownership as knowing about know-how.)

As a legal clerk, Xu dedicated attention—as the quote says—to "knowing," noting that for a sound judgement, a thorough scrutiny of all matters was key. As a scholar, Xu dedicated most of his attention to words and meanings, showcasing the main skill that the rulers of the Yuan appreciated about the *Ru*. In combining both roles, Xu contributed to the historical reception of craft knowledge and practical abilities in Chinese history, creating paper trails that survived the bodies that had originally performed such knowledge, and thus continued a nomenclature of skills.

In acknowledging that Xu advertised naming as a skill to achieve social relevance and political power over a group with abilities that were more important, a last point needs to be made about modern scholarship. For a historian of science, fascinating ambiguities lie in research on the historical and contemporary role of China's bureaucracies and property rights in the 1950s to the 1980s—both in the comparisons between Eastern and Western models, and in a world in which the anthropology of Émile Durkheim, organizational sociology, and debates around economy, crafts, and science took place in separate and very different camps and ideological blocks, and were were being applied in different ways in discussions about different regions of the world. Against the background of Cold War politics, Marxist-Maoist historians in China have explored the feudal character of labor relations in their past, while Western historians of science such as Robert Merton have revisited Marxian historiography as a question of how bureaucracy affected sciences in the West. The separation of individuals from their skills, as well as the rationalities and irrationalities of property relations, featured prominently in these debates, which identified the West as the origin of approaching knowledge as property and property as an individual's right, and the East as being ruled by copious bureaucracies that created heritages and pasts, but no legal framework for the protection of individual knowledge and skills. Through such debates, a fine but all-too-unyielding line of separation was drawn that researchers reflect when they approach law as authorizing the social norms for property and see governance defining the—public and private—policy structures that bureaucracy then implemented, while ascribing to bureaucracy the function of identifying the sum of institutional practices that enforce, utilize, or instrumentalize such property rights. Inasmuch as law historians accept such domains, they can pinpoint the emphasis of Chinese law on land and things and otherwise address property of knowledge in China as a subsidiary issue of marriage and inheritance rights.[110] Historical, anthropological, and social science scholarship interferes in an equal way, as it similarly relegates craftsmen and crafts production to a research topic that is mainly relevant to economic or social history, while leaving epistemic issues aside.[111]

In following the practice of rectifying names, however, we can see that Chinese historical actors both during the Yuan and Ming eras, *used* the fact that knowledge and its ownership cannot be broken apart to manipulate the ownership of craft knowledge. As I have attempted to make clear in this chapter, this worked by way of exclusion. Following the Chinese scholarly practice of meaning-making addresses the two major concerns in which historical scholarship sees bureaucracy enacting ownership of knowledge. These are claims to knowledge and rights to use it, and both are asserted in two ways: by managing information—codification—and by institutionalizing structural rules. Yet, as following this practice makes clear, practices can also operate in different domains. By tracing meaning-making, we can see that three different transformative logics define the way in which historical actors have used bureaucratic practices to manipulate the ownership of skills: (1) descriptive modes, (2) decision-making processes, and (3) the installation of procedural concepts. Tracing one practice of rectifying names, in this chapter I have highlighted descriptive modes to provide some insights into the possibilities and limits of knowing and owning crafts in historical China. Along those lines, we can see that accounting produces not only tangible objects but also barriers of knowledge ownership, and that scholars generate a very particular landscape of how knowledge can be owned—for instance, by naming some tasks but ignoring others; or by reinterpreting historical relationships. The procedural concept informing bureaucracy lays out possibilities for the scale and scope of ownership claims that then linger on, as in the case of the household categories installed by the Yuan, which lasted even far beyond Ming times. Today, the names that have survived in historical records can be used to claim ownership in the form of modern regiments such as brands or cultural and regional heritage, such as by determining which locality provided what skill set and since when.

One effect of historical tax and administrative practices is that nowadays crafts and craft categories are more closely linked with localities rather than individuals I have written elsewhere of how Chinese bureaucratic practice caused workers of the state—the craftsmen or the officials—to inscribe their names (*wule gongming* 物勒工名, *wule guanming* 物勒官名) directly onto products to guarantee their quality and verify the delivery of tax.[112] Texts and archives have no such names. They mainly count how many lacquer craftsmen were recruited, verifying Xitang, Jiaxing Prefecture as a center of this trade; the prefecture of Jingdezhen as a porcelain hub or that weavers accumulated in Suzhou and Hangzhou. In combination with the increased importance that tax records in books, and genres such as the local gazetteers, gave to the naming of "crafts" for further appropriation by the state, an idiosyncratic landscape of how crafts were known and owned in imperial China emerges: a landscape in which practitioners' knowledge became visible and relevant as a local resource, but was not attached to individual creativity and family names. Such visibility, though, is only a historical artifact. In the past, as today, the literati's sharpest weapon—for good or bad—was and is the word. But it is also true that the word has to stand its ground against the knowledge contained in the exceptionally adept (*qiao*), splendid, and ingenious artifacts that have survived, and as knowledge in bodies that has to be transmitted over generations and survive in modern times—as a historical as well as an epistemic asset important to recognize for us today, too.

Notes

1. "禁制：私造：謂兵器之類，無故私造者" Xu Yuanrui 徐元瑞, *Lixue zhinan* 吏學指南 [Guidebook for clerks], Yuan keben (Beijing: Beijing Tushuguan Chubanshe, 2004), *juan* 6, 5b. The original Yuan edition has nineteen characters per line, the commentaries are in twenty-four characters in double-lines with black croakers, and it is framed all around. The oldest extant edition is the *Jujia biyong shilei quanji* 居家必用事類全集 [Complete collection of household necessities], preserved in Korea and Japan. In 1969 Wenhai published a version of the "Household necessities," which contained several typos and other mistakes. Most facsimiles today use the 1673 version and correct mistakes with reference to the original classical texts. This incorporates the *Lidai lishi leilu* 歷代吏師類錄 [History of clerks], which chronicled popular historical master clerks up until the Song and included a final chapter of notes on varied warnings, sayings, and a biography, all in prose form. For the paratexts, see Xia Lingwei 夏令伟, "*Quan Song wen Quan Yuan wen* buyi" 《全宋文》《全元文》补遗 [Addenda to the complete collections of Song and Yuan prose], *Jiangsu daxue xuebao (shehui kexue ban)* 19, no. 5 (2017): 16–20.

2. Yang Shitie 杨世铁, "*Lixue zhinan* de cishu xingzhi" 《吏学指南》的辞书性质 [The dictionary nature of *Lixue zhinan*], *Huaibei shifan daxue xuebao (zhexue shehui kexue)* 35, no. 5 (2014): 47–49; Ye Xinmin 叶新民, "Yibu Yuanchao gongwen yongyu cidian—*Lixue zhinan* jianjie"一部元朝公文用语

辞典——《吏学指南》简介 [A dictionary of Yuan dynasty official documents—a brief introduction to *Lixue zhinan*], *Neimenggu shehui kexue (wenshizhe ban)*, no. 6 (1988): 68–71. According to Yang's counting, only 25 of the 1,450 give phonetics. Ye counts 91 categories in 2,109 paragraphs.

3. Not to be confused with the Yuan-dynasty hereditary household category *ruhu* 儒戶 mentioned later, which was a legal identity. Those registered as *ruhu* were predominantly literati of the former Jin and Song, and their offspring; people with other legal identities, however, were allowed to study Confucian classics and take state examinations (when they were held again after 1313), who would be referred to as *Ru* in the sense of a social and cultural identity. See Xiao Qiqing 蕭啟慶, "Yuandai de ruhu: rushi diwei yanjinshi shang de yizhang" 元代的儒戶：儒士地位演進史上的一章 [The Confucian households of the Yuan dynasty: A chapter in the history of Confucians' social status], in *Yuandai shi xintan* 元代史新探 [A new exploration in the history of the Yuan dynasty] (Taipei: Xinwenfeng, 1983), 1–58.

4. Yang, "*Lixue zhinan* de cishu xingzhi," 47–49; Guo Chaoying 郭超颖 and Wang Chenglüe 王承略, "Cong *Lixue zhinan* kan Yuandai liyuan yishi" 从《吏学指南》看元代吏员意识 [On the consciousness of officials in the Yuan dynasty from *Lixue zhinan*], *Jiangxi shehui kexue* 35, no. 2 (2015): 114–119.

5. For a general overview, see Bettine Birge, "Gender, Property, and Law in China," *Journal of the Economic and Social History of the Orient* 44, no. 4 (2001): 575–599. Administrative regulations were promoted in 1297, though. See Xie Hongxing 谢红星, "Dianli falü tixi xingcheng zhi qianye: Yuandai qilüyong geli jiqi falüshi diwei" "典例法律体系"形成之前夜：元代"弃律用格例"及其法律史地位 [The eve before the formation of the "classical legal system": Yuan dynasty "Qilüyong geli" and its status in legal history], *Jiangxi shehui kexue* 40, no. 3 (2020): 136–147.

6. Victor H. Mair, "Persian Scribes (Munshi) and Chinese Literati (Ru): The Power and Prestige of Fine Writing (Adab/Wenzhang)," in *Literacy in the Persianate World: Writing and the Social Order*, ed. Brian Spooner and William L. Hanaway (Philadelphia: University of Pennsylvania Press, 2012), 388–414.

7. Thomas T. Allsen, *Commodity and Exchange in the Mongol Empire: A Cultural History of Islamic Textiles* (Cambridge: Cambridge University Press, 1997), 31–33. For an overview, see Morris Rossabi, ed., *Eurasian Influences on Yuan China* (Singapore: ISEAS–Yusof Ishak Institute, 2013), esp. 200–201.

8. Especially sociolinguists such as Michael Silverstein and Greg Urban, eds., *Natural Histories of Discourse* (Chicago: University of Chicago Press, 1996), were among the first to connect how collectives construe, understand, and make sense of the world to a verbal and semiotic "meaning-making," noting that as collectives made sense of the world, they created words that then again could come to create realities. Organizational sociologists such as Karl E. Weick, *Sensemaking in Organizations*, Foundations for Organizational Science (Thousand Oaks, CA: Sage, 1995), reconnect such notions back to Émile Durkheim, thereby pinpointing the processes of objectification and meaning-making that anticipate decision-making in bureaucracy.

9. Émile Durkheim, *Les formes élémentaires de la vie religieuse: Le système totémique en Australie* (Paris: Puf, 1912), 324.

10. Daniel Harrison Kulp, *Country Life in South China* (New York: Teachers College, Columbia University, 1923), 148–150. Note that in this 1920s discussion, capital played much less a role than social wealth and prosperity and therefore collective ownership responsibilities and rights were discussed in different terms than in the post-WWII debates imbued by political dichotomies of communism-versus-capitalism.

11. For a summary of this literature and the development of the field, I refer to Frank Dobbin, "How Durkheim's Theory of Meaning-Making Influenced Organizational Sociology," in *The Oxford Handbook of Sociology and Organization Studies: Classical Foundations*, ed. Paul S. Adler (New York: Oxford University Press, 2009), 200–222.

12. For a study of Zhu Bishan, see J. Keith Wilson, "The Fine Art of Drinking: The Chinese Silversmith Zhu Bishan and His Sculptural Cups," *Bulletin of the Cleveland Museum of Art* 81, no. 10 (1994): 380–401. Studying the local culture of Suzhou where Xu lives, Michael Marmé notes that "the havoc Mongol rule inflicted on handicrafts has been greatly exaggerated." Marmé, *Suzhou: Where the Goods of All the Provinces Converge* (Stanford, CA: Stanford University Press, 2005), 55. See also James C. Y. Watt, *The World of Khubilai Khan: Chinese Art in the Yuan Dynasty* (New York: Metropolitan Museum of Art, 2010), 295.

13. For the relation of the tax practice to law in the *longue durée*, see Yuan Yuanweiyang 袁远维扬, "Wule gongming zhi de lunli yunhan" "物勒工名"制的伦理蕴含 [The ethical implications of the "Wule Gongming" system], *Hubei jingji xueyuan xuebao (renwen shehui kexue ban)* 15, no. 4 (2018): 22. For a general overview of inscriptions, see Dagmar Schäfer, "Inscribing the Artifact and Inspiring Trust: The Changing Role of Markings in the Ming Era," *East Asian Science, Technology and Society* 5, no. 2 (2011): 239–265. For the *longue durée* view on such practices, see Anthony J. Barbieri-Low, *Artisans in Early Imperial China* (Seattle: University of Washington Press, 2007).

14. "賦役平：謂理財之最，取辦有法，催科不擾者。" Xu, *Lixue zhinan*, juan 1, 6a.

15. Yao Sui 姚燧, "Song Li Maoqin xu" 送李茂卿序 [Preface to Li Maoqing] (Beijing: Renmin Wenxue Chubanshe, 2011), 4:71.

16. Guo and Wang, "Cong *Lixue zhinan* kan Yuandai liyuan yishi," 115.

17. Sukhee Lee, *Negotiated Power: The State, Elites, and Local Governance in Twelfth- to Fourteenth-Century China* (Boston: Harvard University Asia Center, 2014), 220. Lee refers to the neo-Confucian scholar Cheng Duanli 程端禮 (1271–1345), Lu Wengui 陸文圭 (1252–1336), and Xu Qian 許謙 (1270–1337).

18. "貢舉法廢，士無入仕之階，或習刀筆以為吏胥，或執僕役以事官僚，或作技巧販鬻以為工匠商賈。" Song Lian 宋濂, *Yuan shi* 元史 [History of Yuan] (Beijing: Zhonghua Shuju, 1976), 1269.

19. Xu, *Lixue zhinan*, 1a. I have left the term *fa* here untranslated because it can be interpreted as both administrative rules or methods for regulating work and ordering society. The *Guidebook* does not, however, address craft methods themselves.

20. Xu, *Lixue zhinan*, juan 1, 1b. These talents are resonant with the "overseeing" tasks that Xu identified later, such as "calculating materials" (*jiliao* 計料) and "thorough inspection" (*jianhe* 檢覈).

21. On political stance, see Kang Jinsheng 康金声, "Sima Xiangru xinlun" 司马相如新论 [New findings on Sima Xiangru], *Shanxi daxue xuebao (zhexue shehui kexue ban)*, no. 4 (2002): 10–11, and Yang Fuyou 杨富有, "Yuan Shangdu yongshishi de neirong jiqi yiyi fenxi" 元上都咏史诗的内容及其意义分析 [The content and meaning of historical poems in the capital of the Yuan dynasty], *Neimenggu minzu daxue xuebao (shehui kexue ban)* 38, no. 3 (2012): 27–29. For Sima Xiangru's interest in language, see Ding Yiru 丁憶如, "Sima Xiangru fupian zhi yinyun fengge yanjiu" 司馬相如賦篇之音韻風格研究 [The study on the linguistic style of Sima Xiangru's rhaposodies] (master's thesis, Taipei, National Chengchi University, 2007). For the role of *fu* in Song exams, see Martin Kern, "The 'Biography of Sima Xiangru' and the Question of the Fu in Sima Qian's Shiji," *Journal of the American Oriental Society* 123, no. 2 (2003): 304.

22. "以美人爲君子, 以珍寶爲仁義, 以水深雪雰爲小人, 思以道術相報, 貽於時君, 而懼讒邪不得以通。" Zhang Heng 張衡, "Sichou shi xu" 四愁詩序 [Preface to the Four Sorrows], in *Wenxuan* 文選 [Anthology], ed. Xiao Tong 蕭統 (Taipei: Wunan Tushu, 1998), 751; "途之大者謂之道, 小者謂之術 [. . .] 莊周以江湖對道術而言, 則直指爲道路無疑矣。" Sun Yi 孫奕, *Lüzhai shier bian* 履斋示儿編 [Collection of instructions for my sons in the Lüzhai Study] (Beijing: Zhonghua Shuju, 1985), 669.

23. Liu Xie 劉勰, *Wenxin diaolong zhu* 文心雕龍注 [The literary mind and carving of dragons with annotations], ed. Fan Wenlan 范文瀾 (Beijing: Renmin Wenxue Chubanshe, 1962), 727.

24. Yang Shuhong 杨淑红, "Yuandai qiyue wenshu de kanbu yu yanjiu zongshu" 元代契约文书的刊布与研究综述 [A survey of contract documents in the Yuan dynasty], *Zhongguoshi yanjiu dongtai*, no. 1 (2011): 28–34.

25. "良賤：名編戶籍, 素本齊民。" Xu, *Lixue zhinan*, juan 6, 4a.

26. "賦役：謂徵催錢糧, 均當差役也。" Xu, *Lixue zhinan*, juan 3, 2b.

27. "貴賤：身富位尊曰貴, 卑下無位曰賤。《刑統賦釋》曰：貴賤之賤, 君子有時居之。" Xu, *Lixue zhinan*, juan 6, 4a.

28. Xu notes nothing about the tax immunity that Allsen ascribes to these groups for the purpose of coopting them with patronage and tax immunities. *Commodity and Exchange*, 200. Allsen himself quotes Tao-Chung Yao, "Ch'iu Ch'u-Chi and Chinggis Khan," *Harvard Journal of Asiatic Studies* 46, no. 1 (1986): 201–219, who focused in his study on the era of Chinghis Khan.

29. "韃法：一官、二吏、三僧、四道、五醫、六工、七獵、八民、九儒、十丐, 各有所統轄。" Zheng Sixiao 郑思肖, *Zheng Sixiao ji* 郑思肖集 [Collected works of Zheng Sixiao] (Shanghai: Shanghai Guji Chubanshe, 1991), 186; Hu Zhiyu 胡祗遹, *Zishan da quanji* 紫山大全集 [The big collection of Zishan], Siku quanshu (Taipei: Taiwan Shangwu Yinshuguan, 1986), juan 23, 1–40.

30. Hu, *Zishan da quanji*, juan 22, for instance, notes thirty-six different groups.

31. Zhang, "Legal System," 345.

32. Huang Shijian 黃時鑑, ed., *Tongzhi tiaoge* 通制條格 [Statutes from the comphrehensive regulations] (Hangzhou: Zhejiang Guji Chubanshe, 1986), 342.

33. "霆嘗考之, 韃人始初草昧, 百工之事, 無一而有。[. . .] 後來滅回回, 始有物產, 始有工匠, 始有器械。蓋回回百工技藝極精, 攻城之具尤精。後滅金虜, 百工之事, 於是大備。" Xu Ting 徐霆, *Heida shilüe jiaoyhu* 黑韃

事略校注 [Brief account of the black tatars: An annotated edition], ed. Xu Quansheng 许全胜 (Lanzhou: Lanzhou Daxue Chubanshe, 2014), 99. Xu also mentioned that the Yuan "included" (i.e., conquered) the Uighurs (literally: Hui Hui) into their empire and that only then did the Yuan "begin to have craft products (*wuchan* 物產) and also craftsmen (*gongjiang* 工匠) and tools (*qixie* 器械). The techniques of the Hui Hui are refined."

34. Liu Liya 刘莉亚 and Chen Peng 陈鹏, "Yuandai xiguan gongjiang de shenfen diwei" 元代系官工匠的身份地位 [The social status of craftsmen in the Yuan dynasty], *Neimenggu shehui kexue*, no. 3 (2003): 10–16.

35. For a thorough analysis of Mongol crafts based on archeological excavations in Central Asia, see Susanne Reichert, "Imperial Policies towards Handicraft: The Organization of Production in the Old Mongolian Capital Karakorum," in *Craft Production Systems in a Cross-Cultural Perspective*, ed. Martin Bentz and Tobias Helms (Bonn: Verlag Dr. Rudolf Habelt, 2018), 185–208. This paper addresses handicraft production in the Mongolian steppes from the time of the Xiongnu.

36. Ankeney Weitz, "Art and Politics at the Mongol Court of China: Tugh Temür's Collection of Chinese Paintings," *Artibus Asiae* 64, no. 2 (2004): 248. Or see the investment of Yuan court actors into Tibet Buddhist religious art: Anning Jing, "Financial and Material Aspects of Tibetan Art under the Yuan Dynasty," *Artibus Asiae* 64, no. 2 (2004): 213–241.

37. Allsen, *Commodity and Exchange*, 202.

38. Xu, *Lixue zhinan*, *juan* 7, 5a.

39. Chen Gaohua 陈高华, ed., *Yuan dianzhang* 元典章 [Statutes of the Yuan dynasty] (Tianjin: Tianjin Guji Chubanshe, 2011), *juan* 17, 580–591.

40. Hu, *Zishan da quanji*, *juan* 8, 24b. Here Hu explicitly refers to work processes.

41. Hok-Lam Chan, "The Organization and Utilization of Labor Service under the Jurchen Chin Dynasty," *Harvard Journal of Asiatic Studies* 52, no. 2 (1992): 618. For the original source, see Tuo Tuo 脱脱, *Jin shi* 金史 [History of Jin] (Beijing: Zhonghua Shuju, 1975), which was compiled by the Yuan court in 1344. No contemporary Jin records survived.

42. Zheng, *Zheng Sixiao ji*, 186

43. "應管軍民人匠諸色戶計官吏人等，今后毋得將所管戶計私自役使影占。" Chen, *Yuan dianzhang*, *juan* 3, 72.

44. Many of those works also take for granted that the literati gentleman ranked highest and was most sought after. See, e.g., Heinz Friese, "Zum Aufstieg von Handwerkern ins Beamtentum während der Ming-Zeit," *Oriens Extremus* 6, no. 2 (1959): 160–176. Ho Ping-ti, "Aspects of Social Mobility in China, 1368–1911," *Comparative Studies in Society and History* 1, no. 4 (1959): 330–359. Zhu Cishou 祝慈寿, *Zhongguo gudai gongyeshi* 中国古代工业史 [An industrial history of ancient China] (Beijing: Xuelin Chubanshe, 1988).

45. Gao Rongsheng 高荣盛, "Yuandai jianghu sanlun" 元代匠户散论 [On craftsmen in the Yuan dynasty], *Nanjing daxue xuebao*, no. 1 (1997): 123–129.

46. Liu and Chen, "Yuandai xiguan gongjiang de shehui," 14. Several prior studies come to another conclusion. See Wu Wei 吴伟 and Jiang Maofa 姜茂发, "Woguo Yuandai huji fenlei zhidu

yanjiu" 我国元代户籍分类制度研究 [Research on the classification system of household registration in the Yuan dynasty], *Ningxia shehui kexue*, no. 6 (2009): 111. See also Ota Yaichiro 太田彌一郎, "Gendai no juko to juseki" 元代の儒戸と儒籍 [A study of Ru households and Ru registers], *Tōhoku-Daigaku-tōyōshi-rōnshu* 5 (1992): 166–191. On the Yiwei registers, see Matsuda Koichi 松田孝一, "The Number of Military Households in the Yuan," in *Neilu Yazhou lishi wenhua yanjiu. Han Rulin xiansheng jinian wenji* 内陆亚洲历史文化研究——韩儒林先生纪念文集 [Inland Asian history and culture research——Mr. Han Rulin Memorial Collection], ed. Nanjing daxue yuanshi yanjiushi 南京大学元史研究室 (Nanjing: Nanjing Daxue Chubanshe, 1996), 268–295. Matsuda has pointed out that none of these households were military in the sense of serving in warfare.

47. Brian E. McKnight, "Who Gets It When You Go: The Legal Consequences of the Ending of Households (*JUEHU* 絕戶) in the Song Dynasty (960–1279 C.E.)," *Journal of the Economic and Social History of the Orient* 43, no. 3 (2000): 314; for the role of marriage, lineage, and households, see 355.

48. Birge, "Gender, Property, and Law in China."

49. Song, *Yuan shi*, 2639.

50. Francesca Bray, *Technology and Gender: Fabrics of Power in Late Imperial China* (Berkeley: University of California Press, 1997).

51. "令織女為永妻，織帛償"；"懸巢巧婦子，拂水翦刀花"；"旌婦范氏." Yu Xilu 俞希魯, *Zhishun Zhenjiang zhi* 至順鎮江志 [Local gazetteer of Zhenjiang from the Zhishun reign], 1863, *juan* 2, 28a; *juan* 4, 42a; *juan* 19, 17a.

52. Literal translation for a category of small taxpayer households extant since the Song dynasty; Hok-lam Chan translates them as "rats registers." I have chosen the expression "mousetail" to better reflect that the major characteristic and purpose of these registers was to be continuously updated.

53. "縣政要式：軍、民、站、匠、諸色戶計，各鄉保村莊丁口鼠尾簿一扇，各戶留空紙一面于後，凡丁口死亡，或成丁，或產業，孳畜增添、消乏，社長即報官，于各戶下，令掌簿吏人即便標注，凡遇差發、絲銀、稅糧、夫役、車牛、造作、起發、當軍、檢點簿籍，照各家即目增損氣力分數科攤." Hu, *Zishan da quanji*, *juan* 22, 1a.

54. Terms quoted here appear under the generic heading "amounts and substances" (*tiliang* 體量) in Xu, *Lixue zhinan*.

55. "據即目實在丁口、事產、物力符同，給戶貼、造籍冊，仍細注元簽起時屬何屬，再撥屬何屬，目今現屬何屬當役，因而將民籍、站籍、匠籍，諸一切戶籍細目，手持造籍各三本，以為定例諸一切戶籍細目，手持造籍各三本，以為定例." Hu, *Zishan da quanji*, *juan* 22, 1a.

56. "諸壬子年附籍軍民、諸色人等，別無上司改撥充匠明文，雖稱即目入局造作或于各投下送納生活者，仰憑籍收系應當差役." Huang, *Tongzhi tiaoge*, 8.

57. "戶口增：謂生齒之最，民籍增益，進丁入老，批注收落，不失其實，若有流離，而能招誘復業者." Xu, *Lixue zhinan*, *juan* 1, 6a.

58. Huang, *Tongzhi tiaoge*, 8.

59. "放保定工匠楚通等三百四十一戶為民." Song, *Yuan shi*, 338.

60. 190,000 were returned to civil: "今已選定諸色工匠，餘十九萬九千九百餘戶宜縱令為民。" Song, *Yuan shi*, 266.

61. See Huang, *Tongzhi tiaoge*, 3–26.

62. "差官與察司、總府一同磨勘到各戶根腳、氣力、手狀，已是精當類攢冊帳，各路赴部分簡。" Wang Yun 王惲, *Qiujian xiansheng da quanji* 秋澗先生大全集 [Collected works of Mr. Qiujian], Sibu congkan chubian (Shanghai: Shanghai shangwu yinshuguan, 1919), *juan* 22, 10b.

63. "諸投下官員，招佔已籍系官民匠戶計者，沒其家財，所佔戶歸本籍。" Song, *Yuan shi*, 2641.

64. Song Lian 宋濂, *Song xueshi wenji* 宋學士文集 [Collected works of scholar Song], Sibu congkan chubian (Shanghai: Shanghai Shangwu Yinshuguan, 1919), *juan* 5, 3.

65. "管領隨路人匠都提領所; 掌理人匠詞訟。" Song, *Yuan shi*, 2145, 2271.

66. Chen Dezhi 陈得芝, *Yuandai zouyi jilu* 元代奏议集录 [Collected notes of the Yuan dynasty] (Hangzhou: Zhejiang Guji Chubanshe, 1998), 105.

67. "諸局院官輒以微故毆死匠人者，處死。" Song, *Yuan shi*, 2676.

68. Huang, *Tongzhi tiaoge*, 138.

69. Chen, *Yuan dianzhang*, *juan* 34, 1163.

70. Huang, *Tongzhi tiaoge*, 139.

71. Local documents from Heishui offer one possible confirmation for such practices. Such records were produced in multiple languages. I consulted mainly the Chinese volume. See Li Yiyou 李逸友, ed., *Heicheng chutu wenshu* 黑城出土文书（汉文文书卷）[Documents unearthed in Heicheng (volume of Chinese documents)] (Beijing: Kexue Chubanshe, 1991); Wu Chao 吴超, "*Heishuicheng chutu wenshu* suojian Yijinailu nongye jishu tuiguang chutan" 《黑水城出土文书》所见亦集乃路农业技术推广初探 [A preliminary study on the promotion of agricultural technology in Yijinailu in "Documents Unearthed from Heishuicheng"], *Nongye kaogu*, no. 4 (2011): 417–421; Pan Jie 潘洁, "Heishuicheng chutu Yuandai fushui wenshu yanjiu" 黑水城出土元代賦稅文书研究 [Research on Yuan dynasty taxation documents unearthed in Heishuicheng], *Xixiaxue* 4 (2009): 102–124.

72. Huang, *Tongzhi tiaoge*, 337.

73. "禁諸監官不得令人匠私造器物。" Song, *Yuan shi*, 245. See also Chen, *Yuan dianzhang*, *juan* 2, 71–76; Huang, *Tongzhi tiaoge*, 23–24.

74. "梯已出备"; "修补机张什物、风雨箔、人匠夜坐灯油、柴灰、行移文字纸札。" Chen, *Yuan dianzhang*, *juan* 58, 1955.

75. See Song, *Yuan shi*, 127–144, 401–421.

76. "系官諸色原籍正匠並改色人匠，見入局造作者，仰依舊充匠除豁。" Huang, *Tongzhi tiaoge*, 8.

77. "國家初定中夏，制作有程，乃鳩天下之工，聚之京師，分類置局。" Zhao Shiyan 趙世延 and Yu Ji 虞集, *Jingshi dadian jijiao* 經世大典輯校 ["Jingshi dadian": An annotated edition] (Beijing: Zhonghua Shuju, 2020), 869.

78. Chen, *Yuan dianzhang*, *juan* 58, 1972.

79. Song, *Yuan shi*, 3264–3266. Such weavers were Wu Derong 吳德融, who was "good at multi-layered wefts" (*shan duan* 善緞), Song, *Yuan shi*, 1453; or Shi Dao'an 史道安, Su Tianjue 蘇天爵, *Guochao wenlei* 國朝文類, Sibu congkan chubian (Shanghai: Shanghai Shangwu Yinshuguan, 1919), *juan* 42, 17a. See also Yang Qiong 楊琼, a mason ordered to manage the masons of Yannan and other provinces (*lu*). Xue Zengfu 薛增福, *Quyang Beiyue miao* 曲阳北岳庙 [Beiyue temple in Quyang] (Shijiazhuang: Hebei Meishu Chubanshe, 2000), 16.

80. Wang Ao 王鏊, *Zhengde Gusuzhi* 正德姑蘇志 [Local gazetteer of Gusu from the Zhengde reign], 1506, *juan* 56, 21b.

81. "工欲善其事，必先利其器。盖器不利則事不成，雖有良工，無所施其巧。" Hu, *Zishan da quanji*, *juan* 8, 23b.

82. "唐太宗以弓示弓，人曰：'非良弓也，木心不正，脉理皆邪，故發矢不直，以太宗之神射，弓不良則亦不能中的。'" Hu, *Zishan da quanji*, *juan* 8, 23b.

83. While the term itself has become an idiom, I chose the literal translation to make visible the addressed skill set. "箕裘：謂承祖父之業者。《禮》云：良弓之子，必學為箕；良冶之子，必學為裘。" Xu, *Lixue zhinan*, *juan* 5, 5b.

84. "Furs/felts" indicates ritual procedures in which all were asked to cover up the lamb fur coats. See *Liji zhengyi* 禮記正義 [Book of Rites], *Shisanjing zhushu (zhengliben)* 十三經注疏（整理本）, (Beijing: Beijing Daxue Chubanshe, 2000), *juan* 36, 1246.

85. Huang Ying 黄英, *Neisheng waiwang de fasixiang tixi: Qiu Jun "Daxue yanyi bu" tanyan*. 内圣外王的法思想体系：丘濬《大学衍义补》探研 [The legal thoughts system of the internal saints and external kings theory: A case study of supplementation to the explanations to the Great Learning by Qiu Jun] (Beijing: Zhongguo Shehui Kexue Chubanshe, 2020), 24. Huang follows Qiu's notion of *fa*, including, for example, political methods (*zhengfa* 政法) and teaching through persuasion (*xinfa* 心法), and critically engages with the "Western" and modern frameworks of legal studies. His study touches on ritual methods (*lifa* 禮法).

86. Aurelia Campbell, *What the Emperor Built: Architecture and Empire in the Early Ming* (Seattle: University of Washington Press, 2020).

87. According to Zhang Xuan 張鉉, *Zhida Jinling xinzhi* 至大金陵新志 [New gazetteer of Jinling in the Zhida reign], Siku quanshu (Taipei: Taiwan Shangwu Yinshuguan, 1986), *juan* 8, 1–17, Jiangning had 373 craftsmen households, Shangyuan 437, Jurong 1060, Tanshouzhou 524, and Tanyangzhou 963.

88. See Thomas G. Nimick, "Case Files from the Sichuan Provincial: Administration Commission, with Annotated Index," *Ming Studies* 2003, no. 1 (2003): 62–85, who examines a set of rare archival documents about such disputes in Sichuan.

89. This is especially so in the case of dress, as Chen BuYun noted recently as well in BuYun Chen, "Wearing the Hat of Loyalty: Imperial Power and Dress Reform in Ming Dynasty China," in *The Right to Dress*, ed. Giorgio Riello and Ulinka Rublack (Cambridge: Cambridge University Press, 2019), 416–434.

90. The Ming "Work Law" of the "Daming Law" specifically records the legal provisions for the classification of the handicraft. See Shen Shixing 申時行, *Da Ming huidian* 明會典 [Collected statutes of the great Ming], Wanli ed. (Beijing: Zhonghua Shuju, 2007), *juan* 172, 880, which stipulates penalties when standards were not met by either the craftsman or the clerk in construction work, such as "those who lie about the construction materials, the applied funds and goods and the amount of labor, should be flogged 50 times. If the property has been damaged or the labor has been spent, the cost of the damaged goods and the labor shall be calculated together. If the crime is serious, it should be regarded as embezzlement."

91. Chen Xinyu 陈新宇, "Bifu yu leitui zhi bian—cong 'biyin lütiao' chufa," 比附与类推之辨———从 "比引律条"出发, Zhengfa luntan 29 (2011): 113–121.

92. Shen, *Da Ming huidian*, *juan* 9, 55; *juan* 59, 363.

93. For an overview on the military implications of this incident, see David M. Robinson, *Bandits, Eunuchs, and the Son of Heaven: Rebellion and the Economy of Violence in Mid-Ming China* (Honolulu: University of Hawai'i Press, 2001), 75–79. Natural disasters are increasingly reported—this can mainly also be read as a critique of the political disinterest of the Chenghua emperor, who was more interested in military affairs. After his release, the Zhengtong emperor reigned a second time under the reign name "Tianshun," 1457–1464.

94. *Ming Xianzong shilu* 明憲宗實錄 [Veritable records of the emperor Xianzong in the Ming dynasty] (Taipei: Zhongyang Yanjiuyuan Lishi Yuyan Yanjiusuo, 1986), *juan* 2, 53a; Charles O. Hucker, *A Dictionary of Official Titles in Imperial China* (Stanford, CA: Stanford University Press, 1985), 568 (7724), thus offers "crafts institute" as a possible translation.

95. Kuai Xiang 蒯祥 (1398–1481) was originally a Suzhou carpenter before he became Yongle's major adviser supervising the construction of the Beijing Forbidden City. *Ming Xianzong shilu*, *juan* 32, 4b.

96. *Ming Xianzong shilu*, *juan* 64, 7a.

97. Each person paid nine qian of silver a month. See Shen, *Da Ming huidian*, *juan* 189, 950.

98. Shen, *Da Ming huidian*, *juan* 189, 950. Then the scope gradually expanded and the amount of silver also changed. In the forty-first year of Jiajing that culminated in 1562 is the stipulation that "from this autumn it is not allowed to go to the Ministry of Works for service without permission."

99. Edward L. Farmer, ed., *Zhu Yuanzhang and Early Ming Legislation: The Reordering of Chinese Society following the Era of Mongol Rule* (Leiden: Brill, 1995), 42, 33.

100. Zhu Xi 朱熹, *Hui'an xiansheng Zhu wengong wenji* 晦庵先生朱文公文集 [Collected works of Zhu wengong Hui'an], Sibu congkan chubian (Shanghai: Shanghai Shangwu Yinshuguan, 1919), *juan* 65, 3171.

101. Qiu uses the term *zhi*, which can be translated as "duties" or "professions/trades." I translated it as "assignments," as he refers to the *Zhouli* narrating how the king of Zhou disseminated tasks for ritual performances among his people: (1) three forms of agriculture that produce

the nine grains; (2) gardening and growing plants; (3) materials of the woods and marshes; (4) husbandry; (5) the hundred crafts that adorn and transform the eight materials; (6) trade and merchandising; (7) women's work of reeling and weaving silk; (8) civil servants collecting and redistributing materials (as tax); (9) people roaming freely/freelancers.

102. Shisanjing zhushu zhengli weiyuanhui 十三經註疏整理委員會, ed., *Zhouli zhushu* 周禮註疏 [Zhouli: An annotated edition], *Shisanjing zhushu (zhengliben)* 十三經註疏（整理本）, (Beijing: Beijing Daxue Chubanshe, 2000), *juan* 1, 38.

103. *Zhouli zhushu*, *juan* 39, 1241–1242.

104. I choose a literal translation here of the word for scribes or authors to highlight Qiu's play of words, i.e., scribes "produce" (*zuo* 作) like craftsmen "produce" pots or tables.

105. This expression is habitually translated as "a division of labor that eases things." However, the saying literally emphasizes the communal nature of an effort as well as the performance of varied tasks, not a division of labor per se.

106. Wang Zhaoyu's 王昭禹 (fl. 1080) *Zhouli xiangjie* 周禮詳解 [Detailed explanations of the ritual of Zhou] as quoted in Qiu, *Daxue yanyi bu*, *juan* 97, 3a. For a discussion of Wang Zhaoyu's approach to ritual, see also Hiu Yu Cheung, "Sequence of Power Ritual Controversy over the Zhaomu Sequence in Imperial Ancestral Rites in Song China (960–1279)" (PhD diss., Arizona State University, 2015), 232–239.

107. Luo Lixiang 羅麗馨 (Lo Li-hsiang), "Mingdai jiangji renshu zhi kaocha" 明代匠籍人數之考察 [On the number of artisan households in Ming China], *Shihuo yuekan* 17, no. 1–2 (1988): 1–20, gives an overview of tasks (or professions) acknowledged in central state registers based on the "Provisions and Tax" (*shihuo* 食貨) chapters of official historiography and Local Gazetteers of the Jiangnan Region.

108. Peter K. Bol, "The Rise of Local History: History, Geography, and Culture in Southern Song and Yuan Wuzhou," *Harvard Journal of Asiatic Studies* 61, no. 1 (2001): 37–76.

109. See Wang Hong 王红, *Laozihao* 老字号 [Time-honored brand] (Beijing: Beijing Chubanshe, 2018), and Thomas David DuBois, "China's Old Brands: Commercial Heritage and Creative Nostalgia," *International Journal of Asian Studies* 18, no. 1 (2020): 1–15.

110. Song Guohua 宋國畫, *Yuandai fazhi bianqian yanjiu* 元代法製變遷研究 (Beijing: Zhishi Chanquan Chubanshe, 2017), 72–86, shows the range of issues addressed in legal texts.

111. See, for instance, Christine Moll Murata, *State and Crafts in the Qing Dynasty (1644–1911)* (Amsterdam: University of Amsterdam Press, 2018); Richard von Glahn, *Fountain of Fortune: Money and Monetary Policy in China (1000–1700)* (Berkeley: University of California Press, 1996); Fan Jinmin 范金民, *Yibei tianxia: Ming Qing Jiangnan sichou shi yanjiu* 衣被天下：明清江南丝绸史研究 [The world of clothes: A study on the history of Jiangnan Silk in Ming and Qing dynasties] (Nanjing: Jiangsu Renmin Chubanshe, 2016).

112. See Schäfer, "Inscribing the Artifact."

Bibliography

Allsen, Thomas T. *Commodity and Exchange in the Mongol Empire: A Cultural History of Islamic Textiles*. Cambridge: Cambridge University Press, 1997.

Barbieri-Low, Anthony J. *Artisans in Early Imperial China*. Seattle: University of Washington Press, 2007.

Birge, Bettine. "Gender, Property, and Law in China." *Journal of the Economic and Social History of the Orient* 44, no. 4 (2001): 575–599.

Bol, Peter K. "The Rise of Local History: History, Geography, and Culture in Southern Song and Yuan Wuzhou." *Harvard Journal of Asiatic Studies* 61, no. 1 (2001): 37–76.

Bray, Francesca. *Technology and Gender: Fabrics of Power in Late Imperial China*. Berkeley: University of California Press, 1997.

Campbell, Aurelia. *What the Emperor Built: Architecture and Empire in the Early Ming*. Seattle: University of Washington Press, 2020.

Chan, Hok-Lam. "The Organization and Utilization of Labor Service under the Jurchen Chin Dynasty." *Harvard Journal of Asiatic Studies* 52, no. 2 (1992): 613–664.

Chen, BuYun. "Wearing the Hat of Loyalty: Imperial Power and Dress Reform in Ming Dynasty China." In *The Right to Dress*, edited by Giorgio Riello and Ulinka Rublack, 416–434. Cambridge: Cambridge University Press, 2019.

Chen Dezhi 陈得芝. *Yuandai zouyi jilu* 元代奏议集录 [Collected notes of the Yuan dynasty]. Hangzhou: Zhejiang Guji Chubanshe, 1998.

Chen Gaohua 陈高华, ed. *Yuan dianzhang* 元典章 [Statutes of the Yuan dynasty]. Tianjin: Tianjin Guji Chubanshe, 2011.

Cheng Jufu 程鉅夫. *Cheng xuelou wenji* 程雪樓文集 [Collected works of Cheng Xuelou]. Taipei: Guoli Zhongyang Tushuguan, 1970.

Chen Xinyu 陈新宇. "Bifu yu leitui—cong 'biyin lütiao' chufa, zhengfa luntan" 比附与类推之辨——从"比引律条"出发, 政法论坛. *Zhongguo zhengfa daxue xuebao* 2 (2011): 113–121.

Cheung, Hiu Yu. "Sequence of Power Ritual Controversy over the Zhaomu Sequence in Imperial Ancestral Rites in Song China (960–1279)." PhD diss., Arizona State University, 2015.

Da Ming huidian 明會典 [Collected statutes of the great Ming dynasty]. Wanli ed. Beijing: Zhonghua Shuju, 2007.

Ding Yiru 丁憶如. "Sima Xiangru fupian zhi yinyun fengge yanjiu" 司馬相如賦篇之音韻風格研究 [The study on the linguistic style of Sima Xiangru's rhapsodies]. Master's thesis, National Chengchi University, Taipei, 2007.

Dobbin, Frank. "How Durkheim's Theory of Meaning-Making Influenced Organizational Sociology." In *The Oxford Handbook of Sociology and Organization Studies: Classical Foundations*, edited by Paul S. Adler, 200–222. New York: Oxford University Press, 2009.

DuBois, Thomas David. "China's Old Brands: Commercial Heritage and Creative Nostalgia." *International Journal of Asian Studies* 18, no. 1 (2020): 1–15.

Durkheim, Émile. *Les formes élémentaires de la vie religieuse: Le système totémique en Australie*. Paris: Puf, 1912.

Fan Jinmin 范金民. *Yibei tianxia: Ming Qing Jiangnan sichou shi yanjiu* 衣被天下：明清江南丝绸史研究 [The world of clothes: A study on the history of Jiangnan silk in Ming and Qing dynasties]. Nanjing: Jiangsu Renmin Chubanshe, 2016.

Farmer, Edward L., ed. *Zhu Yuanzhang and Early Ming Legislation: The Reordering of Chinese Society following the Era of Mongol Rule*. Leiden: Brill, 1995.

Friese, Heinz. "Zum Aufstieg von Handwerkern ins Beamtentum während der Ming-Zeit." *Oriens Extremus* 6, no. 2 (1959): 160–176.

Gao Rongsheng 高荣盛. "Yuandai jianghu sanlun" 元代匠户散论 [On craftsmen in the Yuan dynasty]. *Nanjing daxue xuebao*, no. 1 (1997): 123–129.

Glahn, Richard von. *Fountain of Fortune: Money and Monetary Policy in China (1000–1700)*. Berkeley: University of California Press, 1996.

Guo Chaoying 郭超颖, and Wang Chenglüe 王承略. "Cong *Lixue zhinan* kan Yuandai liyuan yishi" 从《吏学指南》看元代吏员意识 [On the consciousness of officials in the Yuan dynasty from *Lixue zhinan*]. *Jiangxi shehui kexue* 35, no. 2 (2015): 114–119.

Ho, Ping-ti. "Aspects of Social Mobility in China, 1368–1911." *Comparative Studies in Society and History* 1, no. 4 (1959): 330–359.

Huang Shijian 黃時鑑, ed. *Tongzhi tiaoge* 通制條格 [Statutes from the comprehensive regulations]. Hangzhou: Zhejiang Guji Chubanshe, 1986.

Huang Ying 黄英. *Neisheng waiwang de fasixiang tixi: Qiu Jun "Daxue yanyi bu" tanyan* 内圣外王的法思想体系：丘濬《大学衍义补》探研 [The legal thoughts system of the internal saints and external kings theory: A case study of supplementation to the explanations to the Great Learning by Qiu Jun]. Beijing: Zhongguo Shehui Kexue Chubanshe, 2020.

Hucker, Charles O. *A Dictionary of Official Titles in Imperial China*. Stanford, CA: Stanford University Press, 1985.

Hu Zhiyu 胡祗遹. *Zishan da quanji* 紫山大全集 [The big collection of Zishan]. Siku quanshu. Taipei: Taiwan Shangwu Yinshuguan, 1986.

Jing, Anning. "Financial and Material Aspects of Tibetan Art under the Yuan Dynasty." *Artibus Asiae* 64, no. 2 (2004): 213–241.

Kang Jinsheng 康金声. "Sima Xiangru xinlun" 司马相如新论 [New findings on Sima Xiangru]. *Shanxi daxue xuebao (zhexue shehui kexue ban)*, no. 4 (2002): 10–11.

Kern, Martin. "The 'Biography of Sima Xiangru' and the Question of the Fu in Sima Qian's Shiji." *Journal of the American Oriental Society* 123, no. 2 (2003): 303–316.

Kulp, Daniel Harrison. *Country Life in South China*. New York: Teachers College, Columbia University, 1923.

Lee, Sukhee. *Negotiated Power: The State, Elites, and Local Governance in Twelfth- to Fourteenth-Century China*. Boston: Harvard University Asia Center, 2014.

Liji zhengyi 禮記正義 [Book of Rites]. *Shisanjing zhushu (zhengliben)* 十三經注疏（整理本）. Beijing: Beijing Daxue Chubanshe, 2000.

Liu Liya 刘莉亚 and Chen Peng 陈鹏. "Yuandai xiguan gongjiang de shenfen diwei" 元代系官工匠的身份地位 [The social status of craftsmen in the Yuan dynasty]. *Neimenggu shehui kexue*, no. 3 (2003): 10–16.

Liu Xie 劉勰. *Wenxin diaolong zhu* 文心雕龍注 [The literary mind and carving of dragons with annotations]. Edited by Fan Wenlan 范文瀾. Beijing: Renmin Wenxue Chubanshe, 1962.

Li Yiyou 李逸友, ed. *Heicheng chutu wenshu* 黑城出土文书（汉文文书卷） [Documents unearthed in Heicheng (volume of Chinese documents)]. Beijing: Kexue Chubanshe, 1991.

Luo Lixiang 羅麗馨 (Lo Li-hsiang). "Mingdai jiangji renshu zhi kaocha" 明代匠籍人數之考察 [On the number of artisan households in Ming China]. *Shihuo yuekan* 17, no. 1–2 (1988): 1–20.

Mair, Victor H. "Persian Scribes (Munshi) and Chinese Literati (Ru): The Power and Prestige of Fine Writing (Adab/Wenzhang)." In *Literacy in the Persianate World: Writing and the Social Order*, edited by Brian Spooner and William L. Hanaway, 388–414. Philadelphia: University of Pennsylvania Press, 2012.

Marmé, Michael. *Suzhou: Where the Goods of All the Provinces Converge*. Stanford, CA: Stanford University Press, 2005.

Matsuda Koichi 松田孝一. "The Number of Military Households in the Yuan." In *Neilu Yazhou lishi wenhua yanjiu: Han Rulin xiansheng jinian wenji* 内陆亚洲历史文化研究——韩儒林先生纪念文集 [Inland Asian history and culture research——Mr. Han Rulin Memorial Collection], edited by Nanjing daxue yuanshi yanjiushi 南京大学元史研究室, 268–295. Nanjing: Nanjing Daxue Chubanshe, 1996.

McKnight, Brian E. "Who Gets It When You Go: The Legal Consequences of the Ending of Households (JUEHU 絕戶) in the Song Dynasty (960–1279 C.E.)." *Journal of the Economic and Social History of the Orient* 43, no. 3 (2000): 314–363.

Ming Xianzong shilu 明憲宗實錄 [Veritable records of the emperor Xianzong in the Ming dynasty]. Taipei: Zhongyang Yanjiuyuan Lishi Yuyan Yanjiusuo, 1986.

Moll Murata, Christine. *State and Crafts in the Qing Dynasty (1644–1911)*. Amsterdam: University of Amsterdam Press, 2018.

Nimick, Thomas G. "Case Files from the Sichuan Provincial: Administration Commission, with Annotated Index." *Ming Studies* 2003, no. 1 (2003): 62–85.

Ota Yaichiro 太田彌一郎. "Gendai no juko to juseki" 元代の儒戶と儒籍 [A study of Ru households and Ru registers]. *Tōhoku-Daigaku-tōyōshi-rōnshu* 5 (1992): 166–191.

Pan Jie 潘洁. "Heishuicheng chutu Yuandai fushui wenshu yanjiu" 黑水城出土元代賦稅文書研究 [Research on Yuan dynasty taxation documents unearthed in Heishuicheng]. *Xixiaxue* 4 (2009): 102–124.

Qiu Jun 丘濬. *Daxue yanyi bu* 大學衍義補 [Supplement to the extended meaning of the Great Learning]. Jianningfu ed. 1486.

Reichert, Susanne. "Imperial Policies towards Handicraft: The Organization of Production in the Old Mongolian Capital Karakorum." In *Craft Production Systems in a Cross-Cultural Perspective*, edited by Martin Bentz and Tobias Helms, 185–208. Bonn: Verlag Dr. Rudolf Habelt, 2018.

Robinson, David M. *Bandits, Eunuchs, and the Son of Heaven: Rebellion and the Economy of Violence in Mid-Ming China*. Honolulu: University of Hawai'i Press, 2001.

Rossabi, Morris, ed. *Eurasian Influences on Yuan China*. Singapore: ISEAS–Yusof Ishak Institute, 2013.

Schäfer, Dagmar. "Inscribing the Artifact and Inspiring Trust: The Changing Role of Markings in the Ming Era." *East Asian Science, Technology and Society* 5, no. 2 (2011): 239–265.

Shea, Eiren L. *Mongol Court Dress, Identity Formation, and Global Exchange*. London: Routledge, 2020.

Silverstein Michael, and Greg Urban, eds. *Natural Histories of Discourse*. Chicago: University of Chicago Press, 1996.

Song Guohua 宋國畫. *Yuandai fazhi bianqian yanjiu* 元代法製變遷研究. Beijing: Zhishi Chanquan Chubanshe, 2017.

Song Lian 宋濂. *Song xueshi wenji* 宋學士文集 [Collected works of scholar Song]. Sibu congkan chubian. Shanghai: Shanghai Shangwu Yinshuguan, 1919.

Song Lian 宋濂. *Yuan shi* 元史 [History of Yuan]. Beijing: Zhonghua Shuju, 1976.

Sun Yi 孫奕. *Lüzhai shier bian* 履斋示儿编 [Collection of instructions for my sons in the Lüzhai study]. Beijing: Zhonghua Shuju, 1985.

Su Tianjue 蘇天爵. *Guochao wenlei* 國朝文類 [Guochao wenlei]. Sibu congkan chubian. Shanghai: Shanghai Shangwu Yinshuguan, 1919.

Tuo Tuo 脫脫. *Jin shi* 金史 [History of Jin]. Beijing: Zhonghua Shuju, 1975.

Wang Ao 王鏊. *Zhengde Gusuzhi* 正德姑蘇志 [Local gazetteer of Gusu from the Zhengde reign]. 1506.

Wang Hong 王红. *Laozihao* 老字号 [Time-honored brand]. Beijing: Beijing Chubanshe, 2018.

Wang Yun 王惲. *Qiujian xiansheng da quanji* 秋澗先生大全集 [Collected works of Mr. Qiujian]. Sibu congkan chubian. Shanghai: Shanghai Shangwu Yinshuguan, 1919.

Watt, James C. Y. *The World of Khubilai Khan: Chinese Art in the Yuan Dynasty*. New York: Metropolitan Museum of Art, 2010.

Weick, Karl E. *Sensemaking in Organizations: Foundations for Organizational Science*. Thousand Oaks, CA: Sage, 1995.

Weitz, Ankeney. "Art and Politics at the Mongol Court of China: Tugh Temür's Collection of Chinese Paintings." *Artibus Asiae* 64, no. 2 (2004): 243–280.

Wilson, J. Keith. "The Fine Art of Drinking: The Chinese Silversmith Zhu Bishan and His Sculptural Cups." *Bulletin of the Cleveland Museum of Art* 81, no. 10 (1994): 380–401.

Wu Chao 吴超. "'Heishuicheng chutu wenshu' suojian Yijinailu nongye jishu tuiguang chutan" 《黑水城出土文书》所见亦集乃路农业技术推广初探 [A preliminary study on the promotion of agricultural technology in Yijinailu in "Documents Unearthed from Heishuicheng"]. *Nongye kaogu*, no. 4 (2011): 417–421.

Wu Wei 吴伟 and Jiang Maofa 姜茂发. "Woguo Yuandai huji fenlei zhidu yanjiu" 我国元代户籍分类制度研究 [Research on the classification system of household registration in the Yuan dynasty]. *Ningxia shehui kexue*, no. 6 (2009): 111.

Xia Lingwei 夏令伟. "*Quan Song wen Quan Yuan wen* buyi" 《全宋文》《全元文》补遗 [Addenda to the complete collections of Song and Yuan prose]. *Jiangsu daxue xuebao (shehui kexue ban)* 19, no. 5 (2017): 16–20.

Xiao Qiqing 蕭啟慶. "Yuandai de ruhu: Rushi diwei yanjinshi shang de yizhang" 元代的儒戶：儒士地位演進史上的一章 [The Confucian households of the Yuan dynasty: A chapter in the history of Confucians' social status]. In *Yuandai shi xintan* 元代史新探 [A new exploration in the history of the Yuan dynasty], 1–58. Taipei: Xinwenfeng, 1983.

Xie Hongxing 谢红星. "Dianli falü tixi xingcheng zhi qianye: Yuandai qilüyong geli jiqi falüshi diwei" "典例法律体系"形成之前夜：元代"弃律用格例"及其法律史地位 [The eve before the formation of the "classical legal system": Yuan dynasty "Qilüyong geli" and its status in legal history]. *Jiangxi shehui kexue* 40, no. 3 (2020): 136–147.

Xue Zengfu 薛增福. *Quyang Beiyue miao* 曲阳北岳庙 [Beiyue temple in Quyang]. Shijiazhuang: Hebei Meishu Chubanshe, 2000.

Xu Ting 徐霆. *Heida shilüe jiaoyhu* 黑鞑事略校注 [Brief account of the Black Tatars: An annotated edition]. Edited by Xu Quansheng 许全胜. Lanzhou: Lanzhou Daxue Chubanshe, 2014.

Xu Yuanrui 徐元瑞. *Lixue zhinan* 吏學指南 [Guidebook for clerks]. Yuan keben. Beijing: Beijing Tushuguan Chubanshe, 2004.

Yang Fuyou 杨富有. "Yuan Shangdu yongshishi de neirong jiqi yiyi fenxi" 元上都咏史诗的内容及其意义分析 [The content and meaning of historical poems in Yuan Shangdu]. *Neimenggu minzu daxue xuebao (shehui kexue ban)* 38, no. 3 (2012): 27–29.

Yang Shitie 杨世铁. "*Lixue zhinan* de cishu xingzhi" 《吏学指南》的辞书性质 [The dictionary nature of *Lixue zhinan*]. *Huaibei shifan daxue xuebao (zhexue shehui kexue)* 35, no. 5 (2014): 47–49.

Yang Shuhong 杨淑红. "Yuandai qiyue wenshu de kanbu yu yanjiu zongshu" 元代契约文书的刊布与研究综述 [A survey of contract documents in the Yuan dynasty]. *Zhongguoshi yanjiu dongtai*, no. 1 (2011): 28–34.

Yao Sui 姚燧. "Song Li Maoqin xu" 送李茂卿序 [Preface to Li Maoqing]. Vol. 4. Beijing: Renmin Wenxue Chubanshe, 2011.

Yao, Tao-Chung. "Ch'iu Ch'u-Chi and Chinggis Khan." *Harvard Journal of Asiatic Studies* 46, no. 1 (1986): 201–219.

Ye Xinmin 叶新民. "Yibu yuanchao gongwen yongyu cidian" 一部元朝公文用语辞典—《吏学指南》简介 [A dictionary of Yuan dynasty official documents—A brief introduction to *Lixue zhinan*]. *Neimenggu shehui kexue (wenshizhe ban)*, no. 6 (1988): 68–71.

Yuan Yuanweiyang 袁远维扬. "Wule gongming zhi de lunli yunhan" "物勒工名" 制的伦理蕴含 [The ethical implications of the "Wule Gongming" system]. *Hubei jingji xueyuan xuebao (renwen shehui kexue ban)* 15, no. 4 (2018): 22.

Yu Xilu 俞希鲁. *Zhishun Zhenjiang zhi* 至顺镇江志 [Local gazetteer of Zhenjiang from the Zhishun reign]. 1863.

Zhang Heng 張衡. "Sichou shi xu" 四愁詩序 [Preface to the Four Sorrows]. In *Wenxuan* 文選 [Anthology], edited by Xiao Tong 蕭統, 751–end. Taipei: Wunan Tushu, 1998.

Zhang, Jinfan. "The Legal System of 'Zu Shu Bian Tong' in the Yuan Dynasty: (1279 A.D.–1368 A.D.)." In *The History of Chinese Legal Civilization*, 833–908. Singapore: Springer Singapore, 2020.

Zhang Xuan 張鉉. *Zhida Jinling xinzhi* 至大金陵新志 [New gazetteer of Jinling in the Zhida reign]. Siku quanshu. Taipei: Taiwan Shangwu Yinshuguan, 1986.

Zhao Shiyan 趙世延 and Yu Ji 虞集. *Jingshi dadian jijiao* 經世大典輯校 [Jingshi dadian: An annotated edition]. Beijing: Zhonghua Shuju, 2020.

Zheng Sixiao 郑思肖. *Zheng Sixiao ji* 郑思肖集 [Collected works of Zheng Sixiao]. Shanghai: Shanghai Guji Chubanshe, 1991.

Zhouli zhushu 周禮註疏 [Zhouli: An annotated edition]. *Shisanjing zhushu (zhengliben)* 十三經注疏（整理本）. Beijing: Beijing Daxue Chubanshe, 2000.

Zhu Cishou 祝慈寿. *Zhongguo gudai gongyeshi* 中国古代工业史 [An industrial history of ancient China]. Beijing: Xuelin Chubanshe, 1988.

Zhu Xi 朱熹. *Hui'an xiansheng Zhu wengong wenji* 晦庵先生朱文公文集 [Collected works of Zhu wengong Huian]. Sibu congkan chubian. Shanghai: Shanghai Shangwu Yinshuguan, 1919.

9

OWNABILITY, OWNERSHIP, KNOWLEDGE, AND GENETIC INFORMATION IN THE UNITED STATES

Myles W. Jackson

Gene patenting and genetic information offer us prime examples of the instability of ownership and knowability in the age of biocapitalism. Starting in the 1970s, the United States Patent and Trademark Office (USPTO) began awarding patents for short sequences of DNA, specifically ones that could be used to search for genes. During the early 1980s, when genes were initially being awarded patents by the USPTO, the Japanese Patent Office, and the European Patent Office, the procedure was to isolate a gene, determine the function of its protein product by biochemical assays, and then patent it. The procedure first required knowability—that is, knowledge of what the gene coded for (i.e., its utility)—before ownership could be granted; one needed to know what one owned.

By the 1990s, that had changed. Broad utility patents were being issued for rather vague statements about a gene's utility. That is to say, one could obtain a patent for a gene that coded for a protein whose precise function remained unknown. In this case, one owned something before one precisely knew what one had. With this change in the relationship between ownability and knowability came a change in expertise. Previously, molecular biologists and their "wet" skills, techniques, and practices had been necessary to define the function of the gene product and therefore its utility. By the last decade of the century, their authority had been controversially usurped by a generation of computer scientists who had invented algorithms to find gene sequences with unprecedented rapidity, and who began to model proteins using computer graphics with a view to finding structural homologies and deducing similar functions. The tools and skills of the computer scientist were now seen as being both necessary and sufficient to ascertain knowability. This transformation in expertise raises a number of interesting questions. Does the sequence or genetic information, specifically the genetic code as it appears in the patent description, trump materiality to convey ownership? Do patent owners forfeit their right to ownership if they patent an incorrect sequence, a practice much more common than one might believe? Is the gene sequence sufficient to determine protein function? How can a gene become *visible* before it has been sequenced in a "wet" laboratory? How has the disappearance of materiality with the appearance of bioinformatics changed both intellectual property and the discipline of molecular biology?

I have detailed elsewhere the history of the patent portfolio for the CCR5 gene, which codes for a chemokine receptor.[1] The story raised questions regarding the relationship between property and knowability. In 2000 Human Genome Sciences, Inc. (henceforth HGS), which was subsequently bought out by GlaxoSmithKline, obtained a patent for a gene without knowing the precise function of its protein product. The company maintained that describing the sequence was synonymous with knowability. Thus, much along the lines of Bruno Latour's 2012 work on mapmaking, means of knowing could be understood as description.[2] Using bioinformatic techniques, which compare the sequence with ones whose protein products are well known, they deduced that their gene coded for a chemokine receptor. Using more traditional molecular biological and biochemical techniques, other laboratories determined independently that the receptor was recognized by HIV-1. The sequence HGS listed in the patent specification contained a number of serious errors; it turned out that it would not code for a protein that HIV-1 would recognize. Luckily for the company, its scientists submitted the gene into a public depository. As a result, their patent was not revoked. The key point here, as with this chapter, is that verified, documented materiality is privileged by the law. HGS could show that they possessed the gene by depositing a physical copy in a depository. The material object trumps the written word. The story of the CCR5 gene patents was, in one important way, rather typical of many gene patents of the 1990s and 2000s: sequencing companies submitted literally tens of thousands of DNA sequence applications hoping that subsequent research would render (an albeit small) percentage of the patent material useful. These patent applications had potential financial value.

This chapter analyzes the main legal theories of ownership that have shaped, and have been shaped by, genetic information. These theories have become particularly relevant now that private genomics companies, such as 23andMe and AncestryDNA (the genetic testing company of Ancestry.com), are sharing their clients' genetic information with various interested third parties. While one may naïvely think that these customers *own* their DNA, the situation is actually much more complicated and centers around the question, What does it really mean to own something? Whereas having a certain piece of DNA is a property of the human body, which is independent of knowing what precisely that gene does in that body, the genetic information encoded in that piece of gene is ownable. Indeed, ownership is granted to the person or persons who can glean information about what genetic sequence is and what that gene does. As a result, there is an obvious conflict between possessing a gene and knowing and owning the knowledge of that gene.

This chapter details the ownability of genetic information as property, which can potentially be converted into future financial gain. In short, this chapter demonstrates the *constructedness* and instability of expertise, ownership, ownability, knowability, and privacy.

LEGAL PRECEDENTS AND OWNERSHIP

As any law student in the United States will tell you, property does not necessarily refer to a physical object but rather to an association, or bundle, of rights, which can be—and usually are—enjoyed by more than one person.[3] Despite the economic, political, and historical importance of property, the law is vague, as Roberts outlines:

> But even within law and the legal academy, we have no clear consensus regarding precisely what property *is* or how property rights should be distributed amongst individuals. . . . Consequently, no single set of rights uniformly applies across all situations to all kinds of property.[4]

Since there are critical definitions between various legal regimes worldwide, it is important to stress that the case I am describing in this chapter applies to the United States. Relationships between ownability and knowability with regard to genetic information might look very different outside the United States.

Likewise, ownability is equally difficult to define legally, since different bundles of rights are associated with different types of property.[5] For instance, a 1963 lawsuit declared, "Ownership is not a single concrete entity but a bundle of rights and privileges as well as of obligations."[6] As is the case with ownership of other entities such as land, the various ways of knowability are unstable at the time of ownership, and utility changes and becomes increasingly complex with time. This chapter addresses several fundamental questions about ownership and ownability. Who owns the genetic information given to personal genetic companies? And what can be done with it? How do theories of ownership and ownability necessitate stricter privacy legislation? Questions about ownership, ownability, and property of genetic information are embedded in larger debates about agency. If body parts are property and can be owned, then they become passive agents and a means to an end.[7] The sociopolitical and ethical ramifications of the instabilities of ownership, ownability, possession, and knowability are enormous.

The influence of capital—in these particular cases, biocapital—dictates the power relations in the biotech sector. The means of production are raw materials, natural entities (such as genes), and legal and technical tools—in this case, intellectual property law, which results in the commodification of the so-called subjects of labor. These legal and technical tools structure the social relationships of those in the biotech world. People whose DNA sequences have been patented or used to trace their ancestry are alienated from the means of production, as they do not share in the governance of biomedical research, nor do they reap the profits. Historians are obliged to address the unequal power relations between actors in order to illustrate the politics of knowledge ownership. Waiting until closure is reached prevents historians from playing a role in the outcome of a controversy. Precisely because historians are best placed to illustrate that there always have been alternatives and that nothing is inevitable, they should be at the table during decision-making.

So, who owns your genetic information? Several US court cases have set the legal precedent for the behavior of personal genomics companies.[8] In 1990 the Supreme Court of California ruled that Mr. John Moore of Washington was not entitled to any royalties generated from cells that had been taken from his body. Moore, who was first treated for his hairy cell leukemia by Dr. David W. Golde of UCLA Medical Center back in 1976, had signed a consent form agreeing to have his spleen removed as part of his treatment. For the ensuing seven years, he returned to UCLA for continued treatment, which included the removal of tissue and blood. By August 1979, Golde and Shirley G. Quan had created a cell line from Moore's blood, and seventeen months later the physician and the University of California applied for a patent on the cell line, which was granted by the USPTO on March 20, 1984.[9] The physician and the university received some $440,000 as well as stock options from Genetics Institute, which had hired Golde as a paid consultant.[10] Moore became aware of the patent and decided to sue Golde and UCLA.

The key issue under consideration was whether the plaintiff could argue for conversion, or taking with the intent of exercising over the property a form of ownership that is inconsistent with the owner's right of possession. This was predicated on the patient's ownership over excised cells. Two aspects of the California supreme court majority report are particularly germane to this essay. First, California statutory law drastically limits a patient's control over their excised cells. According to Health and Safety Code 7054.4, "Notwithstanding any other provision of law, recognizable anatomical parts, human tissues, anatomical remains, or infectious waste following conclusion of scientific use shall be disposed of by internment, incineration, or any other method determined by scientific department [of health services] to protect the public health and safety."[11] Clearly, this statute is about public safety: one may not possess a diseased tissue extracted from one's body and then take it home to show friends and family. The majority report, however, felt that one cannot escape the conclusion that the statute's practical effect is to limit, drastically, a patient's control over excised cells. By restricting how excised cells may be used and requiring their eventual destruction, the statute eliminates so many of the rights ordinarily attached to property that one cannot simply assume that what is left amounts to "property" or "ownership" for purposes of conversion law.[12]

Second, the California supreme court opined that the patented product had been altered sufficiently such that Moore could no longer claim ownership. "This is because the patented cell line is both factually and legally distinct from the cells taken from Moore's body."[13] Hence, the majority concluded that "the use of excised human cells in medical research does not amount to a conversion."[14] They felt extending conversion law to biomedical research would drastically restrict access to the necessary raw materials.[15] "The theory of liability that Moore urges us to endorse threatens to destroy the economic incentive to conduct important medical research. If the use of cells in research is a conversion, then with every cell sample a researcher purchases a ticket in a litigation lottery."[16]

Judge Allen E. Broussard, who wrote a report agreeing with some aspects of the majority report and disagreeing with others, also maintained that "a patient may not retain any legal interest in a body part after its removal when he has properly consented to its removal and use for scientific purposes."[17] The key for Broussard, who felt that the plaintiff had indeed established a cause of action for conversion, was not whether a patient retains ownership interest in a body part once it is removed, but rather if a patient has a right to determine, before it is removed, the use to which it will be put. He also criticized the majority's stated concern that extending conversion here would restrict access to existing cell lines. Judge Stanley Mosk offered a much stronger rebuttal to the majority report. He criticized the use of the Health and Safety Code section 7054.4 as a precedent to limit Moore's control over his cells.[18] Contra his colleagues, Mosk felt that the concepts of ownership and property were extremely broad in California state law and that "the limitation or prohibition diminishes the bundle of rights that would otherwise attach to the property, yet what remains is still deemed in law to be a protectible property interest."[19] Under his liberal reading of property and ownership, Moore could legally assert an ownership interest in his cells, thereby warranting his sharing in the commercial rewards of their sale.[20]

The second case relevant to the use and selling of genetic information by personal genomics companies is *Greenberg v. Miami Children's Hospital Research Institute*.[21] Deborah and Daniel Greenberg had two children who suffered from Canavan disease, a degenerative disorder affecting the nerve cells of the brain. Those afflicted usually die by the age of eighteen months. During the early to mid-1980s, they and other families with afflicted children, in conjunction with the nonprofit patient advocacy group National Tay-Sachs and Allied Disease Association, offered Dr. Reuben Matalon of the University of Illinois at Chicago tissue samples so that he could try to find the gene. By 1993 he and his lab, which had relocated to Miami's Children Hospital some three years earlier, had identified the gene and the mutations associated with the disease. The plaintiffs argued that they had provided Matalon with the biological samples and confidential familial information "with the understanding and expectation that such samples and information would be used for the specific purpose of researching Canavan disease and identifying mutations in the Canavan gene which could lead to carrier detection within their families and benefit the population."[22] Apparently, it was their

> understanding that any carrier and prenatal testing developed in connection with the research for which they were providing essential support would be provided on an affordable and accessible basis, and that Matalon's research would remain in the public domain to promote the discovery of more effective prevention techniques and treatments and, eventually, to effectuate a cure for Canavan disease.[23]

Unbeknownst to the families, in 1994 the hospital applied for a gene patent, which was granted three years later. It was also granted patents on prenatal testing. Once the hospital had found a marketer, it granted that marketer an exclusive license for the genetic test, which was too expensive for the families who had donated the samples. The plaintiffs also alleged that the Miami Children's Hospital had enviously guarded their intellectual property by sending threatening enforcement letters to other centers offering Canavan testing.[24] In October 2000 the families, along with the National Tay-Sachs and Allied Disease Association and Dor Yeshorim, a nonprofit organization that offers genetic screening to the Jewish community around the globe, sued the hospital on six grounds: lack of informed consent, breach of fiduciary duty, unjust enrichment, fraudulent concealment, conversion, and misappropriation of trade secrets. Relevant to this chapter once again is the notion of conversion, which is defined by Florida state law as "an unauthorized act which deprives another of his property permanently or for an indefinite time."[25] The court ruled that since the body tissue had been donated for research, there was no expectation of return of the physical object and the knowledge derived from it, and therefore conversion was not applicable. They cited *Moore v. Regents of the University of California* as the precedent. In addition, since the Supreme Court of Florida decided not to recognize a property right in the body of another after death, they determined that the property right of the knowledge in blood and tissue samples dissipates once the sample is voluntarily given to a third party. Also taking their cue from that case, they opined that "the patented result of research is 'both factually and legally distinct' from excised material used in research."[26] Finally, much along the lines that the California supreme court had argued, the US District Court for the Southern District of Florida ruled that "if adopted, the expansive theory [of conversion] championed by Plaintiffs would cripple medical research as it would bestow a continuing right for donors to possess the results of any research conducted by the hospital. At the core, these were donations to research without any contemporaneous expectations of return."[27]

The third and final precedent case for the ownership of biological materials involves the work of Dr. William Catalona, a renowned urologist who, starting in 1983 as an employee of the Washington University Division of Urologic Surgery, created a collection of over thirty thousand prostate tissue samples donated by his own and his colleagues' patients with a view to determining the genetic basis of prostate cancer.[28] In 2003 he resigned his position and accepted a new job offer from Northwestern University. He wished to take along his collection of samples, asserting that the informed consent forms he had obtained transferred ownership of the samples from the university to him personally. Approximately six thousand patients who had donated their biological material signed a document requesting that Washington University release their samples to Catalona.[29] Citing its intellectual property policy, namely that "all intellectual property (including . . . tangible research property) shall be owned by the university if significant university resources were used or if it is created pursuant to a research project," Washington University insisted that the samples were their property and sued Catalona.[30] As the senior district judge stated in the introduction of the opinion, "Central to the several pending summary judgment motions, and preliminary injunction motion(s) is the issue of 'ownership.'"[31] Later in the decision, the importance of ownership was underscored: "The sole issue determinative of this permanent injunction; in fact of this lawsuit; is the issue of ownership."[32] Specifically, the crucial question was whether the research participants retained ownership rights after they had made voluntary donations of their biological samples to a research institution for medical research.

The court turned to the decisions of the two aforementioned cases for their ruling. In March 2006, the US District Court ruled in favor of Washington University. The previous two cases had deemed research participants to be donors who had surrendered their property rights once their biological materials had been excised for research. With that in mind, the Missouri court "finds that W[ashington] U[niversity] has met its burden in establishing ownership of the subject materials and that the R[esearch] P[articipants] have not put forth adequate evidence to challenge WU's ownership claim."[33] The court also ruled that the informed consent forms transferring ownership to Catalona were invalid because they had not received the approval of the university's Human Studies

Committee, nor were they ever submitted to an institutional review board for prior approval.[34] Finally, the court agreed with the earlier cases that

> medical research can only advance if access to these materials to the scientific community is not thwarted by private agendas. If left unregulated and to the whims of an RP, these highly-prized biological materials would become nothing more than chattel going to the highest bidder. . . . Allowing an RP to choose who can have the sample, where the sample will be stored, and/or how the sample can be used is tantamount to a blood donor being able to dictate that his/her blood can only be transfused into a person of a certain ethnic background, or a donated kidney being transplanted only into a woman or man. This kind of "selectiveness" is repugnant to any ethical code which promotes medical research to help all of mankind.

Hence, the court ruled that Washington University owned all of the biological material and that neither Catalona nor any of the research participants had any ownership or proprietary interest in it.[35]

PRECISION MEDICINE, PERSONAL GENOMICS COMPANIES, AND THE OWNABILITY AND OWNERSHIP OF KNOWLEDGE

Perhaps the most important advance in medical treatment over the past twenty-five years, precision medicine is a medical model whereby treatment is tailored to an individual patient, based on her/his genetic makeup. It does not intend to create specific drugs or treatments for each individual; however, it does seek to create categories of individuals, or populations, which will respond in the same way to certain medications or treatment regimes. It involves researching the relationship of genetic markers to certain illnesses within various populations which, in the United States, are controversially—some might argue dubiously—often defined by "race" or "ethnicity."[36] The precision medicine market generated $18 billion in 2017 in the United States alone, and it is expected to be worth nearly $100 billion by 2024.[37] On December 18, 2015, President Barack Obama signed legislation that provided over $200 million for the Precision Medicine Initiative.[38] This initiative unites an impressive array of for-profit, nonprofit, charitable, and federal institutions, such as the Department of Defense, the National Institutes of Health, the Food and Drug Administration, the American Medical Association, Genentech, Pfizer, the Broad Institute, the Bill and Melinda Gates Foundation, Color Genomics, Amazon Web Science, Microsoft, IBM, and the New York Genome Center, to name just a few. The initiative stresses the importance of securing the privacy of the individual from which the data have been collected, on the one hand, while on the other hand ensuring access to the data to all interested biomedical researchers, rather than only those belonging to institutions that are willing and able to pay for such access.

Taking their legal cue from the rulings in the aforementioned three decisions, personal genomics companies share their clients' genetic information with third parties such as Big Pharma.[39] Anyone living in the United States would have noticed a significant increase in the number of television commercials for personal genomics companies such as AncestryDNA, some of which, some of which are quite comical, while others are serious attempts to encourage the notion of biological identity.[40] Some, such as 23andMe, offer both ethnic ancestry and medical testing.[41] Others, such as AncestryDNA, focus exclusively on ethnic ancestry. Thanks in large part to the Food and Drug Administration (FDA)'s decision in 1985 to permit direct-to-consumer advertising, these companies have joined pharmaceutical companies in advertising their wares to a nation hungry to find out where their ancestors hailed from or what medical ailments might await them in the future.[42] Their use of clients' genetic information has recently made headlines, which illustrates how these companies' practices highlight the instability of the legal notions of ownership and privacy.

Personal genomics companies refer to hundreds of thousands of genetic markers known as single nucleotide polymorphisms (SNPs), from populations in Asia, the Americas, Africa, and Europe. SNPs are singular differences of DNA at specific spots along the chromosomes. They occur approximately once in approximately every three hundred base pairs. Hence, there are circa ten million SNPs in the human genome. Since they are passed down from generation to generation, different populations have different SNPs. SNPs that possess markedly different frequencies between populations are referred to as ancestry informative markers (AIMs), which are used to determine a person's ancestry. By comparing their client's DNA with previously collected samples from around the globe stored in their proprietary databanks, these companies offer customers profiles of their genetic heritage for a cost of anywhere between $49.99 and $199. For example, 23andMe's v4 chip, introduced in December 2013, tested for some 602,000 SNPs from 2,329 Y chromosomes and 19,487 X chromosomes, and 3,154 SNPs from mitochondrial DNA. Their newest v5 chip, which was introduced in late 2017, added approximately fifty thousand more SNPs of custom content. As of November 2017, 23andMe was using twenty-four reference populations around the globe in order to report thirty-one population labels.[43] The technology is rapidly changing and—according to the companies—improving. 23andMe boasts that its "innovative machine learning technology under the hood gets better and more precise as we add new customers and refine our technology."[44] This is, after all, the age of bioinformatics. Customers who have used both the v4 and v5 versions have blogged that the percentages of ethnicities, particularly those from East Asia, were wildly different, and that the results from the v5 chip corresponded much more closely with their family history.[45] AncestryDNA tests for some 700,000 SNPs from 885 Y chromosomes and 17,604 X chromosomes. They boast an accuracy rate of 90 percent without actually defining how one measures that, and claim that clients can find out their ancestry going back to the sixteenth century.[46]

For those whose ancestors originate from parts of the world where few samples have been taken, the tests tend to be inaccurate. A number of my students who independently decided to have their DNA tested by both 23andMe and AncestryDNA, and who were from regions of the globe where there is limited genetic information testing, informed me that the results they obtained from the two companies were different—not necessarily in the ethnicities, but in the corresponding percentages. Apparently, this is a common complaint. "People have sent their DNA to several of these companies and found differences in the results—though not necessarily radical differences. So you have to look at the percentages you receive back with skepticism."[47] As Sheldon Krimsky has pointed out, the companies doing the testing do not share their data with other personal genomics companies, their methods have not been tested by any independent group of scientists, and there are no generally accepted standards of accuracy.[48]

While these companies do profit from these tests, other sources of income come from third parties, such as Big Pharma, which are very interested in the genetic information of various human populations. The white male is no longer used to represent all of humankind. We need medical information on women and people of color, as required by the National Institutes of Health Revitalization Act of 1993.[49] In an age of precision medicine, where one size no longer fits all, Big Pharma has a vested interest in ascertaining which populations will respond best to certain medications. And, as mentioned above, many of them have controversially used race as a proxy for human diversity and populations.

Intellectual property is, of course, critical to the survival of these personal genomics companies. Access to data is granted in the likelihood of potential future payoff, and Big Pharma is willing to take that chance. Personal genomics companies own the data, which become scientifically relevant and knowable once their algorithms determine the data's importance. While gene patents seemed to be the best financial way forward for a number of early sequencing companies of the 1990s, including Incyte Genomics, Human Genome Sciences, and Millennium Pharmaceuticals,[50] nowadays personalized genomics companies are far more interested in protecting the patents of their algorithms. For example, 23andMe has patented algorithms related to providing displays with graphic-user interfaces, summarizing an individual's aggregate contribution to a genetic characteristic, processing data from genotyping chips, finding relatives in a database, identifying matrilineal or patrilineal relatives, processing data from genotyping chips, genome sharing, making genetic comparisons between grandparents and grandchildren, trio-based phasing using a dynamic Bayesian network, and correcting errors in ancestry classification.[51] Their patent portfolio more closely resembles a computer information company than a traditional wet-biomedical research company. By and large, they do not seem all that interested in patenting specific gene sequences, perhaps as a result of the USPTO's decision to make it substantially more difficult to patent products of nature after the US Supreme Court ruled in June 2013 to disallow gene patents based on DNA that was merely excised from the genome.[52] One notable exception is the patenting of genetic polymorphisms, which give rise to a particular disease, such as Parkinson's.[53] Similarly, Ancestry.com DNA LLC, the subsidiary of Ancestry.com, owns patents as well, most of which are for algorithms for, for example, identifying ancestral relationships using a continuous stream input, discovering population structure from patterns of identity by descent, identifying family networks using combinations of DNA analysis and genealogical information, a method and system for displaying genetic and genealogical data, and a method for molecular genealogical research. They have also applied for a patent for the computational methods needed to reconstruct the chromosomes and genomes of ancestors based on genetic data.[54]

These algorithm patents are critical for the viability of these companies, as witnessed by the lawsuit 23andMe filed against Ancestry.com DNA LLC on May 11, 2018, in the Northern District of California. In this, 23andMe alleges that Ancestry.com DNA LLC has infringed on the patent for algorithms it uses to find relatives in a database that share a common ancestor (known as "identity by descent," or IBD) within a certain number of generations. The patent claims to determine the IBD by obtaining DNA sequence information of first and second users stored in a database of many users, and to ascertain a degree of relative relationship based on the number of generations within which the two users share a common ancestor. The lawsuit also claims trademark infringements by Ancestry.com DNA LLC by using the word *ancestry* in the company's advertisement.[55] Ancestry.com DNA LLC is banking on the hope that 23andMe's patent on the algorithm for IBD will not hold up in a court of law. The theory behind IBD dates back some seventy years to the pioneering research of the French mathematician Gustave Malécot, and the IBD method has been used by hundreds of scientists over the past decade. On October 29, 2008, the first computational framework for analyzing IBD was published as an open-source software program, GERMLINE. 23andMe filed its patent on the following December 31.[56] As is the case with many patents on algorithms, this patent might not hold up, as it may fail the novelty criterion of patentability.[57] In short, while these companies still depend on traditional forms of intellectual property protection, such as patents and copyright, it turns out that more of their income involves their clients' genetic information, as I will discuss later. Interestingly, these companies do not know the relevance of certain bits of the data until their patented algorithms tell them which ones are scientifically—and therefore commercially—relevant. Ownership precedes knowing exactly what they have.

So, do personal genomics companies *really* own the genetic information of their clients? Well, it depends on what you mean by "own." Over the past several years, the instability of the legal notion of ownership—and the related notion of privacy—has been the subject of legal concerns. When a customer accepts 23andMe's terms of service, they are agreeing to the company's waiver of property rights, which states that "you specifically understand that you will not receive compensation for any research or commercial products that include or result from your Genetic Information or Self-Reported Information."[58] The client does not own any part of the services, which include the genetic information determined by the company's tests and provided by 23andMe. It is important to note that the terms of service refer specifically to "genetic information" and not to the actual DNA material, which will always belong to the donor:

> You acknowledge and agree that 23andMe (or 23andMe's licensors, as applicable) own all legal right, title, and interest in and to the Services, including any intellectual property rights (including but not limited to patents) which subsist in the Services (whether those rights happen to be registered or not, and wherever in the world those rights may exist).[59]

23andMe actually grants their customers a "limited license" to copy and distribute their genetic information freely for noncommercial purposes. The terms of service go on to say that if a customer has granted permission to 23andMe Research to do so, the company may share anonymized and aggregate genetic and self-reported information with third parties who are interested in publishing articles in peer-reviewed journals. Moreover, "23andMe may also include your information in Aggregated Genetic and Self-Reported Information disclosed to third-party non-profit and/or commercial research partners who will not publish that information in a peer-reviewed scientific journal."[60] If you withdraw from 23andMe Research, your data that have already been used in studies cannot be withdrawn, and you must give the company thirty days to withdraw your information.[61] In addition, if you do not give consent to 23andMe Research, "your Genetic Information and Self-Reported Information may still be used for other purposes, as described in our Privacy Statement."[62]

While the company's CEO, Anne Wojcicki, has stressed that 23andMe does not work with insurance companies,[63] the company's terms of service does convey a chilling warning:

> Currently, very few businesses or insurance companies request genetic information, but this could change in the future. While the Genetic Information Nondiscrimination Act [GINA]

was signed into law in the United States in 2008, its protection against discrimination by employers and health insurance companies for employment and coverage issues has not been clearly established. In addition, GINA does not cover life, long-term care, or disability insurance providers. Some, but not all, states and other jurisdictions have laws that protect individuals with regard to their Genetic Information. You may want to consult a lawyer to understand the extent of legal protection of your Genetic Information before you share it with anybody."[64]

However, a number of legal scholars have complained that GINA provides too little consumer protection.[65] Furthermore, the genetic information that you do choose to share with your physician or other healthcare providers may become part of your medical record and, through that route, be accessible to other healthcare providers and/or insurance companies in the future. Genetic information that you share with family, friends, or employers may be used against your interests. Even if you share genetic information that has no or limited meaning today, that information could have greater impact in the future as new discoveries are made. If you are asked by an insurance company whether you have learned genetic information about health conditions and you do not disclose this to them, this may be considered to be fraud.[66]

AncestryDNA's terms and conditions were recently a source of contention. They originally stated that "you grant AncestryDNA and the Ancestry Group Companies a perpetual, royalty-free, world-wide, transferable license to use your DNA . . . to use, host, sublicense and distribute the resulting analysis to the extent and in the form or context we deem appropriate on or through any media and medium and with any technology or devices now known or hereafter developed or discovered."[67]

In May 2017, Joel Winston, a consumer protection lawyer and former deputy attorney general of New Jersey, published a blog entry that sharply criticized AncestryDNA's practices, claiming that "according to its privacy policies Ancestry.com takes ownership of your DNA forever. Your ownership of your DNA, on the other hand, is limited in years."[68] Thus, Winston highlighted how your genetic information can be used, and can be seen, as a form of ownership by means of a proprietary license. The terms and conditions (TaC) go on to say that a client has no right to any commercial gain from their sample.[69] Consumers may, of course, withdraw consent; however, just as is the case with 23andMe, the company can continue to use your information for thirty days from the date of your request. In addition, any research, including research published online, that already uses your data cannot be withdrawn. The terms and conditions insist that Ancestry.com "will not share your Genetic Information with insurance companies, employers, or third-party marketers *without* your express consent."[70] While DNA as a material substance is being replicated and translated into mRNA to produce proteins within a person's body, it still remains owned by that person. The information encoded in someone's DNA, however, can be licensed to, and therefore owned by, another entity. Recall that under US law, property is a bundle of rights, which can be owned by numerous entities simultaneously.

Winston's blog entry ignited a firestorm on the internet. Ancestry.com's chief privacy officer Eric Hearth called Winston's remarks "inflammatory and inaccurate."[71] His defense of his company's policies was based in large part on the fuzziness of the legal concept of ownership:

> The consumer maintains ownership of their data. This is actually why we need a license in order to conduct our analysis, display their results, and so on. Not only do they own their own data, but we allow them to download their raw data and they can ask us to destroy the data at any time.[72]

Winston countered by differentiating between various legal notions of ownership:

> A license is a contractual form of ownership. At its most basic, a license is defined as "a permit from an authority to own or use something." Ancestry.com does not have "exclusive ownership" because customers still retain ownership of their own DNA. Ancestry.com does not have "absolute ownership" because customers can revoke the license. But, Ancestry .com irrefutably takes ownership of customers' genetic data by contractual license granted in the Terms and Conditions.[73]

Personal genomics companies do not own your DNA—the physical, material object—and it would not be in their interest to argue that they do. An Ancestry.com spokesperson was adamant about that, and the terms and conditions explicitly state that. In addition, the terms and conditions make it clear that "you always maintain ownership of your data." The company immediately continues by explaining, "but we need the ability to use your data for the purposes set out in our Privacy Statement and these Terms, and, if you agree to it, in our Informed Consent to Research."[74] The same spokesperson did concede that it is "broadly correct" that the license allows the company the privileges of ownership: "We couldn't send samples to the lab to be analyzed, transmit the results, etc. if we didn't have a license."[75] Ancestry.com decided to alter its terms and conditions slightly in 2017 in direct response to Winston's critiques. The company dropped the word *perpetual* to describe the license, and removed the phrase "to the extent and in the form and context we deem appropriate."[76]

Finally, AncestryDNA.com also warns its customers that their DNA may be used against them or a genetic relative in a court of law. This might seem farfetched, but it has already happened in the United States. On April 24, 2018, Joseph James DeAngelo was arrested in Sacramento on suspicion of being the so-called Golden State Killer, accused of murdering a dozen people, raping at least fifty women, and committing over a hundred burglaries from 1974 to 1986. The case had gone cold decades ago. Then in 2017 Paul Holes, who had previously been an investigator with the California's Contra Costa County District Attorney's Office, took the genetic information that had been left at one of the crime scenes and entered it into GEDmatch, a small DNA analysis company that possesses an online database filled with genetic information. This Florida-based website pools raw genetic data from anyone who uploads their genetic information online, often those who have their genetic profiles determined by 23andMe and AncestryDNA. Since they are publicly shared, no court order was necessary. More than a hundred users had corresponding matches at specific SNPs, possibly representing distant cousins. Holes contacted "one or two people," which subsequently led to DeAngelo's arrest. Curtis Rogers, the cocreator and operator of GEDmatch, was quick to point out that

> it has always been GEDmatch's policy to inform users that the database could be used for other uses, as set forth in the site policy. While the database was created for genealogical research, it is important that GEDmatch participants understand the possible uses of their DNA, including identification of relatives that have committed crimes or were victims of crimes.[77]

Less than two months later, Gary Hartman of Tacoma, Washington, was arrested and charged with raping and murdering a twelve-year-old girl some thirty-two years earlier. DNA evidence from a discarded restaurant napkin was used in identifying the suspect. In 2016 police began working with a genetic genealogist who was able to track down two brothers, using the DNA databases on the internet. They were then put under surveillance, and the police collected the napkin and sent it to the Washington State Patrol Crime Laboratory, where the sample was found to be a match.[78]

And on July 17, 2018, John D. Miller was arrested in Indiana for the abduction, rape, and murder of eight-year-old April Tinsley in 1988. Comparing the DNA from the scene of the crime with the DNA in a genealogical database, police found matches with the sequences which led them to two men, Miller and his brother. DNA found in the Millers' garbage led to John's arrest and confession.[79] In mid-October 2018, an article in the *New York Times* warned its readers that "already, 60% of Americans of Northern European descent—the primary group using these sites—can be identified through databases whether or not they've joined one themselves." Researchers say it will soon be possible [within two to three years] to identify up to 90 percent of white Americans from genealogical databases.[80] Privacy, it turns out, is just as unstable as ownership and property.

In addition, there is always the threat of data breaches, and such a case occurred in the summer of 2020. The DNA analysis site used to catch the Golden State Killer, GEDmatch, was the victim of hackers on July 19 and 10. The DNA profile data of over one million clients could be viewed by law enforcement agencies, even though many had opted not to have their data made available.[81]

The importance of ownership over genetic information for biotech companies became evident in the 1990s in Oregon, which was the first state to enact a law relevant to such information.[82] In 1995 Senate Bill 276, sponsored by then state senator Neil Bryant, declared that genetic information was the property of the individual from which it was obtained. While the bill was passed, it was revised six years later. In 1997 SmithKline Beecham (now GlaxoSmithKline) sponsored a bill that would repeal the property provision; the company feared that if individuals owned their genetic information, then the company would have to share the royalties generated by gene patents or, even worse, the transfer of rights to their biomedical researchers might not be enforceable. During this time, Oregon wanted to lure biotech companies to the Greater Portland area. After much debate over the ensuing four years, in 2001 the Oregon legislature overhauled its genetic privacy law. It deleted the provision that genetic information and DNA samples are the property of the individual, and in exchange, tightened up the privacy requirements surrounding such information.[83]

Despite Oregon's valiant efforts to create more stringent privacy laws, many feel that more needs to be done, as preexisting federal privacy laws as stipulated by GINA are considered insufficient. A number of politicians have recently expressed concerns that tighter restrictions should be placed on personal genomics companies. For example, in late November 2017, Democratic US senator from New York, Chuck Schumer, warned that "many consumers don't realize that their sensitive information can end up in the hands of unknown third-party companies. There are no prohibitions, and many companies say that they can still sell your information to other companies."[84] He called on the Federal Trade Commission to "take a serious look at this relatively new kind of service and ensure that these companies can have clear, fair privacy policies."[85]

CONCLUSION

In conclusion, this chapter has discussed an example of the related constructedness and instabilities of genetic information, ownership, knowability, expertise, and privacy. The sharing with third parties of genetic information that companies hold on their clients—who are seeking to gain an understanding of their ancestry and genetic predispositions—illustrates the legally contentious notion of ownership. The expertise of such ancestry testing is also being questioned, as the results are often inconsistent. Yet, the data do give these companies a good idea of the possible afflictions from which certain populations suffer. Big Pharma can now focus their efforts on treating certain genetic ailments of these groups, while insurance companies potentially can gauge how to move forward in setting their premiums. On July 25, 2018, GlaxoSmithKline (GSK) announced that they had just entered into a four-year collaboration with 23andMe to develop novel forms of medicine. GSK issued a statement in an attempt to assuage any fears of people who had had their DNA sequenced by 23andMe:

> 23andMe customers are in control of their data. Participating in 23andMe's research is always voluntary and requires customers to affirmatively consent to participate. For those who do consent, their information will be de-identified, so no individual will be identifiable to GSK. The continued protection of customers' data and privacy is the highest priority for both GSK and 23andMe. Both companies have stringent security protections in place when it comes to collecting, storing and transferring information about research participants. 23andMe employs software, hardware and physical security measures to protect the computers where data is stored and information will only be transferred using encryption to offer maximum security.[86]

The press release—tellingly, and wisely—did not mention ownership.

Intellectual property still plays a major role in biomedical research. With some notable exceptions, gene patents are not now considered to be as profitable as they used to be. The intellectual property portfolios of personal genomics companies are instead filled with patents on algorithms, which make their databases both knowable and scientifically and economically relevant. That said, there seems to be a trend among private companies in the biotech sector to increase their financial viability by means of their proprietary databases. This chapter has explored the instability of ownership and knowability of genetic information and the moral implications of such instability. Going forward, it is clear that renewed efforts to ensure privacy are critical during an age when our most intimate data, and our very identities, are becoming a coveted financial entity.

Notes

1. For the complete story, see Myles W. Jackson, *The Genealogy of a Gene: Race, Patents, and HIV/AIDS* (Cambridge, MA: MIT Press, 2015). For a recent work on patenting biological materials in general, see Shobita Parthasarathy, *Patent Politics: Life Forms, Markets, and the Public Interest in the United States and Europe* (Chicago: University of Chicago Press, 2017).

2. I would like to thank the anonymous reader for pointing this out to me.

3. For the classic study on property, see Tony M. Honoré, "Ownership," in *Oxford Essays in Jurisprudence*, ed. Anthony G. Guest (New York: Oxford University Press, 1961). See also Pilar N. Ossorio, "Property Rights and Human Bodies," in *Who Owns Life?*, ed. David Magnus, Arthur Caplan, and Glenn McGee (New York: Prometheus Books, 2002), 224–225.

4. Jessica L. Roberts, "Theories of Genetic Ownership" (working paper, September 9, 2015), 8–9, accessed February 14, 2019, http://petrieflom.law.harvard.edu/assets/publications/Roberts_Genetic_Ownership_Draft.pdf.

5. Ossorio, "Property Rights," 225.

6. Union Oil Co. v. State Board of Equal., 60 Cal.2d 441, 447 (Cal. 1963), https://www.courtlistener.com/opinion/1165762/union-oil-co-v-state-bd-of-equal/.

7. Ossorio, "Property Rights," 226.

8. Christopher Heaney et al., "The Perils of Taking Property Too Far," *Stanford Journal of Law and Science Policy* 46, no. 1 (2009): 46–64. See also Linda L. McCabe and Edward R. B. McCabe, "Gene Patents: Perspective from the Clinic and the Laboratory," in "Gene Patents," ed. Myles W. Jackson, special issue, *Perspectives on Science* 23, no. 1 (2015): 67–69; Lori Andrews and Dorothy Nelkin, "Propriety and Property: The Tissue Market Meets the Courts," in *Who Owns Life?*, ed. David Magnus, Arthur L. Caplan, and Glenn McGee (New York: Prometheus Books, 2002), 200–201; and Margaret Everett, "The Social Life of Genes: Privacy, Property and the New Genetics," in *Information Ethics: Privacy, Property, and Power*, ed. Adam D. Moore (Seattle: University of Washington Press, 2005), 233–234.

9. Moore v. Regents of the University of California, 51 Cal.3d 120, 127–128 (Cal. 1990); 271 Cal. Reptr. 146; 793 P.2d 479.

10. *Moore*, 51 Cal.3d at 128–129.

11. *Moore*, 51 Cal.3d at 141.

12. *Moore*, 51 Cal.3d at 141–142.

13. *Moore*, 51 Cal.3d at 137, 141. See also McCabe and McCabe, "Gene Patents," 67–69; Andrews and Nelkin, "Propriety and Property," 200–201; and Everett, "Social Life of Genes," 233–234.

14. *Moore*, 51 Cal.3d at 144.

15. *Moore*, 51 Cal.3d at 145–146.

16. *Moore*, 51 Cal.3d at 147.

17. *Moore*, 51 Cal.3d at 152. Please note the gendered language in the quotation. Women have also had bodily cells and tissues removed for scientific research.

18. *Moore*, 51 Cal.3d at 164.

19. *Moore*, 51 Cal.3d at 167.

20. *Moore*, 51 Cal.3d at 170.

21. Greenberg v. Miami Children's Hospital Research Institute, 208 F. Supp. 2d 918 (2002). The ruling of the Florida district court, which was final, can be found at Greenberg v. Miami Children's Hospital Research Institute, 264 F. Supp. 2d 1064 (2003).

22. *Greenberg*, 208 F. Supp. 2d at 922.

23. *Greenberg*, 208 F. Supp. 2d at 922.

24. *Greenberg*, 208 F. Supp. 2d at 922.

25. *Greenberg*, 264 F. Supp. 2d at 1075.

26. *Greenberg*, 264 F. Supp. 2d at 1075.

27. *Greenberg*, 264 F. Supp. 2d at 1076.

28. Washington University v. William J. Catalona, M.D., 437 F. Supp. 2d 985 (8th Cir. 2006).

29. *Catalona*, 437 F. Supp. 2d at 993.

30. *Catalona*, 437 F. Supp. 2d at 990.

31. *Catalona*, 437 F. Supp. 2d at 988.

32. *Catalona*, 437 F. Supp. 2d at 994.

33. *Catalona*, 437 F. Supp. 2d at 997.

34. *Catalona*, 437 F. Supp. 2d at 997 and 1001.

35. *Catalona*, 437 F. Supp. 2d at 1002.

36. Biomedical researchers often use these terms interchangeably.

37. "Precision Medicine Will Surpass USD 96600 Million by 2024," Pharmaphorum (website), April 30, 2019, https://pharmaphorum.com/partner-content/precision-medicine-market-will-surpass-usd-96600-million-by-2024/.

38. "Fact Sheet: Obama Administration Announces Key Actions to Accelerate Precision Medicine Initiative," Obama White House archives, The White House Office of the Press Secretary, February 25, 2016, https://obamawhitehouse.archives.gov/the-press-office/2016/02/25/fact-sheet-obama-administration-announces-key-actions-accelerate.

39. It should be noted that FamilyTreeDNA has not sold its customers' DNA or data to third parties.

40. See, for example, "23andMe TV Commercials," iSpot.tv, accessed February 7, 2019, https://www.ispot.tv/brands/Ias/23andme; and "Ancestry TV Commercials," iSpot.tv, accessed February 7, 2019, https://www.ispot.tv/brands/dhv/ancestry.

41. As of November 2017, 23andMe's database had 2 million users, Ancestry's had 5 million users, FamilyTreeDNA's had 1.5 million users, African Ancestry had 33,000 lineages across 43 African nations, and National Geographic had a database of 830,000 users. See Natasha Stokes, "What You Need to Know Before Buying a Home DNA Test," Techlicious, November 7, 2017, https://www.techlicious.com/guide/dna-home-test-kit-what-you-should-know-our-picks/. Here are some of those commercials, which have been aired in the US: Ancestry, "Testimonial: Kyle," 2015, commercial, accessed October 15, 2020, https://www.ispot.tv/ad/wppp/ancestrydna-testimonial-kyle; Ancestry, "Ancestry Stories: Anthem," 2018, commercial, accessed October 15, 2020, https://www.ispot.tv/ad/dwkp/ancestry-ancestry-stories-anthem; 23andMe, "Getting to Know You," 2018, commercial, accessed October 15, 2020, https://www.ispot.tv/ad/wvMS/23andme-getting-to-know-you; and Ancestry, "Lyn Discovers Her Ethnicity Discoveries," 2016, commercial, accessed October 15, 2020, https://www.youtube.com/watch?v=0l0_ttMidII.

42. While the FDA does not regulate ancestry testing, they are responsible for the genetic testing of diseases, which a number of these companies also offer.

43. "23andMe Revs Up Ancestry Composition Feature," *23andMeBlog* (blog), November 16, 2017, https://blog.23andme.com/ancestry/23andme-revs-ancestry-composition-feature/.

44. "23andMe Revs Up."

45. "23andMe Revs Up." One customer, whose father's ancestry was Italian on both sides, claimed that the percentage rose from 0.4% to 21% Italian when using the v5 chip.

46. "AncestryDNA DNA Test Review (2020 Update): How Does It Compare?," *myFamilyDNA* (blog), accessed February 19, 2020, https://www.myfamilydnatest.com/ancestrydna-review/.

47. Genevieve Rajewski, "Pulling Back the Curtain on DNA Ancestry Tests," Tufts Now, January 26, 2018, https://now.tufts.edu/articles/pulling-back-curtain-dna-ancestry-tests.

48. Rajewski.

49. Steven Epstein, *Inclusion: The Politics of Difference in Medical Research* (Chicago: University of Chicago Press, 2007), 95–122.

50. Note that these companies are not the same as personal genomics companies.

51. "Patent Information," 23andMe, accessed January 22, 2019, https://www.23andme.com/patents/.

52. Association for Molecular Pathology v. Myriad Genetics, Inc., 569 U.S. 576 (2013). For the USPTO's requirements for patent eligibility of objects of nature, see "2105 Patent Eligible Subject Matter—Living Subject Matter," United States Patent and Trademark Office, accessed January 22, 2019, https://www.uspto.gov/web/offices/pac/mpep/s2105.html. It should be noted that the United States Supreme Court's decision did not outright prohibit the patenting of products of nature; it insisted that patented products of nature have attributes that distinguish them for their naturally occurring homologs. See Jackson, *Genealogy of a Gene*, 175–187.

53. See, for example, 23andMe's US Patent 8197811: "The application provides nucleic acid sequences that may be used to determine the presence or absence of nucleotides at polymorphic sites in an individual's RNA or genomic DNA that are associated with susceptibility to or protection from P[arkinson's] D[isease]. In another aspect, the application provides a method for identifying a human subject having an increased or decreased susceptibility to PD." 23andMe, Polymorphisms Associated with Parkinson's Disease, US Patent 8197811, filed November 30, 2010, issued May 29, 2012, https://patents.google.com/patent/US8187811.

54. "Patents Assigned to Ancestry.com DNA LLC," Justia Patents, accessed January 23, 2019, https://patents.justia.com/assignee/ancestry-com-dna-llc.

55. Steve Brachmann, "23andMe Sues Ancestry.com over DNA Genetic Testing Kits," IPWatchdog.com, May 15, 2018, https://www.ipwatchdog.com/2018/05/15/23andme-sues-ancestry-com-dna-genetic-testing-kits/id=97269/.

56. Megan Molteni, "23andMe Is Suing Ancestry over Some Pretty Ancient IP," *Wired*, May 30, 2018, https://www.wired.com/story/23andme-sues-ancestry/.

57. In re Bilski, 545 F.3d 943, 88 U.S.P.Q.2d 1385 (Fed. Cir. 2008), it was ruled that, while a pure algorithm is not patentable subject matter, if the algorithm is embodied in a machine and results in a transformation of some kind, then it may indeed be patentable subject material.

58. "Terms of Service," 23andMe, accessed February 9, 2019, https://www.23andme.com/about/tos/. See also Roberta Estes, "23ndMe, Ancestry and Selling Your DNA Information," *DNAeXplained—Genetic Genealogy* (blog), revised May 25, 2018, https://dna-explained.com/2015/12/30/23andme-ancestry-and-selling-your-dna-information/.

59. "Terms of Service."

60. "Terms of Service."

61. "Research Content Document," 23andMe, accessed February 9, 2019, https://www.23andme.com/about/consent/.

62. "Research Content Document."

63. Eric Johnson, "Is It Safe to Give Your Genetic Data to 23andMe?," Vox, September 22, 2017, https://www.vox.com/2017/9/22/16347728/23andme-dna-testing-genomics-privacy-anne-wojcicki-kara-swisher-lauren-goode-too-embarrassed-podcast.

64. "Terms of Service."

65. Kristen V. Brown, "What DNA Testing Companies' Terrifying Privacy Policies Actually Mean," Gizmodo, October 18, 2017, https://gizmodo.com/what-dna-testing-companies-terrifying-privacy-policies-1819158337.

66. "Terms of Service."

67. As quoted in Joel Winston, "Ancestry.com Takes DNA Ownership Rights from Customers and Their Relatives," ThinkProgress, May 17, 2017, https://archive.thinkprogress.org/ancestry-com-takes-dna-ownership-rights-from-customers-and-their-relatives-dbafeed02b9e/.

68. Winston.

69. "Ancestry Terms and Conditions, Effective 5 June 2018," Ancestry, accessed February 7, 2019, https://www.ancestry.com/cs/legal/termsandconditions/.

70. "Privacy Statement," Ancestry, accessed February 7, 2019, https://www.ancestry.com/cs/legal/privacystatement; emphasis in original.

71. Dan MacGuill, "Can Ancestry.com Take Ownership of Your DNA Data? An Experienced Attorney's Blog Post Has Caused Widespread Concern," Snopes, May 22, 2017, https://www.snopes.com/fact-check/ancestry-dna-steal-own/.

72. MacGuill.

73. MacGuill.

74. Ancestry, "Ancestry Terms and Conditions." See also Brown, "DNA Testing Companies' Privacy Policies."

75. As also quoted in Brown, "DNA Testing Companies' Privacy Policies."

76. MacGuill, "Ownership of DNA."

77. As quoted in Keith Allen, Jason Hanna, and Cheri Mossburg, "Police Used Free Genealogy Database to Track Golden State Killer Suspect, Investigator Says," CNN (website), April 27, 2018, https://www.cnn.com/2018/04/26/us/golden-state-killer-dna-report/index.html.

78. Ralph Ellis, "DNA on Napkin Used to Crack 32-Year-Old Cold Case, Police Say," CNN (website), June 23, 2018, https://www.cnn.com/2018/06/22/us/cold-case-killing-1986/index.html.

79. Eric Levenson and Amanda Watts, "Child-Killer Taunted Investigators for 30 Years with Disturbing Notes. DNA Ends the Mystery of Who Did It, Police Say," CNN (website), July 17, 2018, https://www.cnn.com/2018/07/16/us/cold-case-april-tinsley-dna-trnd/index.html.

80. Heather Murphy, "Your DNA, Identified by the DNA of Others," *New York Times*, October 12, 2018.

81. Peter Aldhous, "A Security Breach Exposed More Than One Million DNA Profiles on a Major Genealogy Website," BuzzFeedNews, July 22, 2020, https://www.buzzfeednews.com/article/peteraldhous/hackers-gedmatch-dna-privacy.

82. Everett, "Social Life," 238–245.

83. Oregon Health Authority, "History of Oregon's Genetic Privacy Law," 3–4, accessed February 7, 2019, http://www.oregon.gov/oha/ph/DiseasesConditions/GeneticConditions/Documents/LAW_ORHxPrivacy.pdf.

84. Daniella Silva, "Senator Calls for More Scrutiny of Home DNA Test Industry," NBC News (website), November 26, 2017, https://www.nbcnews.com/news/us-news/senator-calls-more-scrutiny-home-dna-test-industry-n824031.

85. Silva.

86. "GSK and 23andMe Sign Agreement to Leverage Genetic Insights for the Development of Novel Medicines," GSK (website), Press Releases, July 25, 2018, https://www.gsk.com/en-gb/media/press-releases/gsk-and-23andme-sign-agreement-to-leverage-genetic-insights-for-the-development-of-novel-medicines/.

Bibliography

Aldhous, Peter. "A Security Breach Exposed More Than One Million DNA Profiles on a Major Genealogy Website." BuzzFeedNews, July 22, 2020. https://www.buzzfeednews.com/article/peteraldhous/hackers-gedmatch-dna-privacy.

Allen, Keith, Jason Hanna, and Cheri Mossburg. "Police Used Free Genealogy Database to Track Gold State Killer Suspect, Investigator Says." CNN (website), April 27, 2018. https://www.cnn.com/2018/04/26/us/golden-state-killer-dna-report/index.html.

Andrews, Lori, and Dorothy Nelkin. "Propriety and Property: The Tissue Market Meets the Courts." In *Who Owns Life?*, edited by David Magnus, Arthur L. Caplan, and Glenn McGee, 197–222. New York: Prometheus Books, 2002.

Brachmann, Steve. "23andMe Sues Ancestry.com over DNA Genetic Testing Kits." IPWatchdog.com, May 15, 2018. https://www.ipwatchdog.com/2018/05/15/23andme-sues-ancestry-com-dna-genetic-testing-kits/id=97269.

Brown, Kristin V. "What DNA Testing Companies' Terrifying Privacy Policies Actually Mean." Gizmodo, October 18, 2017. https://gizmodo.com/what-dna-testing-companies-terrifying-privacy-policies-1819158337.

Conley, John M., Robert Cook-Deegan, and Gabriel Lázaro-Muñoz. "Myriad after *Myriad*: The Proprietary Data Dilemma." *North Carolina Journal of Law & Technology* 15, no. 4 (2014): 597–637.

Ellis, Ralph. "DNA on Napkin Used to Crack 32-Year-Old Cold Case, Police Say." CNN (website), June 23, 2018. https://www.cnn.com/2018/06/22/us/cold-case-killing-1986/index.html.

Epstein, Steven. *Inclusion: The Politics of Difference in Medical Research*. Chicago: University of Chicago Press, 2007.

Estes, Roberta. "23ndMe, Ancestry and Selling Your DNA Information." *DNAeXplained—Genetic Genealogy* (blog), revised May 25, 2018. https://dna-explained.com/2015/12/30/23andme-ancestry-and-selling-your-dna-information/.

Everett, Margaret. "The Social Life of Genes: Privacy, Property and the New Genetics." In *Information Ethics: Privacy, Property, and Power*, edited by Adam D. Moore, 226–250. Seattle: University of Washington Press, 2005.

Heaney, Christopher, Julia Carbone, Richard Gold, Tania Bubela, Christopher M. Holman, Alessandra Colaianni, Tracy R. Lewis, and Bob Cook-Deegan. "The Perils of Taking Property Too Far." *Stanford Journal of Law and Science Policy* 46, no. 1 (2009): 46–64.

Honoré, Tony M. "Ownership." In *Oxford Essays in Jurisprudence*, edited by Anthony G. Guest, 107–147. New York: Oxford University Press, 1961.

Jackson, Myles W. *The Genealogy of a Gene: Race, Patents, and HIV/AIDS*. Cambridge, MA: MIT Press, 2015.

Johnson, Eric. "Is It Safe to Give Your Genetic Data to 23andMe?" Vox, September 22, 2017. https://www.vox.com/2017/9/22/16347728/23andme-dna-testing-genomics-privacy-anne-wojcicki-kara-swisher-lauren-goode-too-embarrassed-podcast.

Latour, Bruno. "Visualization and Cognition: Drawing Things Together." *Avant: Trends in Interdisciplinary Studies* 3 (2012): 207–260.

Levenson, Eric, and Amanda Watts. "Child-Killer Taunted Investigators for 30 Years with Disturbing Notes. DNA Ends the Mystery of Who Did It, Police Say." CNN (website), July 17, 2018. https://www.cnn.com/2018/07/16/us/cold-case-april-tinsley-dna-trnd/index.html.

Lewis, Ricki. "Dueling BRCA Databases: What about the Patient?" *PLOS Blogs* (blog), April 20, 2017. https://blogs.plos.org/dnascience/2017/04/20/dueling-brca-databases-what-about-the-patient/.

MacGuill, Dan. "Can Ancestry.com Take Ownership of Your DNA Data? An Experienced Attorney's Blog Post Has Caused Widespread Concern." Snopes, May 22, 2017. https://www.snopes.com/fact-check/ancestry-dna-steal-own/.

McCabe, Linda L., and Edward R. B. McCabe. "Gene Patents: Perspective from the Clinic and the Laboratory." In *Gene Patents: Perspectives on Science*, edited by Myles W. Jackson, 66–79. Cambridge, MA: MIT Press, 2015.

Molteni, Megan. "23andMe Is Suing Ancestry over Some Pretty Ancient IP." *Wired*, May 30, 2018. https://www.wired.com/story/23andme-sues-ancestry/.

Murphy, Heather. "Your DNA, Identified by the DNA of Others." *New York Times*, October 12, 2018.

Oregon Health Authority. "History of Oregon's Genetic Privacy Law." Last updated April 2, 2010. http://www.oregon.gov/oha/ph/DiseasesConditions/GeneticConditions/Documents/LAW_ORHxPrivacy.pdf.

Ossorio, Pilar N. "Property Rights and Human Bodies." In *Who Owns Life?*, edited by David Magnus, Arthur Caplan, and Glenn McGee, 223–242. New York: Prometheus Books, 2002.

Parthasarathy, Shobita. *Patent Politics: Life Forms, Markets, and the Public Interest in the United States and Europe*. Chicago: University of Chicago Press, 2017.

Rajewski, Gene. "Pulling Back the Curtain on DNA Ancestry Tests." Tufts Now, January 26, 2018. https://now.tufts.edu/articles/pulling-back-curtain-dna-ancestry-tests.

Roberts, Jessica L. "Theories of Genetic Ownership." Working paper, September 9, 2015. http://petrieflom.law.harvard.edu/assets/publications/Roberts_Genetic_Ownership_Draft.pdf.

Silva, Daniella. "Senator Calls for More Scrutiny of Home DNA Test Industry." NBC News (website), November 26, 2017. https://www.nbcnews.com/news/us-news/senator-calls-more-scrutiny-home-dna-test-industry-n824031.

Stokes, Natasha. "What You Need to Know Before Buying a Home DNA Test." Techlicious, November 7, 2017. https://www.techlicious.com/guide/dna-home-test-kit-what-you-should-know-our-picks/.

Winston, Joel. "Ancestry.com Takes DNA Ownership Rights from Customers and Their Relatives." Thinkprogress, May 17, 2017. https://thinkprogress.org/ancestry-com-takes-dna-ownership-rights-from-customers-and-their-relatives-dbafeed02b9e/.

IV THE ROLE OF SCHOLARSHIP

10

OBJECTS, KNOWLEDGE, AND MUSEUMS: REFLECTIONS ON THE ENDANGERED MATERIAL KNOWLEDGE PROGRAMME

Lissant Bolton

In 2018 the British Museum launched a grant program to support ethnographic research, the Endangered Material Knowledge Programme.[1] In this chapter I use this program as a point of entry into questions around knowledge and ownership that are increasingly crucial for ethnographic museums. I explore the politics around museum anthropology collections as they impact on material anthropology—that is, on the documentation and description of knowledge associated with objects. These politics relate to ideas about the ownership and deployment of knowledge—in other words, to epistemology. I explore these ideas with specific reference to Vanuatu, where much of my own research has been focused. In doing so, I pay attention to what I consider to be the significant use of ignorance and forgetting in these arenas.

The Endangered Material Knowledge Programme (EMKP) has been established to enable the documentation of knowledge associated with objects and the built environment, and to make it available through an open-access digital repository. The program is supported by a private charitable foundation, the Arcadia Fund (described below). By focusing on objects, by including collected objects in its remit, and by being based in a museum, EMKP invokes some current issues concerning museums, especially those that hold ethnographic collections. These issues are to do with the kinds of moral work that museums are often now required to do on behalf of wider society. By "moral work," I mean the work of reparation and restitution for past wrongs, especially wrongs committed in the colonial era. EMKP also draws attention to aspects of how people understand and control knowledge in and around museums. The program provides a lens through which to consider some of the questions of knowledge and ownership faced by anthropology museums today.

In recent decades, museum anthropology has been caught up in the politics of colonialism and postcolonialism. This politics has been made particularly public in France in recent years. In November 2017, President Emmanuel Macron gave a speech at the

University of Ouagadougou in Burkina Faso, in which he said that he could not accept that a large part of cultural heritage from several African countries is in France, commenting further that African heritage should not be just in European private collections and museums but also in Dakar, in Lagos, and in Cotonou. Later, the Élysée tweeted: "African heritage can no longer be the prisoner of European museums."[2] Macron followed up that speech in March 2018 by appointing two people to make plans to repatriate African artifacts held in French museums: the Senegalese writer and economist Felwine Sarr and the French historian of art Bénédicte Savoy. Savoy was already leading a major research project on issues of provenance and on how the meanings of objects change when they are being transferred into museums often far away,[3] and she had made the politics that can be invested in museum collections very explicit. As was reported in an article in the UK *Art Newspaper*, she told the German newspaper *Der Tagesspiegel*, "I want to know how much blood is dripping from each artwork."[4] The Sarr and Savoy report, presented in November 2018, has by 2022 resulted in a handful of objects being returned to Africa from France.

In the agreement with the British Museum to establish EMKP, Arcadia stipulated that the cultural knowledge the program would record must be endangered and should be anchored by the made world, by objects and the built environment. In fact, there are few forms of knowledge that are not linked to the material world in one way or another. Studying knowledge around objects does not mean just considering how things are made and used, but also addressing nearly every aspect of social life. The study of objects can focus on knowledge of the properties of materials, and documenting materials can lead to recording knowledge about landscape, plants, animals, and weather. Studying craftsmanship and skill leads to a focus on how skills are transmitted from person to person and through different modes of communication. Learning how objects are made and used leads to a consideration of the habitual practices of daily life, and equally of rituals and other special occasions. Researching objects can also lead to considerations of language, of concepts of design and aesthetics, and so on. Investigating the built environment involves a similar range of areas of enquiry. It is not just a matter of knowledge of materials and construction techniques, but also of knowledge of and adaptation to the environment; of forms of decoration and their significance; of the management of light and darkness; of the different types of buildings people make and how they use them; of how space is divided within and between buildings, and what that says about both social organization and the creation and use of outside spaces such as plazas. All of these areas touch on issues of ownership—not only of objects, but also always of the knowledge that surrounds them.

Anthropology, as a discipline, operates on the principle that the knowledge societies hold, and the way they manage and organize that knowledge, is valuable and should be both documented and analytically described. As Thomas Kirsch and Roy Dilley sum it up, "Cultural and social anthropology has long been driven by an encyclopaedic desire to identify, document, classify and archive to the greatest possible extent what was previously unknown."[5] Indeed, to quote Jonathan Mair et al., "knowledge is the value that justifies all aspects of academic activity, whether it is desired as a means of promoting other goods (health, happiness, wealth, well-being) or as an end in itself."[6] This approach characterizes knowledge as open and free to access; it also constitutes ignorance—not knowing—as a negative. The ignorance that comes from *forgetting* is a particularly bad thing, especially as discussed in the context of cultural loss.[7]

Museum anthropology—which focuses on the knowledge that objects reveal and embody—has long been part of that academic project. Anthropology museums collect research and display objects as a way to illuminate the knowledge and practice of the different societies they represent. Objects stand in relation to knowledge in interesting ways because although people can invest new meanings in an object, at the same time, objects embody knowledge in their material form. That knowledge is muted when someone who cannot recognize the materials or appreciate the skills involved views the object, but it is present nonetheless. Of course, what can only be imagined by looking at an object is the nature of the social context around the specific object as it was used—in all the different contexts in which it was used.

DOCUMENTING OBJECTS

Museums are thus based on the principle that objects illustrate knowledge and practice; that knowledge can be gained by looking at an object itself, and knowledge obtained elsewhere can be illuminated and expanded by studying objects. Before the twentieth century, the knowledge that objects illustrate and illuminate was assumed to be quite straightforward. Objects represented a place or a time, and the societies that produced them. Each object's very materiality formed a link to that place and time because the object had been physically there, then. Especially before the advent of photography and film, museums provided visitors with an insight into those other places and other times by collecting and displaying objects from them. Most early collections, such as those made during Captain Cook's explorations of the Pacific (1768–1780), were barely documented at all. The individual object—a club, headdress, or feather cloak—was, in its strangeness and distinctiveness, seen to be sufficient in itself to represent the societies that produced it. It is perhaps partly for this reason that the development of anthropology as a discipline led for many decades to the denigration of museums and material culture as being a less important branch of research. As soon as it was recognized how much there was to be known about those societies, the characterization of objects as emblems may have seemed to trivialize the new project of anthropology.

In particular, the development of the discipline of fieldwork quickly demonstrated the limits to what can be learned by looking at an object in a museum without any reference to the context that produced it. As Malinowski observed in 1922:

> A canoe is an item of material culture, and as such it can be described, photographed and even bodily transported into a museum. But . . . the ethnographic reality of the canoe would not be brought much nearer to a student at home, even by placing a perfect specimen right before him. . . . [Even] the study of . . . its ownership, accounts of who sails in it, and how it is done; information regarding the ceremonies and customs of its construction . . . [do] not touch the most vital reality of a native canoe. . . . [For the sailors] it is . . . a living thing, possessing its own individuality.[8]

I do not need to rehearse here the long history of anthropology's theoretical dalliance with objects, which others have discussed at length. A number of authors have sought to understand how meaning and significance is attached to objects, and how people deploy that meaning in social contexts. Much of this discussion has been focused on the idea of objects as art.[9] In recent decades there has been the recognition that, as Pierre Lemonnier puts it, "the anthropology of objects and techniques allows us to understand aspects of social organisations, cultures, and systems of thought that would be impossible to grasp without studying the most material dimensions of human action."[10] And as Lemonnier also makes clear, it is not only special objects, such as those defined as art, to which people attach meaning, but also ordinary objects—"mundane objects," in Lemonnier's phrasing—such as a garden fence or an eel trap.[11]

If field anthropologists now appreciate the importance of attending to material objects, the recognition of quite how significant objects can be poses challenges for earlier collections. What is now the significance or relevance of an object collected in 1770 or 1860 or even 1920 that is provided with only a geographical provenance or date of collection in museum records? More and more anthropology museums have responded to their ignorance about collections by instituting and developing field research programs that document objects by taking photographs back to the place where those objects were collected or by bringing people from that place to the museum and asking people today what they know about that kind of object. The Endangered Material Knowledge Programme includes a focus on documenting collected objects in this way.

The objective of documenting material knowledge is thus one shared by both Arcadia and the British Museum. EMKP has been set up to document knowledge around objects where there is no other resource available to do so (so, in mostly small-scale societies and often in the global south). Arcadia is a personal charitable fund belonging to Lisbet Rausing and her husband Peter Baldwin; much of their work is achieved through supporting institutions to operate grant programs and related digital repositories. Arcadia approached the British Museum some years ago, seeking to set up a grant program that would document endangered cultural knowledge in digital formats. Arcadia's approach to EMKP, of funding a grant program administered by another institution, is a characteristic mode of operation. For example, they support the Endangered Languages Documentation Programme first set up at the School of African and Oriental Studies, now at the Berlin Brandenburgische Akademie der Wissenschaften, and the Endangered Archives Programme managed by the British Library, both of which award and manage grants for specific documentation projects, and make the results available in online digital repositories.

Rausing and Baldwin set out their objectives for Arcadia in a statement on the fund's website:

> Arcadia supports work to preserve endangered cultural heritage, protect endangered ecosystems, and promote access to knowledge. Our aim is to defend the complexity of human culture and the natural world, so that coming generations can build a vibrant, resilient and green future.[12]

Their statement expresses an overall project similar to academic research, and specifically to anthropology's objective of documenting and describing other people's knowledge and practice. However, Arcadia's objective is not to support anthropology itself, in the sense that anthropology is the discipline in which ethnographic data—field data—is analyzed and described. Rather, as their website makes clear, they are interested in establishing a way of recording ethnographic data as a form of knowledge. Of course, it is not possible to organize and store ethnographic information without analytical thinking, but the analysis is not their aim. In this sense, the Endangered Languages Documentation Programme, which they founded in 2002, provides an explanatory model for what Arcadia is looking for in EMKP. The Endangered Languages Documentation Programme records language through the established analytical and descriptive categories of vocabulary and grammar.

Arcadia is explicitly unwilling to fund revival projects—bringing knowledge and practice back into currency. Externally driven revival projects often have an accidentally transformative effect; they sometimes create a hybrid form compelled by the outside rationale for revival. Thus, performance to outsiders such as tourists, or the sale of objects as art or souvenirs, turns locally specific knowledge and practice into something different from what it previously was. Documentation, however, may revive interest within the community, and local decisions may bring about the restoration and sometimes the transformation of knowledge and practice, for local purposes.

Arcadia is deeply committed to making knowledge available through accessible digital archiving, and to their conception of knowledge as being freely accessible and not owned. They value open access as a means to making such knowledge available to everyone, not least the communities whose heritage it is. Rausing and Baldwin have declared that

> access to the materials must be a crucial part of any effort to safeguard the knowledge and memory they contain. Minorities, exiles, the displaced and various first nations who have often been denied access to their own heritage as a result struggle to maintain their cultural identity. Who could lay claim to rescuing their heritage if we digitise it without making it accessible to them? . . . Digitisation may help to preserve the archives, but without open access the impact of these efforts will be limited.[13]

The underlying motivation for all Arcadia funding programs is thus the concept of endangerment. The sense of things being lost arises from the characterization of knowledge not as something that is endlessly transmuted, modified, and remade, but as something specific which, if changed, is thus also in part forgotten and lost. Anthropology often focuses on the knowledge that exists in the present moment. But it is true that much local knowledge and practice is actually being forgotten in the world today. I am personally aware of this in relation to my own work in the Republic of Vanuatu.

ENDANGERMENT, IGNORANCE, AND SECRECY

The anchor of my involvement in Vanuatu has been my participation in supporting the Women Fieldworkers Program at the Vanuatu Cultural Centre for the last thirty years. Vanuatu is a small nation in the western Pacific, comprising an archipelago of over eighty islands that spread over more than a thousand kilometers from north to south. The population (300,019 in 2020)[14] speaks more than 130 indigenous languages as well as the languages of colonialism (English and French) and a lingua franca called Bislama. The fieldworker program is designed to address and support this immense cultural diversity. It aims to train and support voluntary ni-Vanuatu[15] researchers working in their own villages and districts to document and revive local knowledge and practice. A group of male fieldworkers was set up in 1989. I have chaired the annual women fieldworkers' workshops since the group was founded in 1984 and have participated in several documentation and revival projects set up by fieldworkers in different parts of the Vanuatu archipelago. My own career has thus involved supporting people in Vanuatu who are seeking to document and revive their own knowledge. In that program, we have found again and again that documentation leads to an increased local interest in both knowledge and practice, and sometimes to reviving old traditions. Indeed, this has happened even when revival was not an intended outcome.

Working with this program for so long, I am also aware of the extent to which local knowledge and practice in Vanuatu is rapidly changing. Vanuatu has experienced significant social change since independence in 1980, brought about by factors including population growth and the growth of urban centers, education for employment, the increasing influence of the media (especially social media), new churches, and labor migration to New Zealand and Australia.

Inevitably, these changes have wrought an alteration in epistemology. As has often been demonstrated, and as discussed below, in Vanuatu both knowledge and practice were traditionally deeply tied to place.[16] Indeed, even now a ni-Vanuatu person's primary identification is based neither on kin ties nor on language, but on the place that person comes from, expressed in the Bislama term *manples*. This sees knowledge as being derived from living constantly in a place, knowing its character and history, absorbing the place into oneself by eating food grown on it. Knowledge also comes from an awareness of the place itself embodied in other mostly invisible beings—like people, but not people—who also inhabit the landscape.

As has been argued for different regions of Vanuatu, it is widely held there that knowledge is available—in the sense of being accessible and retrievable—in the landscape. For north Pentecost island, John Taylor has described a place known as *abanoi* (Raga language), a kind of "invisible parallel dimension layered across or threaded within the lived world of human experience," understood to contain the "true and authentic knowledge of the ancestors."[17] Writing about the southern island of Tanna, Joël Bonnemaison observed that people there consider that they belong less to a social order than to a place, so much so that "if their social fabric were destroyed, the Tannese would lose none of their heritage—provided they kept the memory of their places."[18] In fact, he argued, "in traditional thinking, cultural identity is merely the existential aspect of those places where men live today as their ancestors did from time immemorial."[19] In the terms of these characterizations, someone coming from elsewhere can, by living in a place and eating food from it, become part of the place and absorb its knowledge. In this epistemology, knowledge cannot be forgotten or lost because it is always there, in that place.

Lamont Lindstrom, also writing about Tanna, coined the term *geographic oeuvre* to describe the way in which knowledge is so much linked to place that there are geographically based restrictions on who can speak about what.[20] He defines the contents of such oeuvres as "text-like formulaic statements such as genealogical lists, stories, legends, songs, sets of local names for men, women, and pigs, maps of land plot boundaries, medical recipes, spells and magical technologies."[21] Thus, it might be considered that the kind of knowledge that dwells in a place is not practical information (such as how to plant a yam) but, rather, cultural—with political, intellectual, and spiritual reference.

Education, employment, and long periods of residence in other places all detach people from the knowledge belonging to their place. As people's links to their place change, so does their knowledge. As a result, many ni-Vanuatu no longer understand traditional knowledge as the existential aspect of place, but instead characterize it as deriving from a time—from the past. There is thus a profound transformation in epistemology happening, a transformation that could well be considered as endangering traditional knowledge in the sense of reducing complexity. You might say that the way people now move from their island to the town and sometimes to other places means that people are becoming ignorant of the knowledge that dwells in that place.

A number of further points about knowledge in Vanuatu are relevant for my discussion in this chapter. Commentators have described certain kinds of knowledge as being subject to a form of copyright. Copyrightable knowledge is generally what Lindstrom describes as the content of "geographic oeuvres"—that is, knowledge belonging not so much to practical matters as to the interests of social life.[22] Kirk Huffman characterizes copyrighted knowledge in north-central Vanuatu as "certain items of material culture, visual art, rituals, music, song, dance, myths and ideas."[23] Although understood to belong to certain places, in north-central Vanuatu, such forms of knowledge could be, and are, traded from one group to another. Indeed, Huffman argues that some rituals were themselves "thought to have a power and spirit of their own that urges them to get up, move to other areas, to stay there for a while and then move on. . . . The ritual does not (necessarily) disappear from its place of origin, but expands itself spiritually—through the intermediary of men."[24]

Secrets are disclosed to those who participate in some rituals. The major focus of social life in much of north and central Vanuatu was, at the time Europeans first arrived, a diverse range of status alteration systems, also known as "graded" or secret societies. In most places, men belonged to a central public graded society, and often also belonged to one or more allied secret societies.[25] In all of them, membership involved rising or moving through a series of grades or steps. In most cases, these societies focused on the exchange or killing of pigs. There were a related set of status-alteration societies for women throughout the whole region, sometimes related to pigs, but in other cases focusing on the production or presentation of plaited pandanus textiles. In both men's and women's systems, membership nearly always involved the eating of special and restricted food, and in all of them, key moments of transformation occurred through something being performed, spoken, or revealed in such a way that only the participants experienced its full impact. To know about such a ritual is to have experienced participating in it; this knowledge is absorbed bodily as well as intellectually.

I have personally participated in two such status-alteration systems for women: observing a *huhuru* in the east of Ambae island in 1992, and participating as a grade-taker in a *lengwasa* on the island of Maewo in 1997. Both *huhuru* and *lengwasa* transform the status of the women who take a grade within them. In the *huhuru*, as for so many rituals in Melanesia, the key moment of revelation involves something other than the transmission of information—it involves seeing, experiencing, and eating something spiritually powerful and significant.[26] The rituals are affective rather than informative. In different ways, both involve transformative experiences that are to do with the presence of spiritual power, mediated by physical experiences. By taking part in these rituals, women obtain a new aspect to their identity, a new way in which they are distinguished from other people. The consequence of their new knowledge is ultimately to do with relationships.[27]

A number of recent publications have emphasized the socially constructive work of ignorance and secrecy.[28] Dilley and Kirsch observe that "ignorance is not simply the absence of, or a gap in, knowledge. Ignorance is a social fact . . . [that] has generative social effects, . . . is produced in specific socio-cultural contexts and [has] political consequences."[29] Secrecy is the deployment of ignorance. Certainly, the secrets of both *lengwasa* and *huhuru* rituals facilitate contexts of emotional power that make distinctions between people. Taking part in those rituals creates enduring social differences between women based on their experiential knowledge of their participation, and those differences have permanent political significance. Both *lengwasa* and *huhuru* also make differences between women and men, but in a way that creates parallels between them, creating senior women in the same way as men's status-alteration rituals create senior men.

At the same time, a key characteristic of knowledge in many parts of Vanuatu is the idea of it as something that can be acted on, made evident in habitual use or in special contexts. For women in Ambae, a major focus of their lives is making the complex suite of plaited pandanus textiles used in exchange, as clothing, and as furnishings. As well as the ordinary textiles, there is also a category of special textiles known as *singo* that are used in ritual contexts. *Singo* are used in *huhuru* and they are also very important to the Ambae men's status-alteration society, *huqe*.

Singo are made using special distinctive techniques of plaiting and stencil dyeing. Only women who have formally obtained the right to perform those techniques by paying someone to teach them can use that knowledge. Skilled weavers can work out how to perform those special techniques using their own observation and technical ability, but knowing does not confer a right to perform them. They can make *singo* only when they have earned the right to do so by ritual and payment.[30] Women who know, but do not have the right to act on that knowledge, practice a strategic ignorance. The Ambae women's *huhuru* rituals focus on the making of these special textiles. There are a number of different "grades" in *huhuru*, based on these textiles. Once a woman has performed one *huhuru* grade, she will pretty much know exactly what happens in the others. However, just knowing does not count as performing those other *huhuru*. They still have to be performed and experienced for them to have a transformative impact on that woman's identity.

As I have observed already, across Vanuatu, such rules around how knowledge is communicated and shared apply mostly to intellectual and social knowledge such as songs, myths, genealogies, and ritual practices. By contrast, as Lindstrom remarks about Tanna, technical agricultural and economic knowledge—for example, knowledge of soil types or the productive capacities of various garden sites—is shared widely between people.[31] Ni-Vanuatu have an extensive practical knowledge of this kind, and of plants of all types. This includes knowledge of the plants they grow in their gardens as well as plants that grow wild in the bush. It includes knowledge about everyday matters such as the properties of different kinds of trees for firewood—some burn hot, some give off unpleasant vapors, and so forth. Ni-Vanuatu also hold, as the ethnobotanist Annie Walter has demonstrated, significant knowledge about how to care for fruiting trees.[32]

In the last few years, however, agricultural production in Vanuatu has started falling.[33] A significant number of young people are no longer learning, or are not being taught, the detailed knowledge of agriculture and arboriculture that was held by their parents, grandparents, and great-grandparents. As one ni-Vanuatu woman remarked to me, tartly, these days some young people "don't even know which way up to plant a yam."[34] In other words, a new kind of ignorance is developing in the form of not knowing what was previously well known. Knowledge about growing things is becoming endangered, subject to not being transmitted and thus forgotten.

Ignorance thus takes a number of different forms. There is the ignorance that results from being excluded from knowledge because it is a secret. There is the strategic ignorance of knowing but being unable to act on that knowledge or to reveal it as something one knows. Then there is the ignorance that represents having forgotten something, never having learned it, or never bothering to learn it. In all these contexts, both knowledge and ignorance are almost always generated in, and sustained within, relationships between people.

MUSEUMS AND MORAL ACTION

Questions of knowledge in relation to ethnographic collections are a key issue for museums today. Museums are a kind of technology—a very simple technology, or aggregate of technologies—that can be put to a number of different uses. At their most basic level, museums enable the collection, storage, and display of objects. The ways in which museum technologies are applied varies greatly along a number of scales—the kinds of objects displayed, the size and location of the museum, the budget at its disposal, and the messages it sets out to communicate. Museums can be used to convey not just information but also argument and ideology. The knowledge they hold and communicate can be structured in particular ways to tell different stories and make different arguments.

For ethnography collections, the way objects hold and anchor meanings has become increasingly important, as research into material culture has brought museum collections back into focus. This is partly because of changing anthropological perspectives, but it is also the result of political and social changes internationally. A key influence on the significance of ethnographic collections has been the impact of decolonization, globally.

In the postcolonial era, from the late 1970s, and especially from the 1990s, partly as a result of improving global communications, more and more small-scale communities have begun to connect with museums. In fact, museums have often become a kind of front line for communities, both for those seeking to engage with the ethnographic documentation of their knowledge and practice and for those seeking to pursue issues around their identity and autonomy. Several now-famous exhibitions and publications—for example, the 1984 Te Maori exhibition in New York,[35] or James Clifford's much-cited 1997 essay, "Museums as Contact Zones"[36]—have introduced the idea that the communities from whose predecessors the objects were collected have an investment in those collections and a series of rights in relation to them. Terms such as *traditional owners* and *source community* have gained significant influence inside the museum anthropology profession. These terms have particular importance for indigenous communities in settler states who have lost control of their land, and whose identity is thus fragilely constituted by connections they formerly had to their places. In these contexts, museum collections provide a second ground for identity formation. Through them, a community can become—in some sense—the people of the objects, as well as, or even rather than, the people of the place.

A key text edited by Laura Peers and Alison Brown sets out some of the thinking behind this movement. Peers and Brown observe:

> During the great age of museum collecting which began in the mid-nineteenth century, . . . [the relationship between museums and source communities] was a one-way relationship: objects

and information about them went from peoples all over the world into museums, which then consolidated knowledge as the basis of curatorial and institutional authority.... Within this context, ethnographic collections, in particular, were built up on the premise that the peoples whose material heritage was being collected were dying out, and that the remnants of their cultures should be preserved for future generations.... In recent years, however, the nature of these relationships has shifted to become a much more two-way process, with information about historic artefacts now being returned to source communities, and with community members working with museums to record their perspectives on the continuing meanings of those artefacts.[37]

This kind of revaluation has rarely arisen around anthropological research findings. With some exceptions, communities have not sought access to the field notes made by anthropologists who have worked with them. Objects, unlike texts, have a particular significance and power for source communities. They provide a very immediate connection to the past generations who made them, offer insight into past skills and knowledge of material resources, provide evidence enabling historical reconstructions, and represent cultural identity.[38] Furthermore, objects provide opportunities for emotional engagement, ceremony, speeches, and songs. Information by itself, written down in field notes or even published in ethnographies, does not make such a direct link to past generations.

A lot of the thinking and a lot of the activity around relationships between museums and source communities has developed in postcolonial settler states, such as Australia, New Zealand, Canada, and the United States, where the museums and communities are physically within reach of each other, and where the state and some of its citizens see museums as offering a way to negotiate moral issues around colonial injustices. Strategies of engagement, consultation, reconciliation, and restitution have been developed around collections and exhibitions. In some cases, museums have ceded ownership of certain categories of objects to communities by returning them or by establishing principles of joint ownership. In most museums in these countries, the perspectives of communities are regularly taken into account when decisions about which items can be put on display and what can be said about those objects are made. In this context, "community" is a locally defined category and can refer to a language group, tribe, or coresident group, represented either formally or informally by a member or members of that group.

In settler states, museums have become an important context in which indigenous communities distinguish themselves from the society that surrounds them and affirm their distinctive identity, making use of the collections in which they have a stake as traditional owners or custodians. The connections people make to collections can be very important for them in constituting their own identity, building their sense of connection to their elders and predecessors.

For settler states, museums are often an equally useful context in which to negotiate the historical maltreatment of indigenous peoples, the loss of land and autonomy. In Australia, for example, it is a fairly widespread popular opinion, albeit an inaccurate one, that all the Aboriginal and Torres Strait Islander collections in museums were stolen and should be returned—as their land was stolen and is not being returned. I am not suggesting that museums are exempt from the colonial project and its injustices, but rather that it has become common to make museums carry a significant load of the guilt and assume responsibility for a significant proportion of the reparation for the iniquities of colonialism. Museums have become a venue for a kind of moral action; they are being asked to acknowledge past wrongs, to make recompense or at least to demonstrate recognition of what has previously been denied. This is part of what the British historian Sharon Macdonald has described as "the international difficult histories boom."[39] Bain Attwood, writing about the National Museum of Australia, argues that the international difficult histories boom has been especially marked in settler societies.[40]

Attwood asserts that museums have an important role to play in this because the difficult histories movement places a premium on sentimental feeling. Citing Sharon Macdonald, he observes that museum exhibitions enable affective encounters that "are perceived by many as more authentic than narratives presented in the form of disembodied words."[41] In other words, Attwood does not identify the existence of museum collections in and of themselves, but rather the museological technology of affective communication through exhibitions utilizing objects, as a crucial element of a nation's moral work of reparation. The affective communication he is speaking about here is not dissimilar to that which occurs in rituals such as *huhuru* or *lengwasa*, where communication occurs in the context of an immersive experience.

Three issues, in particular, have stood out in community engagement with museums over the last several decades. The first is the issue of human remains—in just about every settler state context, indigenous communities have actively sought for the human remains kept in museum collections to be given back to them. Having dealt with these issues nationally, there has also been a move to secure the return of human remains that are held internationally. In these cases, repatriation claims have generally been made with the support of the relevant national governments. Australia and New Zealand, for example, both have nationally funded bodies that are charged with the responsibility of negotiating the return of human remains held in collections overseas. These claims are almost always focused on remains held in public collections.

The second issue relates to secret/sacred or restricted objects. Many source communities, again, especially in settler states such as Australia and Canada, have asked museums to restrict public access to certain categories of objects in their collections.

These are objects that only certain people within their communities are permitted to see—usually only senior men. The rationale for such requests and exclusions is that within the source communities' framework of knowledge and practice, certain objects are restricted and should not be seen by anyone except for specific categories of people within those societies. In negotiations with museums, communities have often been able to assert their identity and their connection to collections by reimposing secrecy rules on the objects. In some cases, there have also been sustained campaigns for the return of sacred objects from national museums. Claims for restricted objects held internationally have not received similar levels of governmental support and are much less frequently given priority in national policies. This is a much more complicated issue, as some communities do not want these sacred objects returned.

The third issue is the matter of access to collections overall. Members of source communities take pleasure and interest in visiting collections and often ask for the opportunity to perform small ceremonies of respect to the objects—speaking to them, praying or singing in their presence, making offerings, and sometimes holding ceremonies to address spiritual presences attached to the objects, such as ancestral spirits. Often very deeply felt, these ceremonies also act as an assertion of connection to the objects and an affirmation of identity for the community members. The rituals enable the participants to confirm their identity in contrast to curatorial staff and society at large, in a similar way that participating in a *huhuru* or a *lengwasa* ritual enables women to distinguish themselves in relationship to others. Museum collections thus become a useful context for a form of ritually effected self-definition for communities. Indeed, this is a new use for the objects in question, enabling people to modify or enhance their identity in relation to collections in the same way that women on Ambae modify and enhance their identity by performing the *huhuru* rituals in relation to *singo*—those special plaited pandanus textiles.

Both museums and communities have sought to establish relationships in which these issues can be negotiated. Curators find themselves acting personally, while officially representing not just the museum but also the nation-state. Significantly, this has transformed the locus of knowledge in relation to those collections. In the past, the curator was considered to be the expert who knew the most about a collection, but now curators often practice a strategic ignorance in relation to objects, deferring to the knowledge of community members. This can be a genuine ignorance, of course, but there can also be instances where a curator does not contradict the assertion of a community member, even if the curator holds different information about an object. In other cases, it is possible for both to share the knowledge they have, joining it together.

As practices of community engagement have become established and gained traction in ethnographic museum practice, it has also become common for museums to consult communities about the ways in which their culture is represented in displays. Again, this is especially the case in settler states but also increasingly a practice of museums internationally.

All of these developments, including President Macron's initiative, are operating within a context of moralized behavior. Museums are no longer discussed as if they were morally neutral, but rather are regarded and treated as actors that are subject to moral analysis and criticism. Thus, Annie Coombes and Ruth Phillips, introducing a major edited collection on museum transformations, trenchantly criticize the British Museum's Africa Gallery for political inconsistency, for remaining silent about how the museum acquired objects that are now the subject of repatriation claims, and for other "significant occlusions."[42] In other words, they are criticizing the museum for failing to do the moral work that they consider museums should do. Bénédicte Savoy adopts a similar moral and political perspective when she asks how much blood is dripping from each artwork.

DIFFERENT ATTITUDES TO COLLECTIONS

As Paul Basu observes, many of the case studies used to discuss the relations between museums and source communities have been drawn from contexts where there is a high degree of museum awareness.[43] Basu, introducing his Sierra Leone collection digitization project, characterizes Sierra Leone as a place where "indigenous activism has not politicized cultural heritage . . . and where there is little awareness of the cultural materials dispersed in museum collections throughout the world or the possible connections contemporary communities may have with them."[44] He argues that it is not only the case that Sierra Leone is a place where communities are not preoccupied with museum collections. Sierra Leone is also a place where the concept of source community is not "the most adequate or appropriate" way to conceptualize relationships with dispersed collections, given that in West Africa "ethnic identities and territorial boundaries are . . . highly fluid, situational and in a constant process of renegotiation and change."[45] As Basu's discussion makes clear, not everybody from communities whose objects have been collected is seeking a specific relationship with those collected objects.

In fact, outside of settler states, source communities have demonstrated a wide variety of attitudes to collections. To cite another example, in the early 1980s the Australian Museum in Sydney undertook a collaborative project with members of the Abelam community from Apangai village, north of the Sepik River in Papua New Guinea, mediated by the anthropologist Diane Losche. Losche made a field collection in the Abelam region, and then two senior men from Apangai, Nera and Narikowi, visited Sydney for some weeks in 1982 and helped to build an Abelam men's cult house—a *haus tambaran*—in a large gallery in the Australian Museum. The whole gallery was about the Abelam, and the *haus tambaran* was its main feature. Inside the house, Nera and Narikowi installed a display similar to what would be on display inside a *haus tambaran* to male initiates. Losche reported that when she discussed this project with the Abelam, they said they were happy for women and uninitiated men in Sydney—the general public—to go inside the *haus tambaran*, as long as their own women and uninitiated men could not do so.[46] Thus, the restriction they placed on this material demonstration of their knowledge and understanding of the world was only applied within the context of their relationships with each other, with other Abelam people. They considered it important to control access to knowledge—the specific experience of entering the house—within their community, but did not at that time feel any need to control that access for other people. In other words, they were concerned about controlling knowledge—knowledge-as-experience—only within the matrix of their relationships, not outside of it.

From 2005 to 2010 the British Museum instituted a project seeking to reconnect people in Melanesia with the British Museum Melanesian collections.[47] We brought people to London to visit the collections in the storeroom, and we took photographs and, sometimes, video footage of objects in the collection back to relevant communities across Melanesia. Not everyone was interested in seeing the material we had taken to show people. I have often shown photographs of collected objects to people somewhere in Vanuatu. People were often interested to spend a few hours looking at the photographs and talking to me about the objects represented in them, but they often got up from the encounter to get back to cutting copra or dealing with small children, without developing any further interest. Nevertheless, such occasions did often enable the museum to better document the objects being discussed.

One person who came to London with the Melanesia Project was my ni-Vanuatu colleague Jean Tarisesei, from Ambae. She was deeply pleased to see the plaited pandanus textiles from Ambae in the museum and somewhat amazed that so many of them should belong to the special category I mentioned earlier, *singo*. On seeing the textiles, Tarisesei reflected on the restrictions over making *singo* and recalled a story about those restrictions, which she subsequently included in an essay she wrote for the Melanesia Project. She wrote, "When I was little, my father's adopted mother got sick and died because of her knowledge and great skill in making *singo*," going on to describe how that knowledge had caused someone to be jealous of her and to poison her.[48] She commented, "For Ambaeans, both men and women, *singo* is very important. It gives a person an identity, a place to stand in society."⁴⁹ These comments make it clear that the possession of knowledge, skill, and the right to deploy them was powerful and significant in Ambae in the past. Tarisesei, secure in her place-based Ambaean identity, was not concerned about establishing her identity in relation to the collection, but rather in learning from it.

Tarisesei commented further that today only a few women on Ambae have the right to make *singo*, and that their number is declining. She wanted women on Ambae to see the quality of the *singo* in the British Museum collection, wishing that she could take one back to show them.⁵⁰ She wanted to return the knowledge physically embodied in that object to those people who could by studying it improve their skills in plaiting new textiles. At the same time, she made no objection to the fact that the British Museum had collected those textiles. Tarisesei did not donate an object to the museum, but other visitors to collections sometimes do. They want to add something to sit alongside objects from their community, or sometimes, they want to ensure that their community is represented in the museum. Likewise, people seeing images of objects in the collection sometimes send objects to represent themselves.⁵¹

REFLECTIONS

Even though ethnographic museums and collections are now being used as a vehicle to make reparations for past injustices, there is great diversity in the specific details of each case. A single or easy solution does not exist. More to the point, although connecting source communities to collections is often a rewarding experience for the community, that reconnection does not solve every issue. The Maori curator Paul Tapsell made that point some time ago when he commented that for his tribe, resolving the ownership of their objects in New Zealand museums would not become a priority until other more important issues, such as land ownership, had been sorted out.⁵² Although the repatriation or long-term loan of African collections to the places from which they come might be welcomed in some parts of that continent, returning objects will not ensure a forgetting of past wrongs. It will not necessarily even ensure that people thereafter practice a strategic ignorance of past events. In fact, the moral work of reparation that is so often assigned to museums in contemporary Western practice requires the creation and maintenance of ongoing relationships. These relationships are often sustained by the presence of the objects in the very museums accused of holding them inappropriately.⁵³

It could be said that the move by source communities to lay claim to objects in museums is based on a concept of ownership—not just a moral ownership of the objects, but specifically an ownership of a knowledge about those objects that enables the knowledge owner to claim specific rights over the objects. In fact, interestingly, this process of laying claim is a process of making ownable knowledge that previously was constrained not by ownership per se but often by the right to participate in an experience through which the knowledge—broadly defined—was obtained, often in a place- and status-determined context. Both intellectual and social knowledge as well as everyday practical knowledge was, and is, held unevenly within most small-scale communities, generally framed by a shared understanding of which knowledge may and which knowledge may not be acted on by any individual. Arcadia's emphasis on open access and on making knowledge accessible is made at a time when some communities are formalizing the ownership of knowledge in the broader context of international academic interest, and especially the digital realm. The control of knowledge is, in this sense, sometimes a response to this wider context of knowing, a wider set of relationships, and thus sometimes also relates to the politics of indigenous rights. Open-access digitization is not always welcome in this kind of context of control. This means that restricted knowledge cannot be documented for EMKP, although the program does allow a small proportion of documented information to remain closed within the archive.

The Endangered Material Knowledge Project provides an opportunity to document objects in places where there are ongoing transformations in the ways that people now live, and where the pace of forgetting is often increasing. At the same time, the program draws attention to some of the complexities of knowing and owning objects in museums. If there are movements to use objects to make reparation for past colonial wrongs and to establish identities through ownership, EMKP potentially contributes to those by providing more information about what those objects are and have been. In addition, the project acknowledges and celebrates the richness and diversity of human knowledge and practice, keeping that richness in sight, where appropriate, by recording it. The best outcome, it seems to me, happens when the process of documenting objects reminds people about them and encourages the local transmission of knowledge and the strengthening of local knowledge and practice.

Notes

1. "Endangered Material Knowledge Programme (EMKP)," British Museum, accessed January 27, 2022, https://www.britishmuseum.org/our-work/departments/africa-oceania-and-americas/endangered-material-knowledge-programme and https://www.emkp.org/.

2. Anna Codrea-Rado, "Emmanuel Macron Says Return of African Artifacts Is a Top Priority," *New York Times*, November 29, 2017, https://www.nytimes.com/2017/11/29/arts/emmanuel-macron-africa.html.

3. Bénédicte Savoy's project, "Translocations" was based at the Technische Universität Berlin. The project studied large-scale displacements of cultural assets from antiquity to the twentieth century.

Bénédicte Savoy, "Leibniz-Project Cluster Translocations," December 2016, https://www.kuk.tu-berlin.de/fileadmin/fg309/bilder/Forschungsprojekte/Translocations_ENGLISH_WEISS_FINAL.pdf.

4. Kate Brown, "'The Idea Is Not to Empty Museums': Authors of France's Blockbuster Restitution Report Say Their Work Has Been Misrepresented," ArtNet News, January 24, 2019, https://news.artnet.com/art-world/restitution-report-critics-1446934. See also Nicola Kuhn, "Berlins verfluchte Schätze," *Der Tagesspiegel*, February 15, 2018, https://www.tagesspiegel.de/berlin/koloniale-raubkunst-berlins-verfluchte-schaetze/20944002.html.

5. Roy Dilley and Thomas G. Kirsch, eds., *Regimes of Ignorance: Anthropological Perspectives on the Production and Reproduction of Nonknowledge* (New York: Berghahn Books, 2015), 10.

6. Jonathan Mair, Ann H. Kelly, and Casey High, "Introduction: Making Ignorance an Ethnographic Object," in *The Anthropology of Ignorance: An Ethnographic Approach*, ed. Casey High, Ann H. Kelly, and Jonathan Mair (New York: Palgrave Macmillan, 2012), 1.

7. The United Nations's *Declaration on the Rights of Indigenous Peoples* includes the right to maintain and strengthen cultures and traditions. See United Nations, General Assemby, *Declaration on the Rights of Indigenous Peoples*, 61/295 (September 13, 2007), https://www.un.org/development/desa/indigenouspeoples/wp-content/uploads/sites/19/2018/11/UNDRIP_E_web.pdf.

8. Bronislaw Malinowski, *Argonauts of the Western Pacific: An Account of Native Enterprise and Adventure in the Archipelagos of Melanesian New Guinea* (London: Routledge, 1922), 105.

9. Jeremy Coote and Anthony Shelton, *Anthropology and Aesthetics* (Oxford: Clarendon Press, 1992); Alfred Gell, *Art and Agency: An Anthropological Theory* (Oxford: Clarendon Press, 1998); Pierre Lemonnier, *Mundane Objects: Materiality and Non-Verbal Communication* (Walnut Creek, CA: Left Coast Press, 2012).

10. Lemonnier, *Mundane Objects*, 19.

11. Lemonnier, 21–62.

12. Arcadia Fund, accessed September 29, 2022, https://www.arcadiafund.org.uk/.

13. Lisbet Rausing and Peter Baldwin, introduction to *From Dust to Digital: Ten Years of the Endangered Archives Programme*, ed. Maja Kominko (Cambridge: Open Book Publishers, 2015), xxxviii. There is a series of questions to ask about the effect of preserving knowledge in a digital environment, which are very real, but which I do not address in this chapter.

14. Vanuatu 2020 Population & Housing Census Key Indicators Table 1, Vanuatu National Statistics Office, November 17, 2021, https://vnso.gov.vu/index.php/en/census-and-surveys/census/2020populationhousingcensus.

15. *Ni-Vanuatu*, a term meaning "of Vanuatu," is used to refer to the nation's citizens. It is mostly used to refer to the majority of the population, descended from precolonial inhabitants, and is rarely used to refer to the handful of naturalized citizens in the country.

16. Lamont Lindstrom, *Knowledge and Power in a South Pacific Society* (Washington, DC: Smithsonian Institution, 1990), 79–81; John Patrick Taylor, "Ways of the Place: History, Cosmology and

Material Culture in North Pentecost, Vanuatu" (PhD diss., Australian National University, 2003); Lissant Bolton, *Unfolding the Moon: Enacting Women's "Kastom" in Vanuatu* (Honolulu: University of Hawai'i Press, 2003), 67–77; Lissant Bolton, "Describing Knowledge and Practice in Vanuatu," in *Social Movements, Cultural Heritage and the State in Oceania*, ed. Edvard Hviding and Knut M. Rio (Oxford: Sean Kingston, 2011), 301–319.

17. Taylor, "Ways of the Place," 114–115.

18. Joël Bonnemaison, *The Tree and the Canoe: History and Ethnogeography of Tanna*, trans. and ed. Josée Pénot-Demetry (Honolulu: University of Hawai'i Press, 1994), 323.

19. Joël Bonnemaison, "The Tree and the Canoe: Roots and Mobility in Vanuatu Societies," *Pacific Viewpoint* 25, no. 2 (1984): 118.

20. Lindstrom, *Knowledge and Power*, 80.

21. Lindstrom, 80.

22. Lindstrom, 64.

23. Kirk W. Huffman, "Trading, Cultural Exchange and Copyright: Important Aspects of Vanuatu Arts," in *Arts of Vanuatu*, ed. Joël Bonnemaison, Kirk Huffman, and Darrell Tryon (Bathurst, Australia: Crawford House, 1996), 182.

24. Huffman, 190.

25. For example, Peter Blackwood has published a survey of male status alteration systems. See Peter Blackwood, "Rank, Exchange and Leadership in Four Vanuatu Societies," in *Vanuatu: Politics, Economics and Ritual in Island Melanesia*, ed. Michael R. Allen (Sydney: Academic Press, 1981).

26. Lissant Bolton, "Classifying the Material: Food, Textile and Status in North Vanuatu," *Journal of Material Culture* 6, no. 3 (2001): 251–268.

27. See also James Leach, "An Aesthetics of Knowledge: Relations and the Documentation of Traditional Knowledge in Papua New Guinea" (chapter 7 in this volume).

28. Ilana Gershon and Dhooleka Sarhadi Raj, "Introduction: The Symbolic Capital of Ignorance," *Social Analysis: The International Journal of Social and Cultural Practice* 44, no. 2 (2000): 3–14; Casey High, Ann H. Kelly, and Jonathan Mair, eds., *The Anthropology of Ignorance: An Ethnographic Approach* (New York: Palgrave Macmillan, 2012).

29. Dilley and Kirsch, *Regimes of Ignorance*, 15.

30. Bolton, *Unfolding the Moon*, 137.

31. Lindstrom, *Knowledge and Power*, 58, 64.

32. Annie Walter, "Knowledge for Survival: Traditional Tree Farming in Vanuatu," in *Science of Pacific Island Peoples*, vol. 3, *Fauna, Flora, Food and Medicine*, ed. John Morrison, Paul A. Geraghty, and Linda Crowl (Suva, Fiji: Institute of Pacific Studies, the University of the South Pacific, 1994),

33. Ralph Regenvanu, personal communication with author, 2015.

34. Marta Yamsiu, personal communication with author, 2015.

35. See Peter Brunt, "Decolonisation, Independence and Cultural Revivial 1945–89," in *Art in Oceania: A New History*, ed. Peter Brunt et al., (London: Thames and Hudson, 2012), 383; Nicholas Thomas, *The Return of Curiosity: What Museums Are Good for in the 21st Century* (London: Reaktion Books, 2016), 136.

36. James Clifford, "Museums as Contact Zones," in *Routes: Travel and Translation in the Late Twentieth Century* (Cambridge, MA: Harvard University Press, 1997), 188–219.

37. Laura Peers and Alison K. Brown, eds., *Museums and Source Communities: A Routledge Reader* (London: Routledge, 2003), 2.

38. Peers and Brown, 6.

39. Sharon Macdonald, "Post-national Museums?" (Paper presented at the National Museums in a Transnational Age Conference, Monash University, Prato, Italy, November 1–4, 2009), 1, quoted in Bain Attwood, "The International Difficult Histories Boom, the Democratization of History, and the National Museum of Australia," in *International Handbooks of Museum Studies*, vol. 4, *Museum Transformations*, ed. Annie E. Coombes and Ruth Phillips (Oxford: Blackwell, 2015), 61.

40. Attwood, "Difficult Histories Boom," 62.

41. Attwood, 62.

42. Annie E. Coombes and Ruth B. Philips, "Introduction: Museums in Transformation: Dynamics of Democratization and Decolonization," in Coombes and Phillips, *Museum Transformations*, xxxvii.

43. Paul Basu, "Reanimating Cultural Heritage: Digital Curatorship, Knowledge Networks and Social Transformation in Sierra Leone," in Coombes and Phillips, *Museum Transformations*, 337.

44. Basu, 338.

45. Basu, 347, 346.

46. Diane Losche, personal communication, ca. 1982.

47. This was not an internal British Museum project, but was undertaken with Professor Nicholas Thomas of the Cambridge Museum of Archaeology and Anthropology. See Lissant Bolton et al., eds., *Melanesia: Art and Encounter* (London: British Museum Press, 2013).

48. Jean Tarisesei, "Singo: Textiles from Our Island, Ambae," in Bolton et al., *Melanesia*, 280.

49. Tarisesei, 280.

50. Tarisesei, 280.

51. Paraka of Mondika Tribe, "The Apron Is the Sister of the Bilum," in Bolton et al., *Melanesia*, 111.

52. Paul Tapsell, "Afterword: Beyond the Frame," in Peers and Brown, *Museums and Source Communities*, 246.

53. Lissant Bolton, "An Ethnography of Repatriation: Engagements with Erromango, Vanuatu," in Coombes and Phillips, *Museum Transformations*, 229–248.

Bibliography

Attwood, Bain. "The International Difficult Histories Boom, the Democratization of History, and the National Museum of Australia." In Coombes and Phillips, *Museum Transformations*, 61–83.

Basu, Paul. "Reanimating Cultural Heritage: Digital Curatorship, Knowledge Networks and Social Transformation in Sierra Leone." In Coombes and Phillips, *Museum Transformations*, 337–364.

Blackwood, Peter. "Rank, Exchange and Leadership in Four Vanuatu Societies." In *Vanuatu: Politics, Economics and Ritual in Island Melanesia*, edited by Michael R. Allen, 35–84. Sydney: Academic Press, 1981.

Bolton, Lissant. "Classifying the Material: Food, Textile and Status in North Vanuatu." *Journal of Material Culture* 6, no. 3 (2001): 251–268.

Bolton, Lissant. "Describing Knowledge and Practice in Vanuatu." In *Social Movements, Cultural Heritage and the State in Oceania*, edited by Edvard Hviding and Knut M. Rio, 301–320. Oxford: Sean Kingston, 2011.

Bolton, Lissant. "An Ethnography of Repatriation: Engagements with Erromango, Vanuatu." In Coombes and Phillips, *Museum Transformations*, 229–248.

Bolton, Lissant. *Unfolding the Moon: Enacting Women's "Kastom" in Vanuatu*. Honolulu: University of Hawai'i Press, 2003.

Bolton, Lissant, Nicholas Thomas, Elizabeth Bonshek, Julie Adams, and Ben Burt, eds. *Melanesia: Art and Encounter*. London: British Museum Press, 2013.

Bonnemaison, Joël. *The Tree and the Canoe: History and Ethnogeography of Tanna*. Translated and adapted by Josée Pénot-Demetry. Honolulu: University of Hawai'i Press, 1994.

Bonnemaison, Joël. "The Tree and the Canoe: Roots and Mobility in Vanuatu Societies." *Pacific Viewpoint* 25, no. 2 (1984): 117–151.

Brown, Kate. "'The Idea Is Not to Empty Museums': Authors of France's Blockbuster Restitution Report Say Their Work Has Been Misrepresented." ArtNet News, January 24, 2019. https://news.artnet.com/art-world/restitution-report-critics-1446934.

Brunt, Peter. "Decolonisation, Independence and Cultural Revival 1945–89." In *Art in Oceania: A New History*, edited by Peter Brunt, Nicholas Thomas, Sean Mallon, Lissant Bolton, Diedre Brown, Damian Skinner, and Susanne Kucheler, 348–383. London: Thames and Hudson, 2012.

Clifford, James. "Museums as Contact Zones." In *Routes: Travel and Translation in the Late Twentieth Century*, 188–219. Cambridge, MA: Harvard University Press, 1997.

Coombes, Annie E., and Ruth Phillips. "Introduction: Museums in Transformation: Dynamics of Democratization and Decolonization." In Coombes and Phillips, *Museum Transformations*, 21–51.

Coombes, Annie E., and Ruth Phillips, eds. *Museum Transformations*. Vol. 4 of *The International Handbooks of Museum Studies*, edited by Sharon Macdonald and Helen Rees Leahy. Oxford: Blackwell, 2015.

Coote, Jeremy, and Anthony Shelton. *Anthropology and Aesthetics*. Oxford: Clarendon Press, 1992.

Codrea-Rado, Anna. "Emmanuel Macron Says Return of African Artifacts Is a Top Priority." *New York Times*, November 29, 2017. https://www.nytimes.com/2017/11/29/arts/emmanuel-macron-africa.html.

Dilley, Roy, and Thomas G. Kirsch, eds. *Regimes of Ignorance: Anthropological Perspectives on the Production and Reproduction of Nonknowledge*. New York: Berghahn Books, 2015.

Gell, Alfred. *Art and Agency: An Anthropological Theory*. Oxford: Clarendon Press, 1998.

Gershon, Ilana, and Dhooleka Sarhadi Raj. "Introduction: The Symbolic Capital of Ignorance." *Social Analysis: The International Journal of Social and Cultural Practice* 44, no. 2 (2000): 3–14.

High, Casey, Ann H. Kelly, and Jonathan Mair, eds. *The Anthropology of Ignorance: An Ethnographic Approach*. New York: Palgrave Macmillan, 2012.

Huffman, Kirk W. "Trading, Cultural Exchange and Copyright: Important Aspects of Vanuatu Arts." In *Arts of Vanuatu*, edited by Joël Bonnemaison, Kirk Huffman, and Darrell Tryon, 182–194. Bathurst, Australia: Crawford House, 1996.

Kirsch, T. G., and Roy Dilley. "Regimes of Ignorance: An Introduction." In Dilley and Kirsch, *Regimes of Ignorance*, 1–29.

Kuhn, Nicola. "Berlins verfluchte Schätze." *Der Tagesspiegel*, February 15, 2018. https://www.tagesspiegel.de/berlin/koloniale-raubkunst-berlins-verfluchte-schaetze/20944002.html.

Lemonnier, Pierre *Mundane Objects: Materiality and Non-Verbal Communication*. Walnut Creek, CA: Left Coast Press, 2012.

Lindstrom, Lamont. *Knowledge and Power in a South Pacific Society*. Washington, DC: Smithsonian Institution, 1990.

Macdonald, Sharon. "Post-national Museums." Paper presented at the National Museums in a Transnational Age Conference, Monash University, Prato, Italy, November 1–4, 2009.

Mair, Jonathan, Ann H. Kelly, and Casey High. "Introduction: Making Ignorance an Ethnographic Object." In High, Kelly, and Mair, *Anthropology of Ignorance*, 1–32.

Malinowski, Bronislaw. *Argonauts of the Western Pacific: An Account of Native Enterprise and Adventure in the Archipelagos of Melanesian New Guinea*. London: Routledge, 1922.

Paraka of Mondika Tribe. "The Apron Is the Sister of the Bilum." In Bolton et al., *Melanesia*, 111.

Peers, Laura, and Alison K. Brown, eds. *Museums and Source Communities: A Routledge Reader*. London: Routledge, 2003.

Rausing, Lisbet, and Peter Baldwin. Introduction to *From Dust to Digital: Ten Years of the Endangered Archives Programme*, edited by Maja Kominko, xxxvii–xxxviii. Cambridge: Open Book Publishers, 2015.

Tapsell, Paul. "Afterword: Beyond the Frame." In Peers and Brown, *Museums and Source Communities*, 242–251.

Tarisesei, Jean. "Singo: Textiles from Our Island, Ambae." In Bolton et al., *Melanesia*, 278–280.

Taylor, John Patrick. "Ways of the Place: History, Cosmology and Material Culture in North Pentecost, Vanuatu." PhD diss., Australian National University, 2003.

Thomas, Nicholas. *The Return of Curiosity: What Museums Are Good for in the 21st Century*. London: Reaktion Books, 2016.

Walter, Annie. "Knowledge for Survival: Traditional Tree Farming in Vanuatu." In *Science of Pacific Island Peoples*. Vol. 3, *Fauna, Flora, Food and Medicine*, edited by John Morrison, Paul A. Geraghty, and Linda Crowl, 189–200. Suva, Fiji: Institute of Pacific Studies, the University of the South Pacific, 1994.

11

A READER'S GUIDE TO *OWNERSHIP OF KNOWLEDGE*: DIAGRAMMATIC CHAPTER

Vivek S. Oak, Jörn Oeder, and Annapurna Mamidipudi

This chapter aims to clarify the processes of manipulations that are involved in splitting and fixing the kn/own/able. In the first part, we trace the sequence of movements in time and space that allows the *illusion* of the separation of knowables from ownables to exist, both temporally and spatially—that is, the splitting and fixing. Next, we show the step-by-step construction of a new framework, *the grid*, as an analytical tool to study the cases in this book. Finally, we offer a how-to guide to unveil efforts to fix this split, which we show through the application of these analytical methods to four distinct cases—as seen in the chapters by Jackson, Leach, Bolton, and Slaton. This step-by-step guide unveils the major characteristic of our modern regime of knowledge ownership—in which science and technology define the highest echelon of reliable knowing, and law defines the dominant form of rightful owning. This regime is a fragmentation of knowing and owning carried out by actors to carve out knowables and ownables in the domains of epistemology, society, and economy.

SPLITTING

From the viewpoint of a complex reality of kn/own/ables, the fragmentation—or *false* pluralization—of the kn/own/able is a supplementary but *illusory* reality generated by "splitting" the kn/own/able into the knowable and the ownable. The split allows actors to set events on a causal, unidirectional time. This is different from the reality of a plural, nonlinear, and multidimensional universe of ownership of knowledge. What is illusory about this reality is not the actual power of exerting ownership of knowledge through science and law; it is the notion that applying property rights to knowledge could be the prime, or even *only* method in our modern world that can tie knowing and owning together in a "fair" and "just" way, and that it can do this while sustaining and expressing the plurality and selfhood of the knower and enabling society at the same time to access this plurality. This illusion is necessary to obscure the underlying paradoxical nature of owning knowledge—namely, that it is inseparable from knowing—and of the kn/own/able that exists in a continuous and dynamic "back and forth" with the un-kn/own/able.

THE OPERATION OF SPLITTING

The three triads Different practices, instantiations, and domains of the kn/own/able are identified separately (see the triads at the top of figure 11.1), so that actors can activate one or the other. Once treated as separate, they can be made to work hierarchically rather than on equal terms. This is different from Cook Ding's world, where he can legitimately assert having—or owning—knowledge because he is able to employ all three practices simultaneously. Any ownership claim that consists of fewer practices not only is a minor form of knowing or not-knowing, it also markedly results in power hierarchies.

Temporal split (step 1) Time comes in. Owing to the fact that the distinct practices of naming, performance, and use can be actuated asynchronously (nonsimultaneously) on words, bodies, and objects, it is possible for actors to create a linear timeline of distinct moments when each practice acts on a material instantiation as a practice of either knowing or owning (as indicated by moments T_1, T_2, and T_3 in figure 11.1)—for example, defining the moment when an actor invents a formula as a moment of knowing, and the moment of receiving the patent as the moment of owning that knowledge. This is different from the situation of Cook Ding, whose acquiring of the knowledge of "the Way" is not fixed in distinct moments of singular practice, but always fluidly moving between all three and always needing all three to be valid and legitimate.

Spatial split (step 2) Space and materiality come in. As words, bodies, and objects are in fact separate entities, actors can identify particular instantiations exclusively as either a knowable or an ownable. The spatial split of words, bodies, objects is made into a splitting of knowing and owning. For example, the formula is knowable and the patent is ownable.

Actors employ this spatial [or material] split, now abstracted along the timeline, by privileging different moments—as moments of owning or knowing a particular material instantiation—to make claims or manipulate knowledge ownership.

FIXING

We use the word *fixing* to indicate a process in which an attempt is made to staple the seemingly irreversibly fragmented kn/own/able [separated as knowable and ownable]

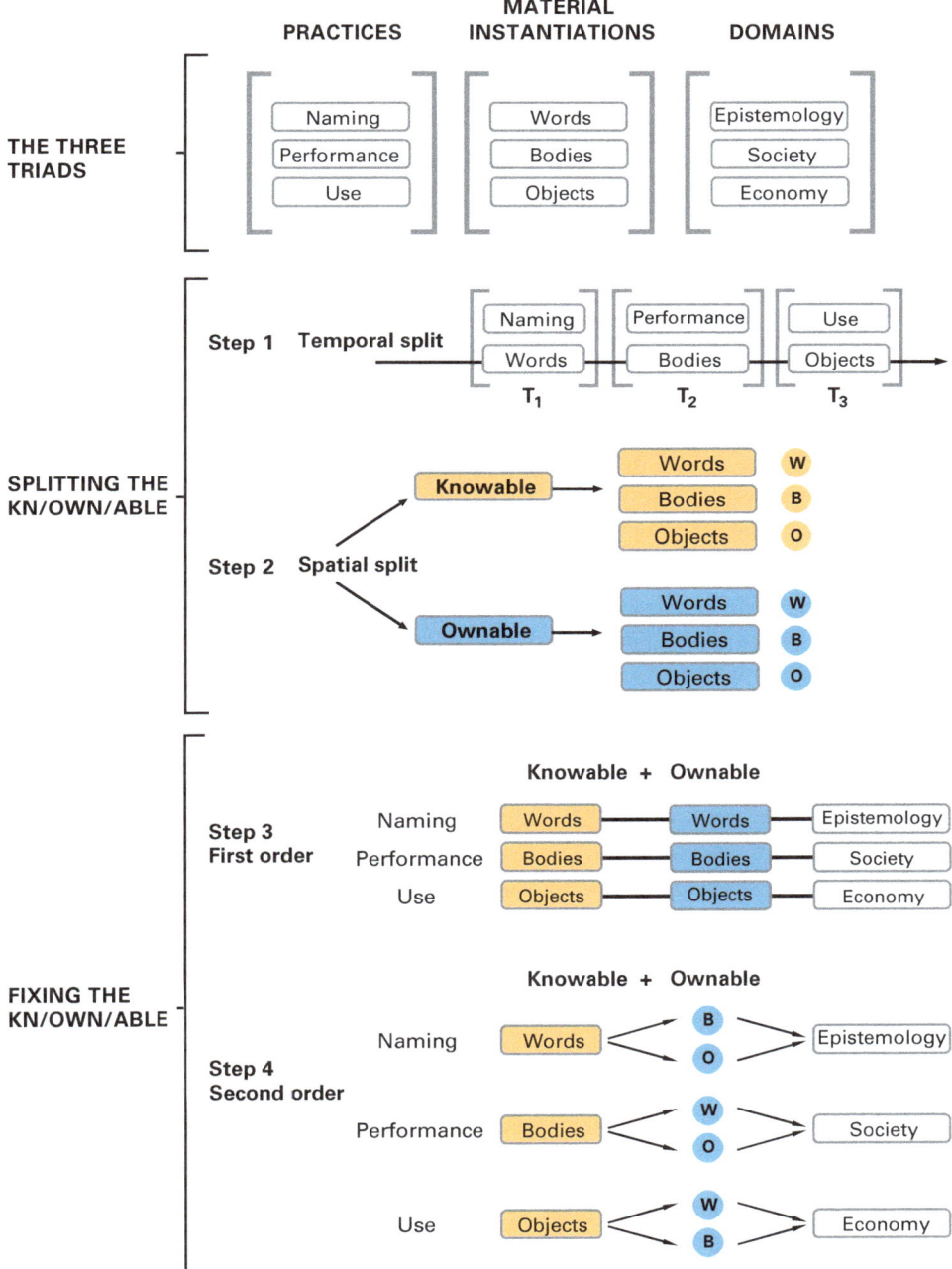

Figure 11.1
The spatial and temporal splits and how they are fixed.

back together again. In order to do so, the actor will need to once again fix knowable and ownable together in material instantiations that they can use to claim knowledge ownership. The emphasis on "seemingly" is important simply because actors identify ownership in social, material, and epistemic domains as irreversibly fragmented because it is this operation that enables them to manipulate the ownership of knowledge.

Between the three practices of naming, performance, and use and the three material instantiations, nine possibilities of fixing the ownership of knowledge can exist.

The first order is about the three primary relations between practices and materials—that is, in a first-order relation, an actor can employ a single practice in order to assert ownership and will only need one (see figure 11.2). Words can be named, bodies can perform, and objects can be used in order to claim ownership. This first order affords the most stable fixing of knowing and owning, and among the first-order relations of material practice-instantiations to claim knowledge ownership, the performing body is the most important one because there cannot be any human action without a performing body. The stability of the first order is primarily a result of practice and material instantiation for both knowing and owning being the same in these cases; and to a lesser degree, the affinity of the practice to the material instantiation is defined or perceived as exclusive, so that naming concerns only words and performing concerns only bodies. The fixing addresses the most obvious manipulation of combining knowing and owning to fix ownership domains—that is, through naming words to own knowledge in the domain of epistemology, through performing bodies to own in the domain of society, and through using objects to own in the domain of economy.

We introduce the remaining six relations as second-order relations. In each of these, actors assert ownership by employing more than one practice—for example, naming a performing body "inventor" or "tanner." Finally, the domain of legitimate ownership is determined by the material instantiation that is employed by the actor to assert ownership; for example, if a word is chosen as the material instantiation, then ownership can be claimed epistemically.

THE OPERATION OF FIXING

The first order (step 3) The initial set of triads is realigned to a new set. We call this a realignment because previously the triads were ordered to keep the practices clustered together. The material instantiations of word, body, and object are always at play, so that knowledge ownership could be claimed across all three domains, not just epistemology. This is modified to establish a new set of triads, each one a singular material-practice-domain. This is one of the effects of the process of manipulation. It is also one

First-order relations	Second-order relations
Naming in WORDS	Naming (in words) the performing BODY
	Naming (in words) the used OBJECT
Performing BODIES	Performing the named WORD
	Performing the used OBJECT
Using the OBJECTS	Using the named WORD
	Using the performing BODY

Figure 11.2
Defining first- and second-order relations through practice-material instantiation relations.

of the significant components in the construction of the illusion of the separation of knowing and owning.

The second order (step 4) Second-order manipulations are the key to reclaiming ownership of knowledge in bodies and objects shattered by the fragmentation. When the practice and the material instantiation are not of the same nature, we can see how knowledge ownership is "fixed" through second-order manipulation. In second-order manipulation, two practices and two material instantiations are activated in fixing knowledge ownership. Here it becomes possible for one practice to act on the second material instantiation; that is, it becomes possible to name [in words performing] bodies, or to name [in words the use of] objects. The domain of ownership is based on the knowing practice; in the first case, through performance of naming words, knowledge is owned in society, and in the second case, through use of named words, knowledge ownership is fixed in the domain of economy.

Exploiting the power of the split by *stapling* together the fragmented knowable and ownable to fix knowledge ownership, we now arrive at the final formulation of a regime of knowledge ownership. From a formerly indivisible, coherent, and pluralist universe of kn/own/ables in which the ownership of knowledge applies syncretically across practices, domains, and materials, the exploitation of splits and their subsequent manipulation via inversion of the triads and fixing leads us to an enclosed, linear, and homogenous regime of knowledge ownership. This regime is carved up into hegemonic fiefdoms by the actors who exert power and authority in particular domains, thus creating multiple new ones in which knowledge can be owned differently, such as science, technology, philosophy, law, and ritual, among others. This leads to the creation of epistemic, social, and economic spheres of knowledge ownership that are amenable to manipulation by powerful actors within those domains. For example, most people would agree that they own their bodies and the knowledge that their bodies bear. However, when scientists extract genetic information from bodies and corporations convert that into transactable forms of intellectual property, which is then recognized as the only valid ownership of such knowledge, the formerly indivisible ownership that had previously existed across all domains is reduced solely to the economic domain, and thus alienated from the body that bears it.

The natural result of these enclosures is the creation of actors who primarily identify themselves in terms of the domains that they acquire power in. Thus, as legitimate forms of knowledge ownership, we have scholarship that is primarily epistemic, performance that is primarily social, and finally, use that is primarily economic, as legitimate forms of knowledge ownership.

THE GRID: WHAT IT IS AND HOW TO USE IT

We propose the grid as an analytical tool to dissect the process of manipulations involved in the assertion of knowledge ownership and thus make it visible. This is a framework in the form of a grid composed of a set of columns and rows. The columns are formed by the three legitimate domains of ownership claims. The rows are formed by the three primary practices that are attached to each of the domains. The columns and rows intersect to form a set of nine cells, with each cell representing a specific combination of material instantiations (that make up a particular case). This set of nine cells, together with the row and column titles, makes up the entirety of this grid (figure 11.3).

The structure of the grid makes obvious when other terrains, such as ethics or environment, present themselves as domains although they are not. We then see that these terrains are in fact merely the outcome of complex manipulations of the kn/own/able on the level of domain-practice relation. For example, the environment is made into a not-ownable domain that can be known by naming but cannot be known by (human) use or through performance.

It is crucial to keep in mind that as soon as the perspective (i.e., the actor under observation in step 1) changes, the outcome of the analysis will be different. But this is

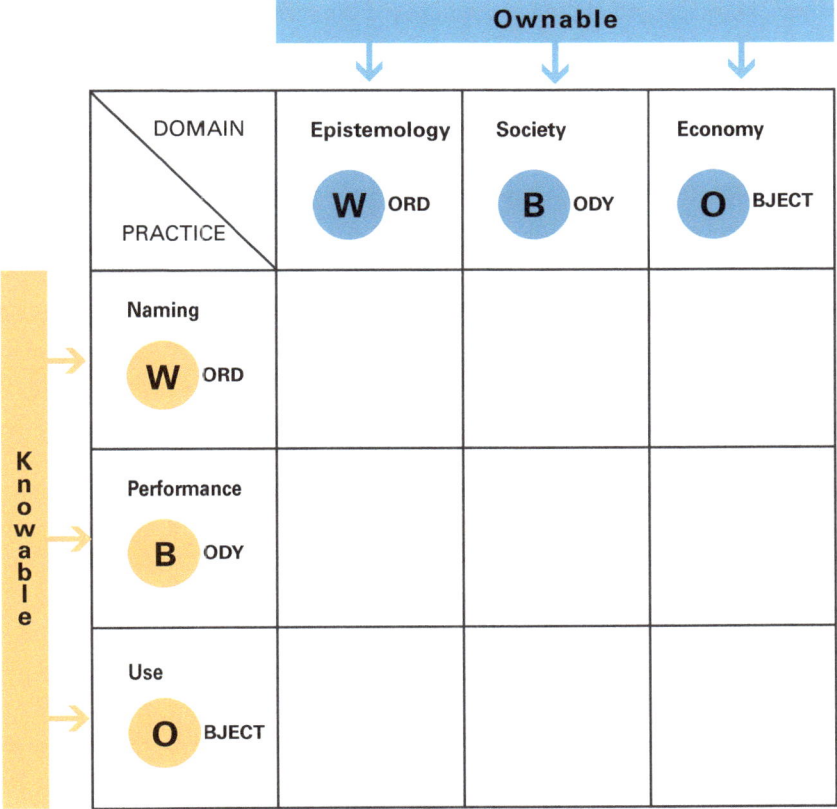

Figure 11.3
The grid for splitting.

a good reminder of why conflicts occur in the first place, and we hope this will encourage readers to use this framework and method beyond the boundaries of this book.

The grid serves two purposes:

1. It offers a sequential method to the reader to examine each case, and an alternative view of the manipulations involved in the assertion of knowledge ownership.
2. Once we define the domains and practices, particular combinations of material instantiations (each corresponding to a distinct case) emerge. These show us the nature of the kn/own/able or knowable+ownable in its specific location, and thus the reader can see the positions that these cases occupy relative to each other. For example, in moving along the diagonal from Slaton's case to Brokaw's case, a decline in the acceptance of the knowledge ownership claim is observed.

Splitting Knowing and owning are split in the dimensions of time and material instantiations as knowables and ownables. We draw them along the two axes. As a result of the spatial split, word, body, and object are separated along the lines of naming, performance, and use, which we indicate as rows in the table (indicating practices of knowing). Along the columns, we show how they are separated into the domains of ownership of epistemology, society, and economy. The downward arrows indicate the direction in which the ownable material instantiation propagates through the grid, and the horizontal arrows indicate the direction in which the knowable material instantiation propagates.

Fixing The authoritative instantiations of ownership in their respective domains propagate vertically through the columns into individual instances. Similarly, the authoritative instantiations of knowledge in their respective domains propagate horizontally through the rows into individual instances, thus forming the individual cells above. In each of the nondiagonal cases, knowing and owning is split, thus demonstrating the operation of second-order relationships that are always defined by two distinct material instantiations. The "+" indicates the act of fixing, where actors are compelled to create the illusion of attaching the knowable material instantiations to the ownable material instantiations to bring legitimacy to knowledge ownership (see figure 11.4). The violence of this act lies in the fact that this creation of knowledge ownership only works to serve the interests of its creators, while it denies the nature of the kn/own/able.

To complete the triad (and because word-body-object are always working together), we add the subordinated material instantiations for each case and arrive at the final iteration of the grid that introduces the cases in this volume. This shows the respective roles (as owning, knowing, and subsidiary) of each material instantiation for each of the nine cases (see figure 11.5).

Finally, when we analyze these cases through the conceptual two-dimensional axes of tacitness and alienability (as discussed in chapter 1, where they are shown to misrepresent how knowledge is and can be owned), we can see that they are ordered in the following way: first, knowledge is classified as tacit or explicit; then, ownership can be alienable or inalienable—that is, knowledge ownership is treated on the same terms as property ownership.

All of the cases presented in the book cluster along these principles into one of the four quadrants of the resulting grid. The majority of the cases belong to the category of inalienable ownership and tacit knowledge.

Here we introduce the axis of tacitness along knowing practices and the axis of alienability along owning in domains to the grid. These axes divide the grid into four quarters, as shown in figure 11.6.

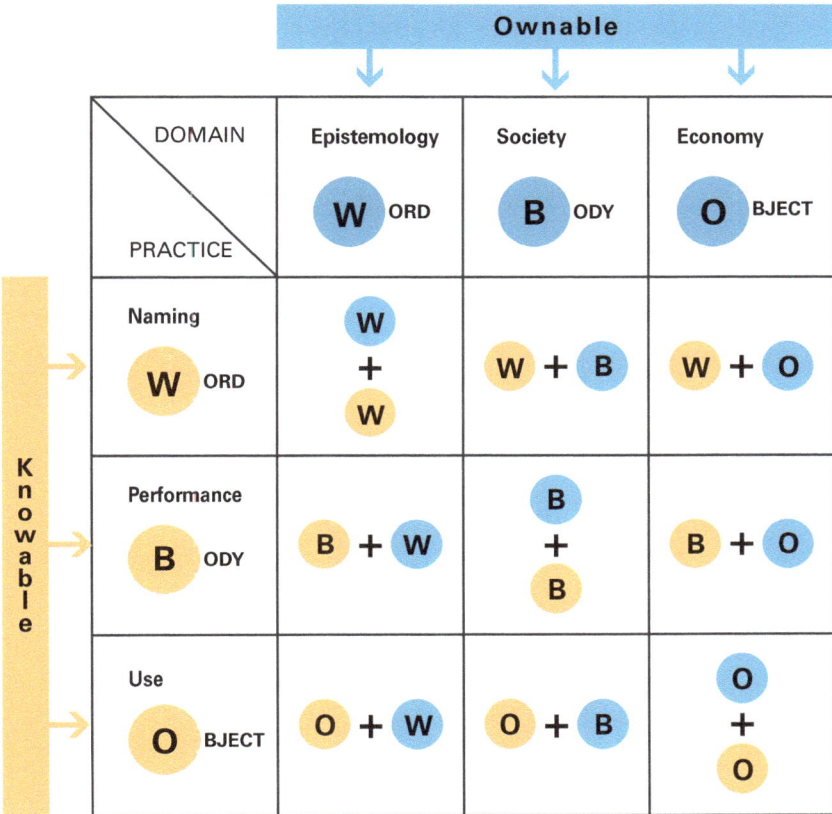

Figure 11.4
Fixing the split.

Science distinguishes between words, bodies, and objects as knowables, the first as bearing explicit knowledge and the other two as bearing tacit knowledge. The top half of the grid is made of knowables that are epistemic, and the bottom half is made of knowables that can perform and be used, but are not epistemic.

Law distinguishes between words, bodies, and objects as ownables. In the first case, top left corner, we find knowledge as alienable property that is ownable (words) and knowledge as inalienable property that is not ownable (bodies and objects). Hence, the left side of the grid is made of ownables that are ownable in the domain of epistemology, and the right side of the grid is made of ownables not ownable in the domain of epistemology.

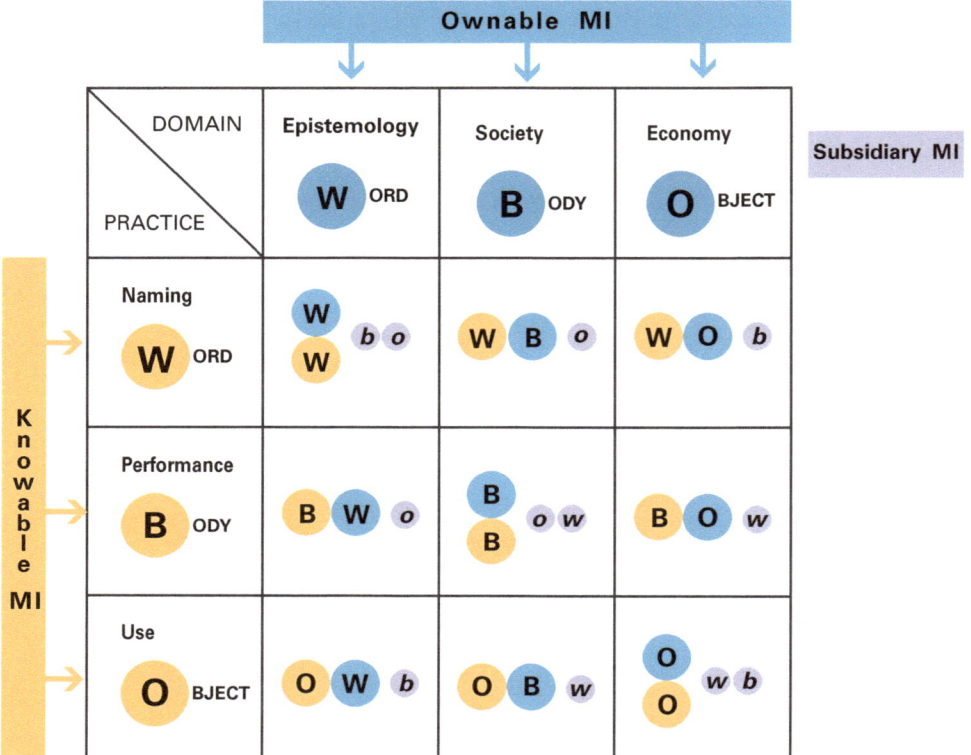

Figure 11.5
The combinations of material instantiations (MI) for each case.

The grid is now divided into four quadrants based on the interaction of the tacitness of knowledge and the alienability of knowledge ownership. We now propagate the grid with the material instantiations notated as WBO to show how knowables and ownables are fixed through actors fixing the domain-practice relations. In the top left quadrant, for example, the practice of knowing is naming, so the knowable is the word (W)]), in orange. The domain of owning knowledge is epistemology, so again, the ownable is the word (W), in blue. The subordinated material instantiations are then body and object (bo), in light purple.

We now reintroduce our cases to the grid (see figure 11.7), to show how actors *fix* knowables and ownables to assert ownership of knowledge. Explicit knowledge as knowable, fixed to alienable property as ownable, is the benchmark for establishing

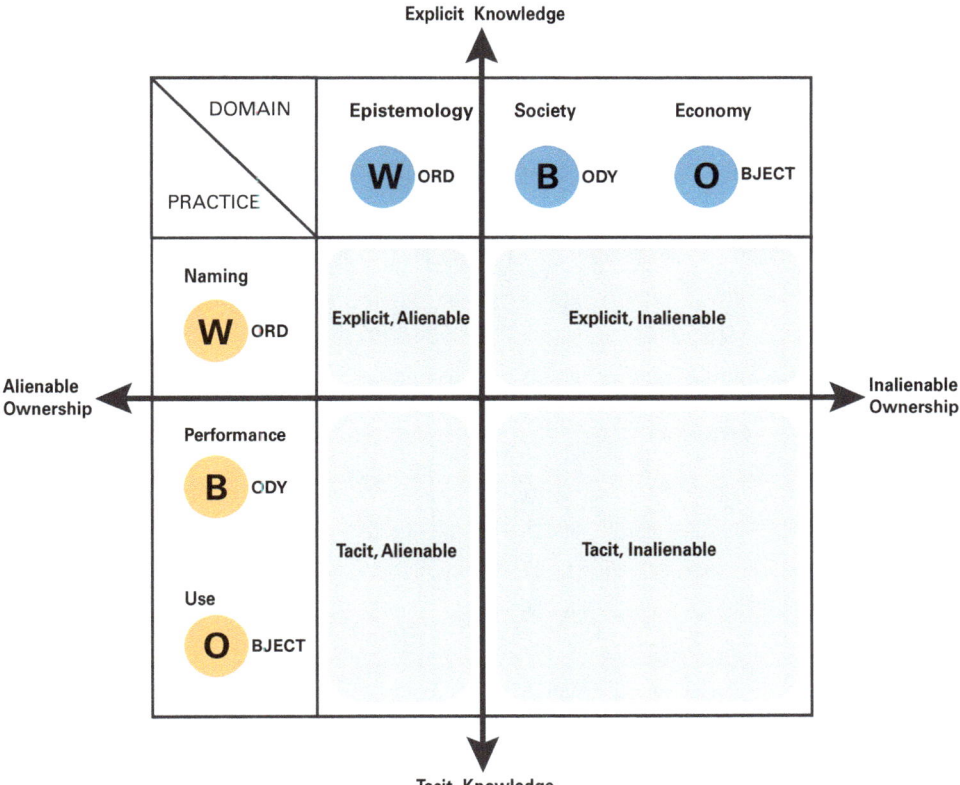

Figure 11.6
Axes of tacit/explicit knowledge, and alienable/inalienable ownership imposed by the science-law relation on the grid.

ownership of knowledge in the domain of epistemology. In the grid, it is the top left quadrant. Where either the knowable is tacit or the ownable is inalienable, as in the other three quadrants of the grid, the fixing of knowable and ownable becomes a form of ownership of knowledge subordinate to that benchmark.

Figure 11.8 represents a "constructed" reality of power relations where knowledge and its ownership appear as separate and divisible and can be structurally analyzed as an inversion of nature, where originally knowledge and ownership are always present together and indivisible as kn/own/ables. In reality, though, each of these material instantiations, practices, or domains can be known or owned because of the mutual

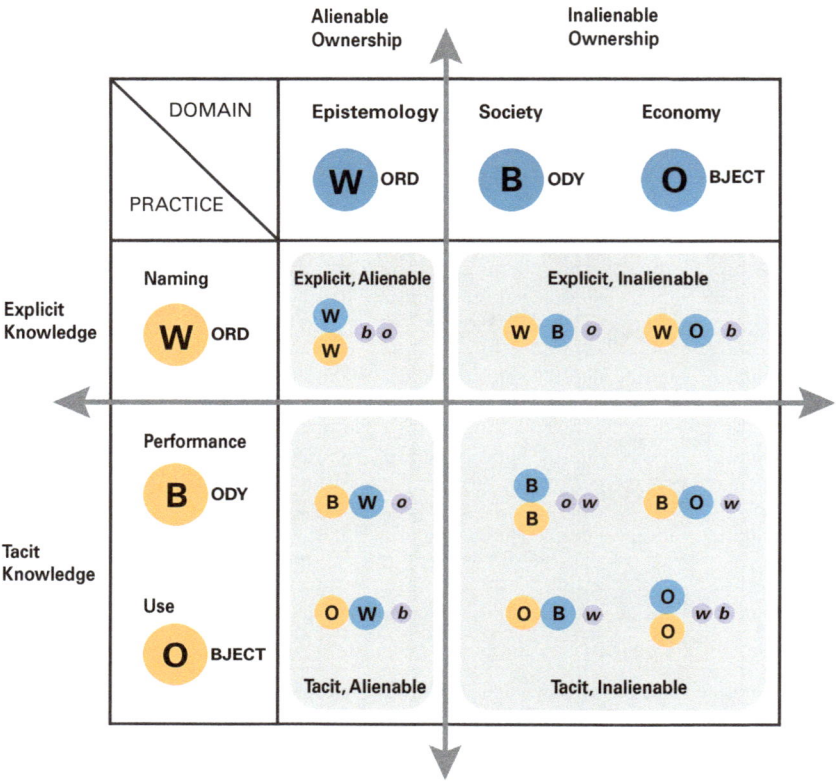

Figure 11.7
Proliferating the quadrants with the WBO notations signifying fixing of ownables and knowables.

conditioning—that is, there is a predetermined reality of kn/own/ables because the gene's product, while yet unknown, is defined as ownable in the domain of economy, or because Carnatic musicians can only own their knowledge through performance in the domain of society, the possibilities for ownership are limited.

This final analytical step leads to the conclusion that the modern regime of knowledge ownership creates an inverted view of the kn/own/ables, by splitting a complex reality into fragments that have to be fixed through the construction of an illusion. Thus, we arrive at an inversion of nature where the illusion holds more power than the reality that it claims to represent. When this inversion is studied as an issue of discourse, as it is by major sociologists of knowledge, the illusion is maintained rather than exposed.

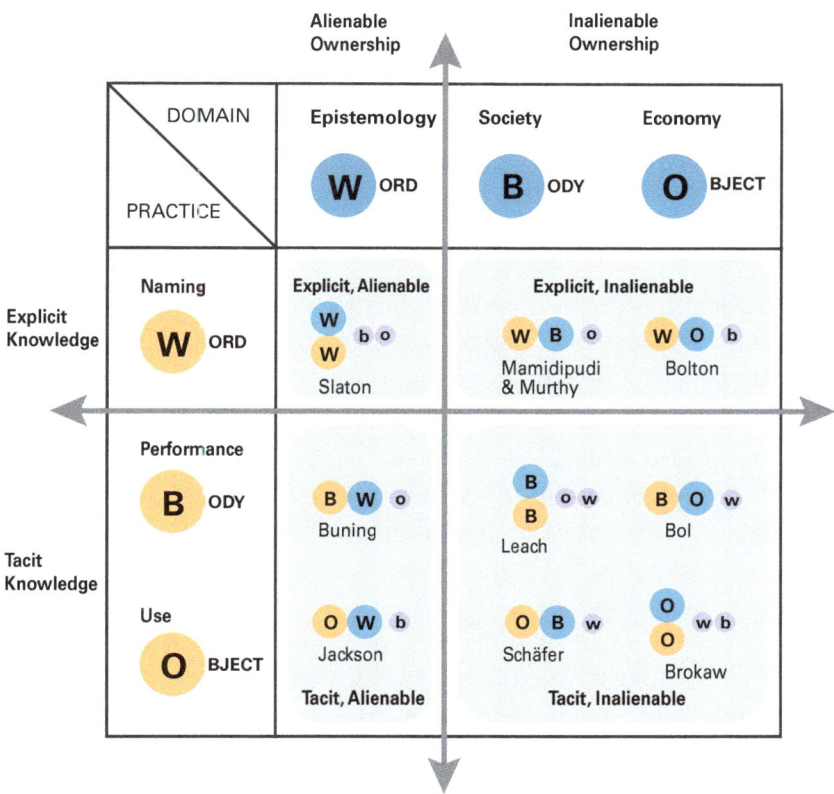

Figure 11.8
Overlaying the quadrants and notation with cases to analyze hierarchy of knowledge ownership.

USING THE GRID: FOUR CASES

We now show how the grid provides a heuristic technique to analyze the cases described in the chapters by Leach, Jackson, Bolton, and Slaton. These cases exemplify how domains and practices are operationalized in conjunction to provide legitimacy for ways of knowing and owning. The reader may want to keep in mind that in our framework, a particular area of human activity qualifies as a domain of knowledge ownership under specific criteria. First, it can be uniquely attached to a human practice that can be used to assert ownership in that domain. Second, domains and practices are constructed as pairs—for example, the practice of performing is employed by actors to establish ownership in society. Third, for a set of domains to be usefully employed in analyzing the ownership of knowledge, they have to be analytically mutually exclusive and nonoverlapping. The domains we have identified span the entirety of human experience and thus are—analytically speaking—sufficiently distinct, expansive, and comprehensive to explain the ownership of knowledge. Finally, in every case, we find that each of the three domains is primarily anchored in a specific material instantiation that alone is capable of bearing the knowledge and its ownership. For example, in the domain of epistemology, words are always present as the sole material instantiation that can bear the ownership of knowledge.

We use shading to indicate the two distinct movements—horizontally, the knowable material instance is colored orange, and vertically, the ownable instance is colored blue. The intersection of both defines the complete set of material instantiations that make up the knowledge ownership claim for that case.

How to use the grid to analyze a case of knowledge ownership:

1. *Identify the actor* who is making a persistent and legitimate claim of knowledge ownership. In Myles Jackson's case, it is the computer scientist. In James Leach's case, it is the Reite community. In Lissant Bolton's case, it is the museum curator. In Amy Slaton's case, it is the educational system.

2. *Identify the material instantiation* through which knowledge ownership is claimed—that is, which material instantiation is known and what is owned. For example, in Myles Jackson's case, the knowable for the computer scientist is the gene, the object, and the ownable is the patented current and future uses of the gene (naming of the use of the object). In James Leach's case, the kn/own/able is the performing bodies of the Reite people. In Lissant Bolton's case, the knowable is the named (labeled) object, as word, and the ownable is the exhibit itself, the object. In Amy Slaton's case, the knowable is the knowledge explicated as words in the classroom and the ownable, again, is words in the form of grades or certificates received by the student.

3. *Allocate the practice as per the knowable material instantiation*—that is, if it is word, it is naming; if it is body, it is performance; and if it is object, it is use. In Myles Jackson's case, the knowable is the applications of the gene, hence the practice of use is deployed. In James Leach's case, since the kn/own/able is a collective body, the practice is performance. In Lissant Bolton's case, the knowable is the label of the Vanuatu object, and thus the practice of naming is deployed. In Amy Slaton's case, the practice of naming is deployed by the student in acquiring the knowledge.

4. *Finally, identify the legitimate domain of ownership* based on the ownable material instantiation—words indicate epistemic ownership, body indicates social ownership. and objects indicate economic ownership. In Myles Jackson's case, the ownable is

the sequence itself, thus the legitimate domain of ownership is one of epistemology, the gene; in James Leach's case, the kn/own/able is the collective Reite body, and hence the ownership is in the social domain; in Lissant Bolton's case, the ownable is the Vanuatu object, which is owned in the domain of economy; in Amy Slaton's case, the ownable is the grades received that permit the student to claim epistemic ownership of their knowledge.

In Myles Jackson's case (see figure 11.9), mapping the gene as a sequence using computer science is sufficient to claim ownership of the future potential use of the gene as

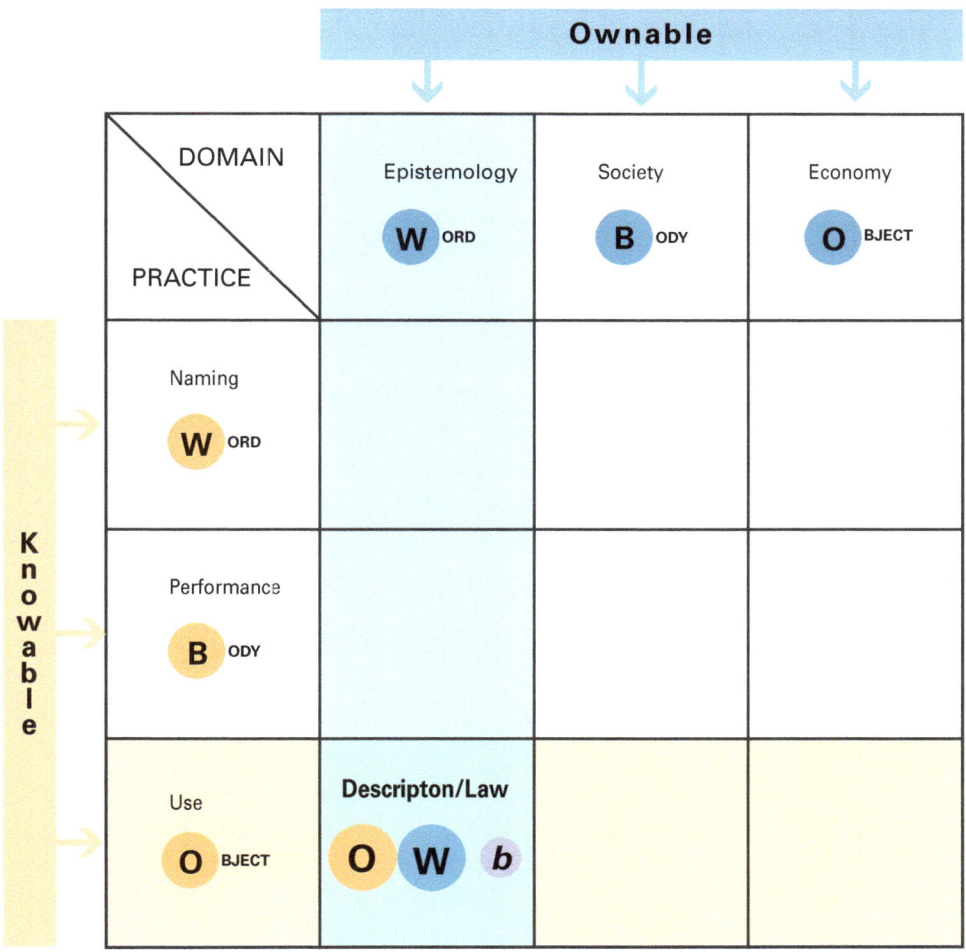

Figure 11.9
Locating Myles Jackson's case in the grid.

patented and legitimized in law, where all current and future uses are made ownable via the patent, which is owned via the material instantiation of the word (i.e., the patent). The knowable, however, is the gene itself, as an object that is being mapped or sequenced. This also illustrates a different kind of manipulation—a movement from scientific ownership that is exercised via named knowledge that has been discovered to legal ownership that is exercised via an assertion of all possible use, both the discovered and the yet to be known.

In James Leach's case (see figure 11.10), we have an intact kn/own/able. It is embodied in a people that both know and own their knowledge as relationships. By virtue of these relationships, individuals, as members of the community, have a legitimate right to demand access to the knowledge that only becomes emergent through these relationships. The knowledge itself is held by the bodies but it can be legitimately acted on only by activating the relationship. The true performance of the relationship enforces a *joint* activation of the knowledge. That is why when knowledge is illegitimately acted on, the result cannot be owned in the social domain (although the results of such illegitimate acts may be irreversible), which is the legitimate domain of ownership in Reite society. Those who do act on knowledge illegitimately are disowned. This case represents the only unfragmented operation of the kn/own/able in this book and thus occupies the center of the grid.

Working backwards from the location, we can see that if we were to analyze the kn/own/able as knowable and ownable in our analysis, knowing is always performed, and social relations are owned mutually and reciprocally. Thus, when this regime interacts with the science-law regime, the top left cell where science and law dominate, naming and property ownership are imposed on this regime. Then, in this encounter, since the collective body of the Reite people neither name nor own their knowledge as property, they become completely disenfranchised as knowledge owners in a modern regime that operates on a different notion of science and law. Here it becomes very clear why the grid is useful for disentangling illegitimate ownership claims.

In Lissant Bolton's case (see figure 11.11), the curator is the actor, so the grid is applied from their point of view. Because the case takes place in the domain of economy, the ownable is the museum exhibit, since one way or another, the museum has acquired the object. The knowable is the labeled use of the exhibit. Now the second order is activated. The practice of naming is one of labeling the objects as museum exhibits—labeling being used as a pointer to a reductionist act of naming. Distorting the naming practice to one of labeling reduces the possible multitude of uses of the object to the singular one of being a museum exhibit, which is done by the actor, the curator. The activation of the second order via the practice of naming (distorted

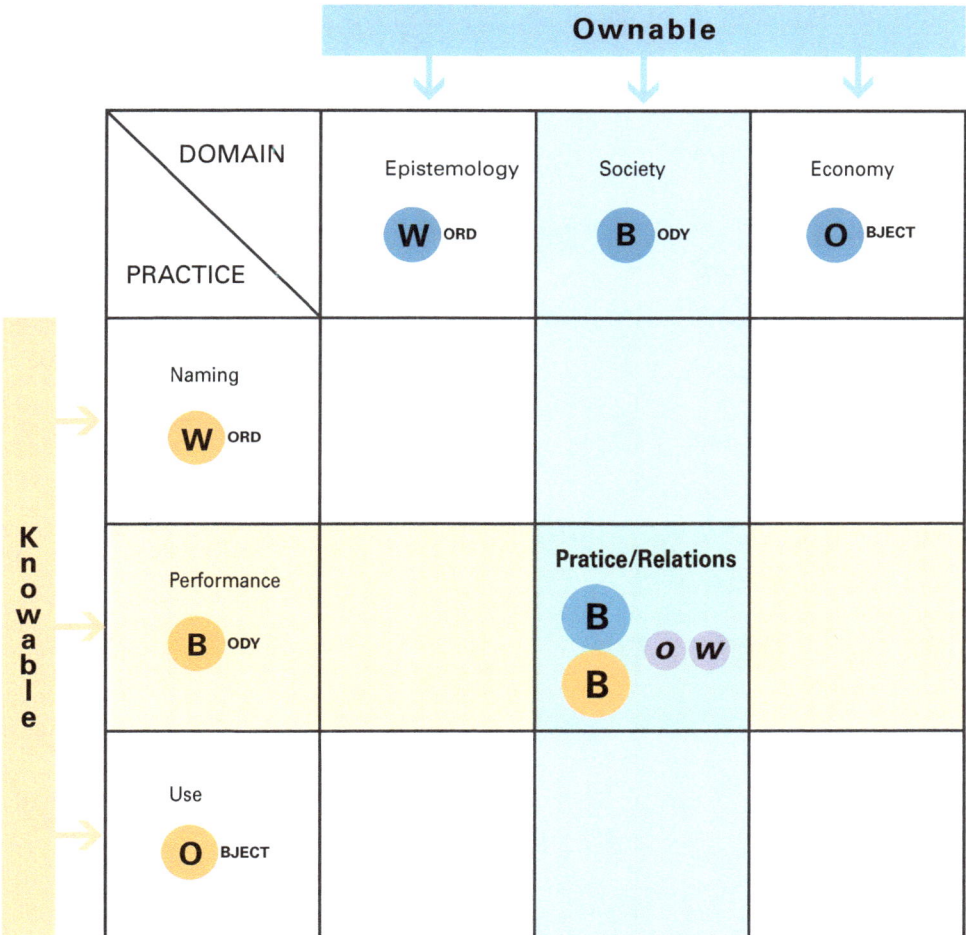

Figure 11.10
Locating James Leach's case in the grid.

as labeling) outright annihilates all other possible uses of the object. So, one could say, the object is colonized by this annihilation of any other possible use, especially the uses it originally had.

In this case, the knowable is the named object drawn on the graph paper at the conclusion of the experiment. The ownable is the name of the student affixed to this graph paper. On the basis of this ownable, the educational system sorts the student into

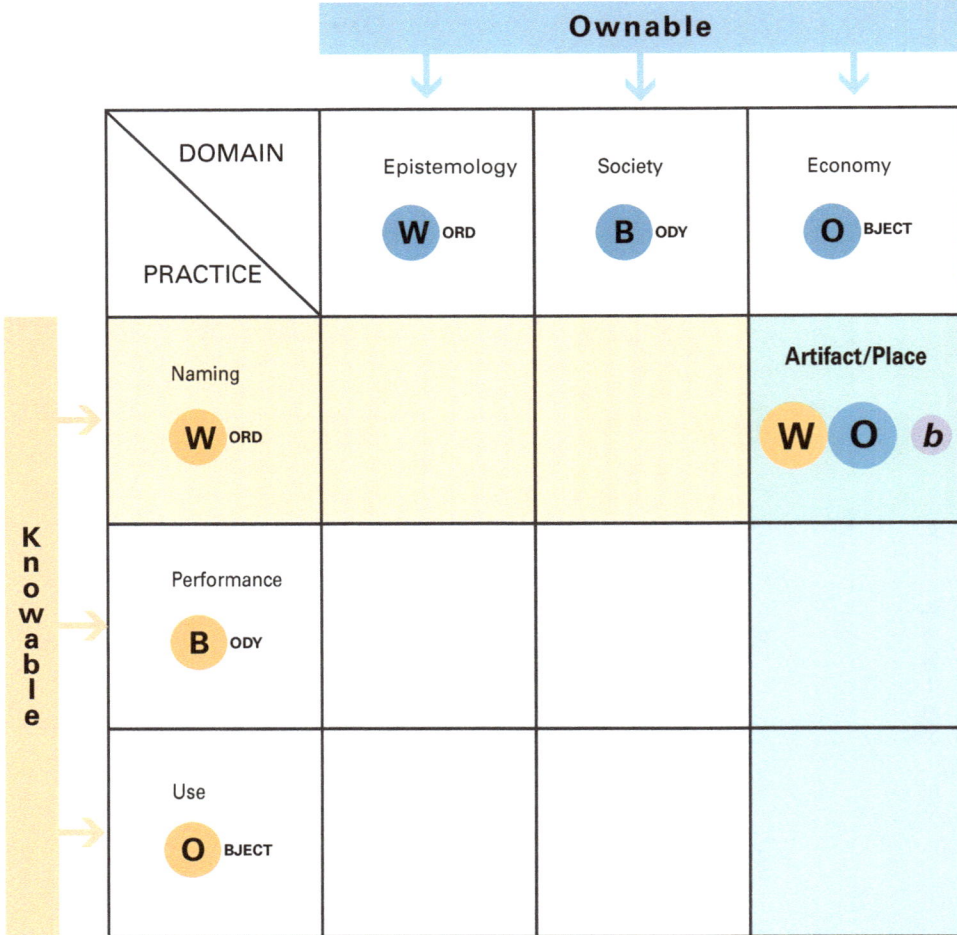

Figure 11.11
Locating Lissant Bolton's case in the grid.

one of two classes (knower, not-knower). The sorting step then determines the future trajectory of the student through the subsequent educational levels.

Amy Slaton's case (see figure 11.12) is unique because it deals with the negative case of knowing and owning—that is, the systematic production (and subsequent labeling) of students as not-knowers, which disowns them from knowledge. This case serves to reinforce a unique point that actors can use the space of the not-kn/own/able that exists outside the grid to further alienate knowledge from its ownership.

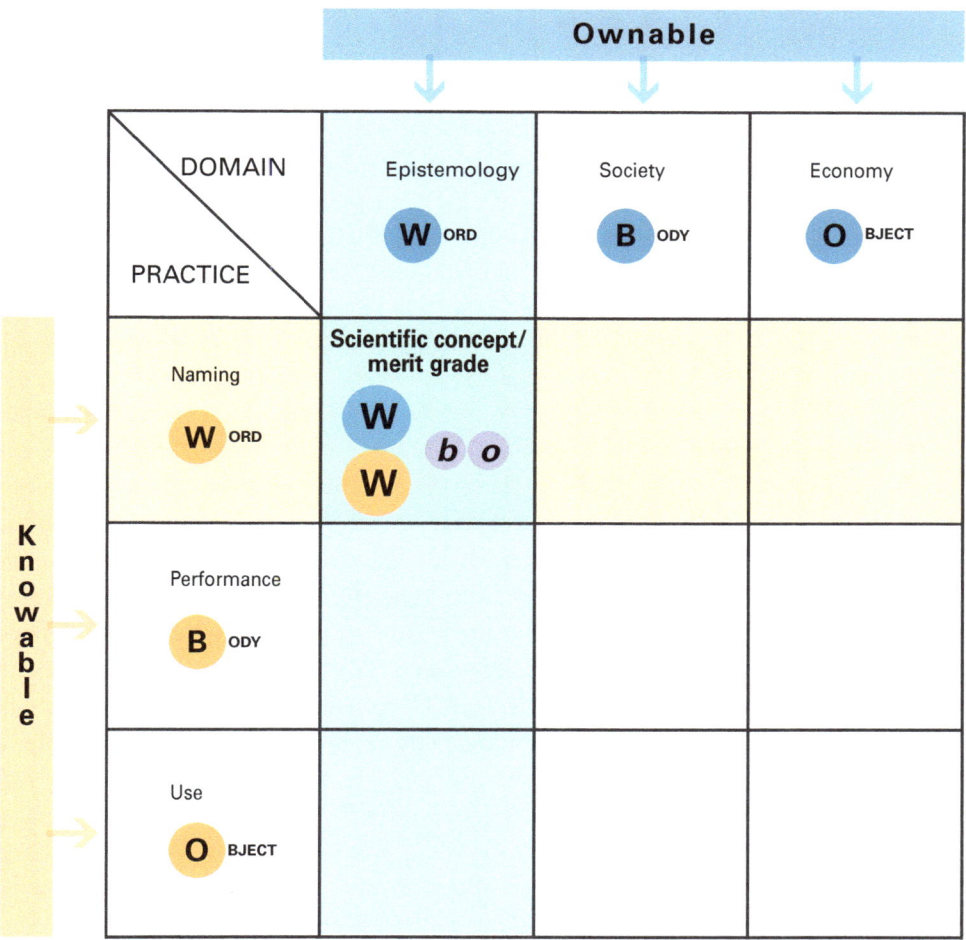

Figure 11.12
Locating Amy Slaton's case in the grid.

CONCLUSION

We have concluded this book with this diagrammatic analysis to help the reader see the effects of mutual conditioning that occur because the reality of the kn/own/able is denied. As a last step, we wish to emphasize that the grid is also helpful in demonstrating that in addition to the domains we have presented, there are other terrains that seemingly present themselves as domains, but either they do not cover an entire practice or they leap across domain boundaries (e.g., ethics or environment).

Assuming that knowledge can be owned purely as property, or that ownership can only follow from naming, often results in the creation of hierarchical orders of ownership claims. Thus, for example, epistemic claims are often thought to be foundational even when social or economic claims of knowledge ownership are made. We live in a world of science and law, in which knowledge is primarily owned by naming. What cannot be named, cannot be known and thus cannot be owned. The exception, such as James Leach's case in this book, proves the rule; or rather, it reveals the operation of the rule, as it makes the kn/own/able in its indivisible state visible to those who have forgotten this possibility. When we live in the world of the kn/own/able, we can accept that Reite gardeners, who perform their knowledge, enjoy a universally valid ownership of knowledge.

ACKNOWLEDGMENTS

The European Research Council project PENELOPE (ERC funding HORIZON 2020 number 682711) at the Deutsches Museum in Munich and the Deutsche Forschungsgemeinschaft (DFG) project 435681850 at the Technische Universität Berlin have supported Annapurna Mamidipudi's research into ownership of knowledge in craft and hand weaving. She is grateful to the PIs of both projects, Ellen Harlizius-Klück and Friedrich Steinle for taking a leap of faith into the world of traditional Indian crafts and immersing themselves in understanding its naming, performance, and use in modern times.

CONTRIBUTORS

Marjolijn Bol is associate professor in technical art history at Utrecht University. Bol's research intersects with historical studies of craft, technology, and knowledge, with a special focus on performative methods (reconstruction) and written sources on art technology. She has published widely on the history of imitative practices and durability in art and continues to explore these topics in *The Varnish and the Glaze* (University of Chicago Press) and the edited volume *The Matter of Mimesis* (Brill), which will be published in 2023. Her research has been supported by grants from the Dutch Research Council (NWO) and European Research Council (ERC). She is on the editorial board of *Studies in Art & Materiality* (Brill) and a member of the Utrecht Young Academy and *De Jonge Akademie* of the Royal Academy of Arts and Sciences (KNAW).

Lissant Bolton is keeper—head of the department—of Africa, Oceania and the Americas at the British Museum. She is also the director of the Endangered Material Knowledge Programme, based at the British Museum. She is an anthropologist specializing in the Western Pacific, notably Vanuatu, where she has long supported indigenous programs to document and revive local knowledge and practice. She writes about gender, textiles, and ideas of tradition in the Western Pacific and also about museums, especially about museum anthropological collections and the politics that surround them.

Cynthia Brokaw, a sinologist and historian of the book, is professor of history and East Asian studies and Chen Family Professor of Chinese Studies at Brown University. Her research centers on the expansion, in both geographical scope and social reach, of commercial woodblock publishing in early modern (sixteenth- to nineteenth-century) China. Brokaw's recent publications, drawing on fieldwork in former publishing sites, examine the nature and the cultural and political impacts of the "core" book culture—the group of widely popular best-selling texts—that early modern commercial publishers both generated and disseminated.

Marius Buning is associate professor in early modern history at the University of Oslo, Norway. He has held fellowships at Harvard University, the Netherlands Institute for Advanced Studies, and the Max Planck Institute for the History of Science. His research interests focus on the origins of intellectual property and the role of the state in shaping notions of scientific and technological progress.

Myles W. Jackson is the Albers-Schönberg Professor in the History of Science at the Institute for Advanced Study in Princeton, New Jersey. He is currently working on two projects: one that examines the ways in which intellectual property issues have changed the conduct and content of molecular genetics, and a study of the collaborations between physicists, radio engineers, and musicians that led to the invention of electronic musical instruments.

James Leach is a social anthropologist and directeur de recherché with the CNRS, based in the Centre de Recherche et de Documentation sur l'Océanie (CREDO), Aix-Marseille University, and honorary professor in anthropology at the University of Western Australia. Leach's research draws on long-term fieldwork with Nekgini-speaking people on the Rai Coast of Papua New Guinea and has attended to kinship and place, art and aesthetics, creativity and ownership, and intellectual property. He has also undertaken research in Europe and Australia with artists and scientists engaged in collaborative knowledge exchange, with free software engineers on gender issues, and with contemporary dance artists on creativity and technology.

Annapurna Mamidipudi is a scholar of science, technology, and society studies (STS). She has been awarded a DFG project titled "Epistemologies of Craft: The Role of Material Innovation in Making Color Expertise" at Technische Universität Berlin. She was previously a postdoctoral researcher at Deutsches Museum working on an ERC project on weaving knowledge in ancient Greece. She is a trustee of the Handloom Futures Trust, which works at the intersection of knowledge and livelihood support for traditional craftspeople in India. Her research interests focus on how craftspeople innovate their material practices and how they make knowledge claims to build value for their work in past and contemporary society.

Viren Murthy teaches transnational Asian history at the University of Wisconsin–Madison and researches Chinese and Japanese intellectual history. He is the author of *The Political Philosophy of Zhang Taiyan: The Resistance of Consciousness* (Brill, 2011) and has published several coedited volumes, including *Confronting Capital and Empire: Rethinking Kyoto School Philosophy* (Brill, 2017), as well as articles in *Modern Intellectual History*, *Modern China*, *Frontiers of History in China*, *Positions: Asia Critique*, *Jewish Social Studies*, *Critical Historical Studies*, and *Journal of Labor and Society*. His manuscript *Pan-Asianism and the Legacy of the Chinese Revolution* will be published by the University of Chicago Press in 2023.

Vivek S. Oak is a fellow with the Handloom Futures Trust, researching evidence-based methods for policy-making for handloom weaving in India. He is a member of the steering committee of the National Federation of Handloom and Handicraft Workers, an Indian trade union of artisanal workers. He has studied for a masters in computer engineering at Carnegie Mellon University, and has experience with designing embedded software applications and augmented reality applications for museums. His research interests—namely, combining technology and the arts as well as exploring the idea of the sacred in artisanal practice—inform his work in handloom policy, knowledge politics, and labor movements.

Jörn Oeder is a fellow with the Handloom Futures Trust in India. From 2016 to 2021, he was a student assistant at the Max Planck Institute for the History of Science (MPIWG), Berlin. Before joining Department III, he was the assistant of MPIWG director emeritus Hans-Jörg Rheinberger, with whom he worked on *Experimentalität* (Kadmos, 2017), *Ordnung und Organisation* (Basilisken-Presse,

2021), and *Spalt und Fuge* (Suhrkamp, 2021), among other publications. He is currently pursuing a BA degree in the history of science and technology at Technische Universität Berlin. His research interest is focused on the history of chemistry and the philosophy of knowledge.

Dagmar Schäfer is a sinologist and historian of science and technology, director of Department III (Artifacts, Action, Knowledge) at the Max Planck Institute for the History of Science (MPIWG), Berlin, and honorary professor at Technische Universität Berlin and Freie Universität Berlin. Her research centers on the history and sociology of technology in China. Author of *The Crafting of the 10,000 Things* (University of Chicago Press, 2011), she has published widely on the premodern history of China (Song to Ming) and the changing role of artifacts in the creation, diffusion, and use of scientific and technological knowledge.

Amy E. Slaton is a historian of science, technology, and engineering and professor of history at Drexel University in Philadelphia. She has published on the history of technical education and labor, primarily in respect to historical concepts of human difference such as race, gender, disability, and sexuality. Her research frequently centers on standards, instrumentation, and testing regimes in places of engineering learning and work. She is coeditor with Tiago Saraiva of the journal *History + Technology*.

INDEX

23andMe, 294, 301–303, 304–305, 307, 308–309

Actor-network theory (ANT), 21, 91, 202
Adorno, Theodor, 134
Alford, William, 49, 55–56, 61, 63
Algorithm patents, 293, 302–304
Alienation, 2, 28–32, 350–355, 360
Allman, Norwood, 71, 84n89
Allsen, Thomas T., 253, 278n7, 280n28
Amazon Web Science, 300
American Medical Association, 300
Ancestry.com, 294, 303, 305–307
AncestryDNA, 294, 301–302, 305–307
Anderson, Benedict, 91
Antons, Christoph, 239
Aplin, Tanya, 94, 98, 104
Arcadia Fund, 319, 320, 323–324, 336
Aristotle, 40n39
Artima, Baldarasse, 155
Atomic force microscopes (AFMs), 181–182, 184–185, 186, 187, 194–195, 196, 198, 201, 202
Attwood, Bain, 331
Australia, 330, 331, 333–334

Bainbridge, David I., 93–94, 96–97, 98, 99, 100–101, 104, 106
Balasubramaniam, G. N, 125, 131, 132–136, 137, 143
Baldwin, Peter, 323, 324

Banana in shoebox, 2, 24, 181, 182–187, 189, 193, 194, 195, 196, 198
Bann, Stephen, 91
Barad, Karen, 183, 186
Barker, Philip, 16, 37n3
Basu, Paul, 333
Batik, 24, 37, 154–155, 167–173, 174
Belgium, textiles, 167
Bellido, Jose, 92
Bentham, Jeremy, 98
Bently, Lionel, 94, 97, 98–99, 101–102, 104
Berg, Maxine, 167
Biagioli, Mario, 102
Biesta, Gert, 197
Big Pharma, 301, 302, 308
Bill and Melinda Gates Foundation, 300
Biocapital, 293, 295
Birge, Bettine, 263, 282n48
Bol, Marjolijn, 19, 24, 153–174, 355
Bolton, Lissant, 26, 319–342, 343, 355, 356, 358, 360
Bonnemaison, Joël, 325
Bracha, Oren, 102
Branding, brands, 7, 47, 69, 172, 174, 246n33, 275–76
Bray, Francesca, 263
Brazil, samba, 123
British Library, 323
British Museum, 319, 320, 322–323, 333, 334–335
Broad Institute, 300

Broadwell, Bethany P., 198, 199, 300
Brokaw, Cynthia, 23, 33, 47–89, 349, 355
Brown, Alison, 329–330
Bryant, Neil, 308
Buddhism, 64, 258, 262
Bulmer, Ralph, 240
Buning, Marius, 23, 47, 91–117, 355

Cage, John, 138
Cai Lun, 54
Caley, Earle Radcliffe, 159–160, 175n14
Canada, postcolonialism, 330
Cang Jie, 54
Cao Shiheng, 70
Capitalism. *See* Economy
Carnatic music
 Balasubramaniam, 131, 132–136, 137, 143
 Brahminization, 125, 131, 132, 143
 caste system and, 124
 devadasis, 125, 129–130, 131–132, 136–137, 141, 144
 gender roles, 130, 138, 139
 improvisation, 122, 124, 127–128, 129, 130
 India, 23–24, 26–27, 121–145
 innovation, 125, 127, 132, 133, 141–143
 Krishna, 124, 125, 138–140, 142, 143, 144
 lyrics, 140
 modernization, 138–140
 nationalism and, 124
 performance, 23–24, 122–123, 124, 128–129, 136–141, 143–145, 130, 354
 pop star, 132–136
 raga as knowledge form, 122, 124, 126–129
 social development, 129–132
 Subbalakshmi, 124, 125, 136–138, 143
 training, 128, 130
 Vedavalli, 125, 141–143, 144
 virtuous knower, 136–138
Cawey, Diacinto, 155
Cennini, Cennino, 163, 167
Chan, Hok-Lam, 261
Chandler, Daniel, 105
Chatterton, Alfred, 31

Chen Peng, 259, 262
Chen Xinyu, 271
Cherniack, Susan, 57
China
 Buddhism, 64, 258, 262
 census records, 261–262
 Confucianism, 49, 50, 51, 55, 57, 60–61, 65, 252–260, 263, 268–269, 274–275
 continuities, 252, 253, 257, 270, 274
 Cook Ding (*see* Cook Ding)
 crafts (*see* Chinese crafts)
 Cultural Revolution, 73–75, 76
 Daoism, 64, 258
 historiography, 259, 261–263, 265, 267, 268, 269–270, 272, 275–276
 imperial China, 47, 58–65, 251–277
 intangible cultural heritage, 8, 9
 intellectual property (*see* Chinese intellectual property)
 Jurchen Jin, 261–262
 king's way, 5, 6–7, 20, 50–51, 54, 55
 Legalism, 60–61
 modern music and nationalism, 123
 Mongols, 253, 255, 259, 263, 266, 268
 PRC, 71–75
 print technology, 55, 65–67
 publishing (*see* Chinese publishing)
 Sagely Way, 51
 sumptuary laws, 258
 thirteenth–seventeenth centuries, 251–277
 Yuan and Ming period, 251–277
Chinese crafts
 bans, 251
 continuities, 270–272
 gender roles, 263
 Guidebook for Clerks (Lixue zhinan), 251–277
 historiography, 259, 261–263, 265, 267, 268, 269–270, 272
 imperial China, 58–59, 251–277
 knowledge ownership and, 2
 material mimesis, 154
 naming, 25, 31, 251–255, 257–258, 272–277
 performance, 25, 258, 261, 265–266

registers, 251–252, 261, 263–265, 267, 270, 272–273, 274
regulation in thirteenth–seventeenth centuries, 251–277
status, 265–266
tanning, 8, 9, 21–22, 259, 262, 268
Yuan and Ming period, 251–277
Chinese intellectual property
assertions of ownership, 55–71
characteristics, 59–60
crafts (*see* Chinese crafts)
history, 47–76
imperial China, 49–71
imperial ideology, 60–65
invention concept, 47, 49, 50, 53, 54–55, 58–59, 72–75
non-development issue, 35, 49–50
patents, 72
PRC (1949–1976), 71–75
publishing (*see* Chinese publishing)
Chinese publishing
assertions of ownership, 55–58
authorship concept, 50–54
customary regulations, 68–71
extra-official ownership claims, 67–68
Masha publishers, 58, 63, 65, 70
PRC, 72–75
print technology, 55, 65–67
profitable ownership, 23
publishing communities, 68–71
Qing period, 71–72
Song period, 56–57, 59–60, 61–65, 66, 75, 76
Clark, Grahame, 156
Clifford, James, 329
Collins, Harry, 30
Colonialism, 103, 132, 220, 319–320, 324, 329–331
Color Genomics, 300
Colston, Catherine, 98, 100
Commodity, commodification, 31, 40n32, 121, 144, 146n1, 167, 171, 296. See also *Economy*
Confucianism *(Ru)*, 49, 50, 51, 55, 57, 60–61, 65, 141, 252–277, 278n3

Consequentialism, 99
Cook, Captain, 321
Cook, Nicholas, 122–123
Cook Ding, 5–6, 19, 20, 344
Coombes, Annie, 333
Copinger, *Law of Copyright*, 92
Copyright
capitalist concept, 121
China and, 23, 58, 60, 65, 67, 71–74
copyright knowledge, 326
critics, 48
textbooks, 92–94, 96, 97, 101–103
Cornish, William, 94, 97, 104
Crackled crystals, 154, 155–162
Crafts
China (*see* Chinese crafts)
endangered materials, 320
material mimesis (*see* Material mimesis)
tacit knowledge, 31
tanning (*see* Tanning)
textiles (*see* Textiles)
Creativity mantra, 48, 103, 104
Crook, Tony, 229, 236, 241

Damon, Fred, 242
Daoism, 64, 258
Darder, Antonia, 192
Davenport, Neil, 98
Davis, Jennifer, 94, 98, 104
DeAngelo, Joseph James, 306–307
Deazley, Ronan, 92
Democritus, 157
Dewey, John, 188, 197, 200
Diagrams
cases, 355–360
domains, 345, 349, 351–355, 356–361
first-order relations, 346, 347
fixing, 343, 344–348, 350
grid use, 348–360
material instantiations, 344, 345–353, 356, 356–358
ownership guide, 343–362

Diagrams (cont.)
  practices, 345, 347, 349, 351, 352–355, 358–361
  second-order relations, 346, 347
  semiotics, 105
  splitting, 343–344, 345, 349, 350
  WBO, 349, 351, 352, 353, 354, 355, 358–361
Dilley, Roy, 321, 327
Ding Richang, 66
Di Zhaoying, 56
DNA. *See* Genetic knowledge
Domains
  diagrams, 345, 349, 351–355, 356–361
  economy (*see* Economy)
  epistemology (*see* Epistemology)
  role, 5–7, 18, 24–25
  society, 5–7, 24–25
  third-century China, 5–6
Duan Changwu, 65, 71
Du Fu, 53–54
Durkheim, Émile, 254, 275
Dworkin, Ronald, 91

Economy
  capitalism, 28–29, 272
  Chinese communism, 72
  Chinese empire, 60, 253, 254, 275
  crackled objects, 174
  diagrams, 345, 349, 351, 352, 353, 354, 355, 358, 359, 360, 361
  domain, 5–7, 18–19, 22, 25, 35
  tanning, 8, 10
  traditional knowledge and, 220, 228, 229
  US science education, 184, 191–192
Education. *See* Science education
Ehrlich, Eugen, 91
Endangered material
  ignorance and endangerment, 323, 324–328
  languages, 323
  secrets, 326–328
  Vanuatu, 324–328
Endangered Material Knowledge Programme, 319, 320, 322–323, 336

Epistemology
  Chinese crafts, 253, 263
  diagrams, 345, 349, 351–355, 358–361
  endangered material, 326
  hierarchies, 26
  longue durée, 36
  museums and, 319
  naming, 6, 20
  performance, 21
  role, 5–7, 18, 22, 24–25, 35, 351, 352
  scholarship, 27
  science and, 9, 29, 192–197, 198
  traditional knowledge and, 219, 325–326
  Western epistemologies, 193
Essentialism, 99, 185
Ethnicity/race
  imperial China, 261
  personal genomics companies, 301, 302
  postracialism, 186
  United States, 92, 183, 184, 185, 202, 300
  West Africa, 333
Ethnography. *See* Museums
Eugenics, 183
Eurocentrism, 21, 48, 60, 102–103, 108, 251
Exclusion law, 36, 140
Exclusion politics, 26, 131, 139, 145, 185, 200
Explicit knowledge, 4, 29, 32–34, 162, 351, 352, 353, 354, 355
Eyferth, Jacob, 59

Falvo, Michael R., 198, 199
Family Tree, 301
Farmer, Edward, 272
Feng Dao, 62
Fenwick, Tara, 186, 202
Ferguson, Roderick A., 184
Feudalism, 131, 275
Fieldwork, 322
Firth, Alison, 91, 104
Fixing
  diagrams, 343, 344–348, 350
  first order (step 3), 346–347

meaning, 344–346
second order (step 4), 347–348
Flags, 9, 16–17, 20, 24
Fong, Wen, 50, 55–56, 77n8
Foucault, Michel, 2, 15, 196
France, 102, 319–320, 333
Freire, Paolo, 183, 201
Fuxi, 54

Gandhi, Mahatma, 137
Gao Rongsheng, 262
Gates, Kelly, 93
Gates Foundation, 300
Gaultier, Jean Paul, 172
GEDmatch, 307
Gemstones, 155–162
Gender
  Carnatic music and, 130, 138, 139
  Chinese crafts, 263
Genentech, 300
Genetic knowledge
  algorithm patents, 293, 302–304
  biocapital, 295–296
  CCR5, 294
  consumer protection, 305–306
  criminal law and, 306–307
  database access, 303–304, 306–308
  diagrams, 356–358
  excised cells, 296–297
  identity by descent, 303
  naming, 25
  ownability, 293–309
  ownership case law, 295–300
  patentability, 7, 25, 293–309, 356, 358
  personal genomics companies, 301–308
  precision medicine, 300, 302
  privacy, 303–304, 306–308
  race and, 301, 302
  SNPs, 301–302
  United States, 293–309
Genette, Gérard, 96
Geopolitics, 48
Germany, patents, 102

GERMLINE, 303
Ghandhi, M. K., 29, 31
Gillespie, Tarleton, 93
GlaxoSmithKline, 294, 308–309
Goldstein, Paul, 103
Gourlay, K. A., 228
Greimas, A. J., 96
Guilds, 30, 31
Gu Yanwu, 53

Hacking, Ian, 185–186, 202
Han Wudi, Emperor, 61
Hartman, Gary, 307
Hayden, Cori, 233
Hearth, Eric, 306
Hegel, Georg Wilhelm Friedrich, 133, 135
He Zhaohui, 66
Hirsch, Eric, 222, 229, 239
Holes, Paul, 307
Huang Yin, 264, 282
Huffman, Kirk, 326
Human Genome Sciences (HGS), 294, 302
Human remains, 331
Hunn, Eugene, 240, 241, 242
Hu Qiyu, 261, 263, 264, 268–269, 282

IBM, 300
Imitation. *See* Material mimesis
Incyte Genomics, 302
India
  Carnatic music (*see* Carnatic music)
  caste system, 124
  crafts, 30–31
  Tamil nationalism, 124
  textiles, 166
Indigenous people. *See also* Vanuatu
  IP regimes and, 48
  museums and, 330–335
  rituals (*see* Rituals)
  scholarship and, 9–10
  traditional knowledge (*see* Traditional knowledge)
Individualism, 60, 225

Indonesia, batik, 37, 167, 168
Industrial Revolution, 99, 101
Innovation/invention
  Carnatic music, 125, 127, 132, 133, 141–143
  Chinese concept, 47, 49, 50, 53, 54–55, 58–59
  Chinese concept (PRC), 72–75
  Chinese crafts, 252
  creativity mantra, 48, 103, 104
  European concept, 30
  legal manipulation, 7, 48
  patent system and, 48
  problematic concept, 48, 106–107
  public good and, 58–59
  science, 199
  traditional knowledge and, 239
Intangible cultural heritage (ICH), 8, 31, 32, 37
Intangible property, 30, 33, 66, 71–72, 92, 219
Intellectual property
  Chinese history, 47–76
  concepts, 49
  critiques, 48
  Eurocentrism, 48, 102–103
  semiotics, 93–96
  textbooks (*see* University textbooks)
Invention. *See* Innovation/invention
Isidore of Seville, 161
Islam, 102
Italy, patents, 102
Ivanhoe, Philip, 50, 53, 54
Iyengar, Madurai Srirangam, 141

Jackson, Myles, 19, 25, 26, 293–316, 343, 355, 356–357, 358
Japan, 29, 31, 71, 293
Java, batiks, 24, 167–168, 169, 170–171
Jin Jian, 54–55
Jones, M. Gail, 198, 199

Kang Yongshao, 272
Kant, Immanuel, 17
Kublai Khan, 256
Kimsky, Sheldon, 302
Kirsch, Thomas, 321, 327

Knowledge
  alienation, 2, 28–32, 350–355, 360
  concepts, 185–186
  explicit knowledge, 4, 29, 32–34, 162, 351, 352, 353, 354, 355
  tacit knowledge, 28, 29–32, 34–35, 350, 351, 352, 353, 354, 355
Knowledge ownership
  cases, 22–27, 355–360
  domains (*see* Domains)
  excavation site, 16–27
  flags, 9, 16–17, 20, 24
  legal manipulation, 7–8
  manipulation, 7–8, 20–22, 26
  practices, 3–5, 15, 23–24
  premises, 3–5
  process, 1–3
  property and (*see* Property)
  scholarship, 9–10, 25–27
  separation of knowing and owning, 28–35
Krishna, T. M., 124, 125, 138–140, 142, 143, 144
Krishnamachari, T. T., 139
Kuai Xiang, 272

Lafleur, Robert, 52
Laminated floors, 153
Lane, Giles, 233–235
Languages, endangered languages, 323
Latour, Bruno, 4, 91, 294
Lau, Fred, 123
Law
  genetics (*see* Genetic knowledge)
  IP (*see* Intellectual property)
  law and literature movement, 91–92
  living law, 91
  manipulation, 1, 2–3, 7–8, 10, 33, 34, 36
  nature, 91–92
  textbooks (*see* University textbooks)
Law, John, 201
Lawrence, Peter, 223–224, 226, 228, 230, 231
Leach, James, 19, 24–25, 33, 219–250, 343, 355, 357, 358, 359, 362
Lee, Sheila Jackson, 107

Lemonnier, Pierre, 322
Li Bai, 53–54
Lien, Marianne Elisabeth, 201
Lindstrom, Lamont, 223, 230, 325–326, 328
Liu Liya, 259, 262
Liu Ziming, 69
Li Yu, 57–58, 62, 70, 71
Locke, John, 97, 103, 104
Loeber, Dietrich, 74
Long, Pamela O., 30
Longue durée, 36, 274
Losche, Diane, 334
Luhmann, Niklas, 91
Luo Wenjie, 265
Lü Zuqian, 57

Macdonald, Sharon, 331
MacGuill, Dan, 313n71
Macron, Emmanuel, 319–320, 333
Mair, Jonathan, 321
Mair, Victor, 253
Maitland, Frederic William, 94–95, 97
Majnep, Saem, 240
Malécot, Gustave, 303
Malinowski, Bronislaw, 322
Mamidipudi, Annapurna, 1–13, 15–44, 121–152, 343–362
Maori, 221, 329, 335
Marx, Karl/Marxism, 2, 28, 131, 275–276
Material instantiations
  diagrams, 344, 345–353, 356–357
  practices, 15, 16, 20, 22
Material mimesis
  batik, 24, 37, 154–155, 167–173, 174
  crackled crystals, 154, 155–162
  examples, 153
  meaning, 153
  performance, 153–154, 174
  textiles, 24, 154–155, 162–173
McKnight, Brian E., 262
Meera (film), 137
Merton, Robert K., 28–29, 275
Microsoft, 300

Mijer, Pieter, 172–173
Millennium Pharmaceuticals, 302
Miller, John D., 307
Minerals, material mimesis, 154, 155–162
Ming Taizu, Emperor, 66
Mirowski, Philip, 30
Mongols, 253, 255, 259, 263, 266, 268
Moral rights, 23, 105, 106
Mountbatten, Lord Louis, 137
Mukharji, Projit Bihari, 198
Muni, Matanga, 126
Murthy, Viren, 19, 23–24, 121–152, 355
Museums. *See also* Endangered material
  attitudes to collections, 333–335
  colonialism, 329–331
  community engagement, 331, 332–335, 336
  diagrams, 356, 357–358
  documenting objects, 321–324
  Endangered Material Knowledge Programme, 319, 320, 322–323, 336
  ethnography, 319–336
  funding, 26, 323–324
  human remains, 331
  Melanesia Project, 334–335
  moral work, 319, 329–333, 335
  mundane objects, 322
  sacred/secret objects, 331–332
Muslims, 102
Mutual conditioning
  domains (*see* Domains)
  meaning, 3–7, 23–24, 353–354
  power, 26–27
  practices (*see* Practices)
  process, 15, 22–23

Naidu, Sarojini, 137
Naming
  Chinese crafts, 25, 31, 251–255, 257–258, 272–277
  crackled objects, 174
  diagrams, 344, 345, 346, 347, 348, 349, 351–355, 352, 358–361
  epistemology, 6, 20, 25

Naming (cont.)
  genetic knowledge, 25
  IP law and, 92, 96
  modern scholarship, 34, 35–36
  power and, 2, 34
  practice, 3, 4, 5, 7, 15, 16, 19, 20, 22–24, 35–36
  tanning, 4, 8
  teaching methods, 24
  traditional knowledge, 226, 228, 231
Nanoscience, 188, 190–192, 193–194, 196, 199, 200
National Science Foundation, 190, 191
Natural rights, 49, 60, 97, 105, 106
Navyug, Gill, 131
Needham, Joseph, 29
Nehru, Jawaharlal, 137
Neoliberalism, 184, 192
Netherlands
  batiks, 167, 171–173
  patents, 101
Nevius, John Livingston, 53
New York Genome Center, 300
New Zealand, 330, 331, 335
Nombo, Porer, 232, 233–234
Nüwa, 54

Oak, Vivek S., 17, 343–362
Obama, Barack, 300
Oeder, Jörn, 17, 343–362
Ovington, John, 166
Owen, Stephen, 51
Ownership
  bundle of rights, 295, 306
  genes (*see* Genetic knowledge)
  knowledge (*see* Knowledge ownership)
  power and, 295–296
  regulation of (*see* Branding, brands)
  separation of knowledge from, 28–35
  US gene case law, 295–300

Papua New Guinea
  botany, 232–233, 240
  diagrams, 356, 357
  documenting *Reite Kastom*, 229–242
  extractive industries, 225–226
  myth, 231–237
  ornithology, 240
  performance, 24–25, 362
  relationships, 223–229
  traditional knowledge, 24–25, 219–243, 362
  value and circulation, 237–239
Patel, Leigh, 193, 200–201
Patents. *See also* Germany; Italy; Switzerland
  batik, 1703
  China (PRC), 72
  Chinese knowledge and, 23, 47
  criteria, 7, 103
  critics, 48, 100–101
  diagrams, 344
  genes (*see* Genetic knowledge)
  history, 98, 100–102, 108
  justification, 97–99, 106
  tanning and, 22
  textbooks, 92, 94, 95, 98, 99, 100–101, 102, 106–107
Peers, Laura, 329–330
Peller, Gary, 186, 206n27
Peng Baojun, 255
Peng Zeyi, 272
Performance
  Carnatic music, 23–24, 122–123, 124, 128–129, 130, 136–141, 143–145, 354
  Chinese clerks, 268
  Chinese crafts, 55, 258, 261, 265–266
  Cook Ding, 5, 20
  diagrams, 344, 345, 346, 347, 348, 349, 351–355, 356, 357, 358–361
  IP law and, 97, 100, 101
  material mimesis, 153–154, 174
  ownability, 17, 18, 20, 25, 35, 254, 261
  Papua New Guinea, 24–25, 362
  practice, 3–4, 22, 23–24, 35, 225, 227, 229
  repetition, 19
  role, 7, 9, 15, 16, 21, 33, 237
  science education, 190, 192, 197
  tacit knowledge, 9

tanning, 8, 9
traditional knowledge, 225, 227, 228, 229, 237, 327–328, 362
Personal genomics companies, 301–308
Pfizer, 300
Phillips, Jeremy, 91, 104
Phillips, Ruth, 333
Pickering, Andrew, 183, 204n12
Pliny the Elder, 158
Polanyi, Karl, 28
Polanyi, Michael, 28, 29–30
Posidonius, 157
Positivism, 198
Postcolonialism, 30–31, 319–320, 329, 330, 331
Power
  China, 6–7, 26, 60–65, 253–254, 255–259, 269, 272, 275
  Chinese publishing, 68, 70
  diagrams, 353, 354
  domains, 6–7
  geopolitics, 48
  hierarchies, 34, 344
  intellectual property and, 48
  knowledge ownership and, 2, 15, 18–19
  manipulations, 348
  Marx, 131
  mutual conditioning, 26–27
  naming and, 2, 34
  ownership and, 295–296
  Papua New Guinea, 227, 230, 231
  performance and, 7
  rituals, 327
  scholarship, 30–31, 36
  science education, 197–198
Practices
  diagrams, 345, 347, 349, 351, 352–355, 358–361
  meaning, 3–5, 23–24
  naming (*see* Naming)
  performance (*see* Performance)
  use (*see* Use)
Precision medicine, 300, 302
Precision Medicine Initiative, 309

Prévinaire, Jean-Baptiste Theodore, 170, 171
Property
  crafts and, 253–255 (*see also* Chinese intellectual property; Crafts)
  intellectual (*see* Intellectual property)
  knowledge as, 7, 15–16
  legal (rights), 2–3, 16, 36, 22, 275 (*see also* Branding, brands; Intellectual property; Patents)
  material as, 254–255
  objects and, 18–19
  ownership and, 23, 24, 27–35, 184 (*see also* Knowledge ownership; Marx, Karl; Patents)
  possession and, 38n7
Public domain, IP textbooks and, 103–106
Puett, Michael, 50–51

Qin Shihuangdi, Emperor, 61
Qiu Jun, 254, 269–274

Race. *See* Ethnicity/race
Raffles, Thomas Stamford, 167–168
Rao, T. R. Subbha, *Swan's Song*, 129, 139
Rausing, Lisbet, 323, 324
Reyman, Jessica, 92
Ricoeur, Paul, 91
Riles, Annelise, 237, 241, 242
Rituals
  Carnatic music, 130
  Confucianism, 60, 61, 254, 268–270, 272
  ethnography and, 320, 332
  indigenous people, 220
  Papua New Guinea, 224, 227, 229, 232, 240
  Vanuatu, 326–328, 331
Roberts, Jessica L., 295
Rogers, Curtis, 307
Ru. See Confucianism
Rudolph, John L., 188

Sadhasivam, 136–137
Sarr, Felwine, 320
Savoy, Bénédicte, 320, 333
Schäfer, Dagmar, 1–13, 15–44, 55, 251–292, 355

Scholarship, role, 9–10, 25–27
Schumer, Chuck, 308
Science
  culture and, 32
  education (*see* Science education)
  manipulation of knowledge ownership, 8
  scientific workers, 28–29
  tacit knowledge, 29–32, 33, 34–35
Science education
  diagrams, 356, 357
  epistemic continuities, 192–197
  hands-on learning, 187–190
  historical making of knowers and nonknowers, 187–201
  inequities, 181–203
  knowledge citizens, 197
  LEGO bricks, 181, 187, 193–195, 196, 198, 199
  marginalized communities, 188–189, 192, 198, 200, 201–202
  meritocracy, 182–187
  nanoscience, 188, 190–192, 193–194, 196, 199, 200
  National Nanotechnology Initiative (NNI), 191
  National Research Council, 197, 209n67
  Next Generation Science Standards (NGGS), 187, 190, 195, 198
  not-to-be-knowers, 197–201
  shoebox method, 2, 24, 181, 182–187, 189, 193, 194, 195, 196, 198
  United States, 2, 181–203
Secules, Stephen, 183
Seismoscope, 54
Semiotics, intellectual property, 93–96
Seneca the Younger, 157–158
Settler states, 329, 330–333
Shang Wei, 52
Shen Zhou, 56
Sherman, Brad, 94, 97, 98, 99, 101–102, 104
Shiva, Vandana, 103
Shoebox education method, 2, 24, 181, 182–187, 189, 193, 194, 195, 196, 198
Shonibare, Yinka, 172

Shundi (Yuan emperor), 268
Silbey, Jessica, 92
Sillitoe, Paul, 238–239
Sima Guang, 51–52, 53, 57
Sima Xiangru, 257
Simmel, George, 108
Single nucleotide polymorphisms (SNPs), 301–302
Singo, 327–328, 332, 334–335
Sisau, Pinbin, 233–234
Sitaramayya, Todi, 128
Slaton, Amy, 19, 24, 26, 181–216, 343, 349, 355, 356, 357, 360, 361
Smith, Herchel, 97
Social construction of technology (SCOT), 21
Society domain, 5–7, 24–25
Song Lian, 265, 282n59, 283n60
Song Yingxing, 55
Spivak, G. C., 193
Splitting
  diagrams, 343–344, 349, 350
  spatial split, 344, 345
  temporal split, 344, 345
Stimulating creativity mantra, 103, 104
Stockholm Papyrus, 158–159
Stone, Charles, 51–52
Stones, material mimesis, 154, 155–162
Strathern, Marilyn, 91, 222, 229, 239
Subbalakshmi, M. S., 124, 125, 136–138, 143
Subramanian, Lakshmi, 131–132, 137, 148n28
Su Che, 62
Sun Picheng, 71
Su Shi, 53–54, 56–57
Su Song, 54
Switzerland, patents, 101
Symbolic substances, 156–157
Symbolic value, 121, 122, 131

Tacit knowledge, 28, 29–32, 33, 34–35, 350, 351, 352, 353, 354, 355
Tang Taizong, Emperor, 269
Tang Wenzong, Emperor, 61
Tang Xianzu, 52

Tanning, 1, 2, 3–4, 8, 9, 10, 21–22, 259, 262, 268
Tapsell, Paul, 335
Tarisesei, Jean, 334–335
Taylor, Amy R., 198, 199
Taylor, Ashley, 197, 201, 210n82
Taylor, John Patrick, 325
Teaching
  banking knowledge, 183
  IP (*see* University textbooks)
  meritocracy, 182–187
  methods, naming, 24
  science (*see* Science education)
  semiotics of IP, 93–96
Terrell on the Law of Patents, 92, 98
Textbooks. *See* University textbooks
Textiles
  batik, 24, 37, 154–155, 167–173, 174
  block-printing, 163, 166–167
  Coromandel Coast quilt, 164
  India, 30–31, 166
  lampas weave, 165
  material mimesis, 24, 154–155, 162–173
  singo, 327–328, 332, 334–335
Thorley, Simon, 98
Tinsley, April, 307
Trademarks, 47–48, 69, 94, 95, 97, 99, 101, 303. *See also* Branding, brands
Traditional knowledge
  aesthetics, 221–222, 223, 229, 231–237
  appropriate documentation, 240–242
  diagrams, 356, 357
  ethnography, 223–229
  expression, 220
  kastom documentation, 222, 229–230, 233–242
  meaning, 219
  myth, 231–237
  Papua New Guinea, 24–25, 219–243, 362
  performance, 225, 227, 228, 229, 237, 327–328
  relationality, 219, 220, 221, 223–229, 242–243
  rituals, 224–225, 226, 227, 229, 231, 232, 240
  scholarship and, 9–10
  terminology, 222–223
  TKRN project, 232, 233–239, 242, 243
  university textbooks and, 102–103
  value, 219
  value and circulation, 237–239
  Vanuatu (*see* Vanuatu)

UNESCO Convention on the Protection and Promotion of the Diversity of Cultural Expressions (2005), 219
United Kingdom
  British Museum, 319, 320, 333, 334–335
  Endangered Material Knowledge Programme, 319, 320, 322–323, 336
  IP university textbooks, 91–108
  textiles, 167, 168
  Warwick Castle table, 155
United States
  batiks, 172–173
  copyright, 48
  educational system, 26, 27
  genetic knowledge (*see* Genetic knowledge)
  human capital, 188
  IP law, 48, 92, 102, 107
  National Research Council, 197, 209n67
  National Science Foundation, 190, 191
  ownership case law, 295–300
  personal genomics companies, 301–308
  postcolonialism, 330
  precision medicine, 300
  racism, 92, 183–185, 202, 300
  science education, 2, 24, 181–203
University textbooks
  balance, 106–107
  creativity mantra, 103, 104
  framing narrative, 100–103
  IP history, 91–108
  lists, 94, 95
  paratext, 96–97

University textbooks (cont.)
  patents, 92, 94, 95, 98, 99, 100–101, 102, 106–107
  public domain and, 103–106
  semiotics, 93–96, 104–105
  trademarks, 94, 95, 97, 99, 101
Use
  Carnatic music, 126, 133, 140, 142
  Chinese authorship, 74
  Chinese crafts, 5, 6, 59, 252, 258, 276
  diagrams, 344, 345, 346, 347, 348, 349, 350, 351–355, 356, 358–361
  genetic data, 297, 301, 305, 306, 357
  IP law, 105, 106, 107
  material mimesis, 161, 166–167, 173, 174
  Papua New Guinea, 220, 226–227, 234–236, 238, 240, 243
  practice, 3, 4, 15, 16, 22, 23–24, 35
  repetition, 19
  role, 7, 16–17, 18, 19–20
  science and, 30
  scientific instruments, 186, 194, 198
  traditional knowledge, 320, 327–328, 332

Vaidyanatha Iyer, Maha, 132–133
Van Meijl, Toon, 221
Vanuatu
  agriculture, 328
  diagrams, 356, 357
  endangered material knowledge, 319, 324–328
  independence, 325
  scholarship, 26
  secrets, 326–328
  singo, 327–328, 332, 334–335
  underprivileged students, 27
  Women Fieldworkers Program, 324–325
Vasuki, Kaber, 140
Veblen, Thorsten, 156–157
Vedavalli, Vidushi R., 125, 141–143, 144
Venice, 98
Verbong, G. P. J., 170–171
Viana, Hermano, 123

Viveiros de Castro, Eduardo, 198
Vlisco, 171–172

Wadlow, Christopher, 92
Waelde, Charlotte, 94, 97
Walter, Annie, 328
Wang, Fei-hsien, 71
Wang Ao, 268, 284n80
Wang Cheng, 64–65
Wang E, 256
Wang Yucheng, 64
Wang Zhaoyu, 273–274
Warren, Beth, 195, 210n88
Warwick Castle table, 155
Wenhui, Lord, 5, 6
White, Hayden, 91
White, James Boyd, 91
Wikipedia, 102
Winston, Joel, 305–306
Wojcicki, Anne, 304–305
World Intellectual Property Organization (WIPO), 107, 219, 239
Wu Yinzhi, 266

Xu Ting, 259
Xu Wei, 54
Xu Yuanrui, *Guidebook for Clerks (Lixue zhinan)*, 251–277

Yamamoto, Yuzo, 29
Yang Mao, 255
Yang Shitie, 252
Yao Wang, 272
Yar, Majid, 93
Ye Dehui, 70
Yellow Emperor, 54, 273
Yingzong, Emperor, 272
Yuan Mei, 55, 56
Yu Anqi, 70
Yu Shengkan, 255

Zhang Cheng, 255
Zhang Heng, 54

Zhang Linnan, 255
Zhou Liulang, 255
Zhuang Zhou. *See* Zhuangzi
Zhuangzi, 5, 11n9, 59
Zhu Bishan, 255
Zhu Mu, 56, 58, 70
Zhu Xi, 57, 70, 273
Zhu Yuanzhang, Emperor, 270–271

INSIDE TECHNOLOGY SERIES

Edited by Wiebe E. Bijker and Rebecca Slayton

Dagmar Schäfer, Annapurna Mamidipudi, and Marius Buning, editors, *Ownership of Knowledge: Beyond Intellectual Property*

Tamar Novick, *Milk and Honey: Technologies of Plenty in the Making of a Holy Land*

Kristin Asdal and Tone Huse, *Nature-Made Economy: Cod, Capital, and the Great Transformation of the Ocean*

Cyrus C. M. Mody, *The Squares: US Physical and Engineering Scientists in the Long 1970s*

Brice Laurent, *European Objects: The Troubled Dreams of Harmonization*

Florian Jaton, *The Constitution of Algorithms: Ground-Truthing, Programming, Formulating*

Kean Birch and Fabian Muniesa, *Turning Things into Assets*

David Demortain, *The Science of Bureaucracy: Risk Decision-Making and the US Environmental Protection Agency*

Nancy Campbell, *OD: Naloxone and the Politics of Overdose*

Lukas Engelmann and Christos Lynteris, *Sulphuric Utopias: The History of Maritime Fumigation*

Zara Mirmalek, *Making Time on Mars*

Joeri Bruynincx, *Listening in the Field: Recording and the Science of Birdsong*

Edward Jones-Imhotep, *The Unreliable Nation: Hostile Nature and Technological Failure in the Cold War*

Jennifer L. Lieberman, *Power Lines: Electricity in American Life and Letters, 1882–1952*

Jess Bier, *Mapping Israel, Mapping Palestine: Occupied Landscapes of International Technoscience*

Benoît Godin, *Models of Innovation: The History of an Idea*

Stephen Hilgartner, *Reordering Life: Knowledge and Control in the Genomics Revolution*

Brice Laurent, *Democratic Experiments: Problematizing Nanotechnology and Democracy in Europe and the United States*

Cyrus C. M. Mody, *The Long Arm of Moore's Law: Microelectronics and American Science*

Tiago Saraiva, *Fascist Pigs: Technoscientific Organisms and the History of Fascism*

Teun Zuiderent-Jerak, *Situated Interventions: Sociological Experiments in Healthcare*

Basile Zimmermann, *Technology and Cultural Difference: Electronic Music Devices, Social Networking Sites, and Computer Encodings in Contemporary China*

Andrew J. Nelson, *The Sound of Innovation: Stanford and the Computer Music Revolution*

Sonja D. Schmid, *Producing Power: The Pre-Chernobyl History of the Soviet Nuclear Industry*

Casey O'Donnell, *Developer's Dilemma: The Secret World of Videogame Creators*

Christina Dunbar-Hester, *Low Power to the People: Pirates, Protest, and Politics in FM Radio Activism*

Eden Medina, Ivan da Costa Marques, and Christina Holmes, editors, *Beyond Imported Magic: Essays on Science, Technology, and Society in Latin America*

Anique Hommels, Jessica Mesman, and Wiebe E. Bijker, editors, *Vulnerability in Technological Cultures: New Directions in Research and Governance*

Amit Prasad, *Imperial Technoscience: Transnational Histories of MRI in the United States, Britain, and India*

Charis Thompson, *Good Science: The Ethical Choreography of Stem Cell Research*

Tarleton Gillespie, Pablo J. Boczkowski, and Kirsten A. Foot, editors, *Media Technologies: Essays on Communication, Materiality, and Society*

Catelijne Coopmans, Janet Vertesi, Michael Lynch, and Steve Woolgar, editors, *Representation in Scientific Practice Revisited*

Rebecca Slayton, *Arguments That Count: Physics, Computing, and Missile Defense, 1949–2012*

Stathis Arapostathis and Graeme Gooday, *Patently Contestable: Electrical Technologies and Inventor Identities on Trial in Britain*

Jens Lachmund, *Greening Berlin: The Co-Production of Science, Politics, and Urban Nature*

Chikako Takeshita, *The Global Biopolitics of the IUD: How Science Constructs Contraceptive Users and Women's Bodies*

Cyrus C. M. Mody, *Instrumental Community: Probe Microscopy and the Path to Nanotechnology*

Morana Alač, *Handling Digital Brains: A Laboratory Study of Multimodal Semiotic Interaction in the Age of Computers*

Gabrielle Hecht, editor, *Entangled Geographies: Empire and Technopolitics in the Global Cold War*

Michael E. Gorman, editor, *Trading Zones and Interactional Expertise: Creating New Kinds of Collaboration*

Matthias Gross, *Ignorance and Surprise: Science, Society, and Ecological Design*

Andrew Feenberg, *Between Reason and Experience: Essays in Technology and Modernity*

Wiebe E. Bijker, Roland Bal, and Ruud Hendricks, *The Paradox of Scientific Authority: The Role of Scientific Advice in Democracies*

Park Doing, *Velvet Revolution at the Synchrotron: Biology, Physics, and Change in Science*

Gabrielle Hecht, *The Radiance of France: Nuclear Power and National Identity after World War II*

Richard Rottenburg, *Far-Fetched Facts: A Parable of Development Aid*

Michel Callon, Pierre Lascoumes, and Yannick Barthe, *Acting in an Uncertain World: An Essay on Technical Democracy*

Ruth Oldenziel and Karin Zachmann, editors, *Cold War Kitchen: Americanization, Technology, and European Users*

Deborah G. Johnson and Jameson W. Wetmore, editors, *Technology and Society: Building Our Sociotechnical Future*

Trevor Pinch and Richard Swedberg, editors, *Living in a Material World: Economic Sociology Meets Science and Technology Studies*

Christopher R. Henke, *Cultivating Science, Harvesting Power: Science and Industrial Agriculture in California*

Helga Nowotny, *Insatiable Curiosity: Innovation in a Fragile Future*

Karin Bijsterveld, *Mechanical Sound: Technology, Culture, and Public Problems of Noise in the Twentieth Century*

Peter D. Norton, *Fighting Traffic: The Dawn of the Motor Age in the American City*

Joshua M. Greenberg, *From Betamax to Blockbuster: Video Stores tand the Invention of Movies on Video*

Mikael Hård and Thomas J. Misa, editors, *Urban Machinery: Inside Modern European Cities*

Christine Hine, *Systematics as Cyberscience: Computers, Change, and Continuity in Science*

Wesley Shrum, Joel Genuth, and Ivan Chompalov, *Structures of Scientific Collaboration*

Shobita Parthasarathy, *Building Genetic Medicine: Breast Cancer, Technology, and the Comparative Politics of Health Care*

Kristen Haring, *Ham Radio's Technical Culture*

Atsushi Akera, *Calculating a Natural World: Scientists, Engineers and Computers during the Rise of U.S. Cold War Research*

Donald MacKenzie, *An Engine, Not a Camera: How Financial Models Shape Markets*

Geoffrey C. Bowker, *Memory Practices in the Sciences*

Christophe Lécuyer, *Making Silicon Valley: Innovation and the Growth of High Tech, 1930–1970*

Anique Hommels, *Unbuilding Cities: Obduracy in Urban Sociotechnical Change*

David Kaiser, editor, *Pedagogy and the Practice of Science: Historical and Contemporary Perspectives*

Charis Thompson, *Making Parents: The Ontological Choreography of Reproductive Technology*

Pablo J. Boczkowski, *Digitizing the News: Innovation in Online Newspapers*

Dominique Vinck, editor, *Everyday Engineering: An Ethnography of Design and Innovation*

Nelly Oudshoorn and Trevor Pinch, editors, *How Users Matter: The Co-Construction of Users and Technology*

Peter Keating and Alberto Cambrosio, *Biomedical Platforms: Realigning the Normal and the Pathological in Late-Twentieth-Century Medicine*

Paul Rosen, *Framing Production: Technology, Culture, and Change in the British Bicycle Industry*

Maggie Mort, *Building the Trident Network: A Study of the Enrollment of People, Knowledge, and Machines*

Donald MacKenzie, *Mechanizing Proof: Computing, Risk, and Trust*

Geoffrey C. Bowker and Susan Leigh Star, *Sorting Things Out: Classification and Its Consequences*

Charles Bazerman, *The Languages of Edison's Light*

Janet Abbate, *Inventing the Internet*

Herbert Gottweis, *Governing Molecules: The Discursive Politics of Genetic Engineering in Europe and the United States*

Kathryn Henderson, *On Line and On Paper: Visual Representation, Visual Culture, and Computer Graphics in Design Engineering*

Susanne K. Schmidt and Raymund Werle, *Coordinating Technology: Studies in the International Standardization of Telecommunications*

Marc Berg, *Rationalizing Medical Work: Decision Support Techniques and Medical Practices*

Eda Kranakis, *Constructing a Bridge: An Exploration of Engineering Culture, Design, and Research in Nineteenth-Century France and America*

Paul N. Edwards, *The Closed World: Computers and the Politics of Discourse in Cold War America*

Donald MacKenzie, *Knowing Machines: Essays on Technical Change*

Wiebe E. Bijker, *Of Bicycles, Bakelites, and Bulbs: Toward a Theory of Sociotechnical Change*

Louis L. Bucciarelli, *Designing Engineers*

Geoffrey C. Bowker, *Science on the Run: Information Management and Industrial Geophysics at Schlumberger, 1920–1940*

Wiebe E. Bijker and John Law, editors, *Shaping Technology / Building Society: Studies in Sociotechnical Change*

Stuart Blume, *Insight and Industry: On the Dynamics of Technological Change in Medicine*

Donald MacKenzie, *Inventing Accuracy: A Historical Sociology of Nuclear Missile Guidance*

Pamela E. Mack, *Viewing the Earth: The Social Construction of the Landsat Satellite System*

H. M. Collins, *Artificial Experts: Social Knowledge and Intelligent Machines*

http://mitpress.mit.edu/books/series/inside-technology